Design and Construction
of
Steel Bridges

BALKEMA – Proceedings and Monographs
in Engineering, Water and Earth Sciences

Design and Construction of Steel Bridges

of

Steel Bridges

Utpal K. Ghosh
BE, CEng, FIE, MICE, MIStruct E, PE
Consulting Engineer

Taylor & Francis
Taylor & Francis Group

LONDON/LEIDEN/NEW YORK/PHILADELPHIA/SINGAPORE

Library of Congress Cataloging-in-Publication Data
Applied for

Published by: Taylor & Francis/Balkema
                   P.O. Box 447, 2300 AK Leiden, The Netherlands
                   e-mail: Pub.NL@tandf.co.uk
                   www.balkema.nl, www.tandf.co.uk, www.crcpress.com

ISBN-10: 0-415-41836-4
ISBN-13: 978-0-415-41836-2

To my parents
Late Professor Kamal Krishna Ghosh
and
Late Surama Ghosh

Preface

Technology of steel production was developed by Bessemer and Siemens in the second half of the 19th century. This material, having almost identical strength in tension, compression and shear, had a great potential for use in the construction industry. However, commercial application of steel was not possible until the closing years of the 19th century, when the continuing research and development paved the way for improved technology of steel production, which could control and guarantee the quality of the product. This improvement, coupled with advancements of theoretical analysis of structures, technical skill, and development of mechanical appliances encouraged the designers of Forth railway bridge in Scotland (1890) to use steel for the first time as material for construction of a major bridge.

Since then, more than a hundred years have passed and vast improvement in the production technology of steel has taken place. New generation of steel is now available with significant improvements in the properties, such as: fine-grained weldable quality steel with enhanced mechanical properties, consistent strength, toughness, and homogeneous composition, corrosion resistant steel, self-healing 'weathering' steel, etc. Also, advances in structural steelwork fabrication techniques by the use of automatic welding processes, and computer numerically controlled (CNC) machines have shown significant improvements in the quality of fabrication. With the introduction of computers and increased research work, understanding of the behaviour of the structures has improved vastly, with the result that designers can now design economical steel structures with confidence. The guiding codes of practice have also undergone radical changes worldwide. In recent years, considerable improvement has also been achieved in the area of protective treatment of steel.

These improvements, along with the inherent advantages of steel such as lightness, stiffness, and strength, have made steel one of the most versatile and durable construction materials for bridges today. Its advantages for long span bridges have been well established for some time, but its advantages for short and medium span bridges are also being seriously examined, particularly from considerations of durability and ease in post-construction repair and rehabilitation.

The present work is meant to provide comprehensive guidelines for design and construction of steel bridges, and is expected to meet the needs of practising bridge engineers as well as students. The presentation is somewhat different from normal engineering text. The book should be regarded as a complementary work to the more rigorous and analytical studies provided in existing publications and codes of practice. Consequently, topics which are not usually covered in the textbooks, but are nevertheless important for the understanding of the subject, have found place in the work.

The ideas presented in this book have been drawn mostly from the author's own experiences, or published papers and also from his personal notes. However, there are also instances where the author has taken the liberty of delving into the vast pool of knowledge and experience of his distinguished predecessors and colleagues. As far as possible, references to the published literature have been mentioned at the end of each chapter. The author thankfully acknowledges his indebtedness to these earlier writers. However, in case some ideas of earlier writers have appeared in the book without appropriate acknowledgement, it is quite unintentional, and the author would like to add his apologies to his indebtedness. If such instances are brought to the author's notice, the same will be gratefully acknowledged in the subsequent edition of the book.

In writing this book the author has also gained enormously from interactions with a large number of individuals. Many a time, seemingly small points raised in discussions have led to change in the text, or inclusion of an additional point of explanation. The author gratefully acknowledges his debt to each of them. The author would also like to acknowledge the encouragement received from the entire team of professionals at Stup Consultants P. Ltd. Special thanks are due to the author's long time colleague, Amitabha Ghoshal of Stup Consultants P. Ltd., for his support throughout the preparation of the manuscript and reviewing it and making a number of valuable suggestions. The author is also indebted to his young colleague Pinaki Chakraborty of Stup Consultants P. Ltd., for meticulously going through the manuscript and providing many positive suggestions.

The author is also thankful to the publishers and their team for their effort in bringing out the book.

And last, but by no means the least, the author is grateful to his wife Manjula, son Indranil, and daughter-in-law Supriti for their fullest co-operation, encouragement and support in writing the book.

19 May, 2006   **Utpal K. Ghosh**

Contents

Chapter 7
Selection of Structural System   84

Chapter 8
Deck Systems   94

Chapter 9
Beam and Plate Girder Bridges   117

Chapter 1

Evolution of Steel Bridges

1.1 INTRODUCTION

Evolution of steel bridges presents a fascinating study spreading well over two centuries. Use of iron in manufacturing cannons and machinery was known to man since the 16th century. Dud Dudley's success in smelting cast iron from iron ore in 1619 actually heralded the age of iron as a useful building material. But it took more than another century for cast iron to be smelted economically and in large quantities and to emerge as a competitive building material to the structural engineers. However, because of its inherent brittleness and poor tensile capacity, use of cast iron had to be limited only to compression members in the form of an arch.

Thus the world's first cast iron bridge—Coalbrookdale bridge—was built across the river Severn in the county of Shropshire in England. Designed by Thomas Pritchard, this 30.5-metre span arch bridge was built by Abraham Darby and John Wilkinson in 1777-79 using about 400 tons of ironwork. The bridge was designed with the traditional concept of stone bridges and consisted of a series of semi-circular cast iron arch ribs placed side by side. The bottom semi-circular arch rib in each vertical plane was continuous over the span, while two sets of discontinuous upper ribs acted as props to the deck and also provided stiffness to the structural system. This bridge is still standing, albeit with lower carrying capacity.

The Coalbrookdale bridge was followed by Buildwas bridge, the second cast iron bridge in Coalbrookdale over the Severn. Built by Thomas Telford in 1796, this bridge had a comparatively flatter arch form and used only about half the weight of cast iron of Coalbrookdale bridge. Subsequently, during the next 50 to 60 years, many cast iron bridges were built in Britain and other parts of the world. Amongst these, the Vauxhall and Southwark bridges over the Thames in London and Pont du Louvre over the Seine in Paris deserve mention.

During the first quarter of the 19th century, wrought iron was being developed. This material was ductile and malleable and could carry considerable tension and soon established itself as a good material for bridge construction. In 1807, James Finlay built the first elemental suspension bridge in wrought iron, the chain bridge, over the Potomac in Washington D.C. This was followed by a number of wrought iron bridges of different types both in America and Europe. Notable amongst these was Guinless bridge in England, which was built in 1821 by George Stephenson using wrought iron to form a 'lenticular' girder bridge. In 1826, the world's first suspension bridge for vehicular traffic was built by Thomas Telford over the Menai Straits in Wales. It was also the world's first major bridge over sea water. The bridge had a span of 177 m and used wrought iron eye bar chains to form the suspension system. It had no stiffening girder and no wind bracings. Subsequently the bridge deck had to be replaced in 1839, 1893 and again in 1939.

The success of Menai Straits bridge led the way for many more suspension bridges. Amongst these, the Hungerford footbridge over the Thames deserves special mention. This bridge was built with wrought iron by the brilliant Victorian engineer Isambard Kingdom Brunel in 1841-45 with a central span of 206 m and two side spans each 105 m long. The superstructure of this bridge was subsequently removed to Bristol and re-used to build the famous Clifton suspension bridge across the Avon gorge in 1862-64. This beautiful bridge is still in use, a century and half after the original Hungerford footbridge was built.

Wrought iron was also used by Brunel in the bridges across the river Wye at Chepstow, U.K. (1852) and the Royal Albert bridge over the river Tamar at Saltash, Plymouth, U.K. (1859) to carry railways. The Royal Albert bridge at Saltash had two main spans, each of 139 m span. The trusses used in both the bridges consisted of a combination of a tubular wrought iron arch and two suspension chains braced together with light wrought iron diagonal ties and vertical struts to form a single trussed girder. The outward thrust of the arches is balanced by the inward pull of the suspension chains acting in combination. The tubes of the arch are of oval cross section, 3.7 m high by 5.1 m wide. Considering the period (mid-nineteenth century) the scale of construction of these bridges was quite extraordinary. Also, the concept of ties and struts in web members was precursor of the lattice girder form which was to be used in many future bridges.

The Newcastle high level bridge was built by George Stephenson in 1849 at a height of about 37 m above the water level. The bridge consisted of six bow string cast iron arches, each with a horizontal tie at the spring point level to cater for the horizontal thrust from the arch . The bridge was quite 'modern' in its concept and carried railway on the top, with a roadway suspended underneath by wrought iron rods.

Meanwhile, in 1850, another brilliant engineer, Robert Stephenson (son of George Stephenson) built the Britannia railway bridge over the Menai Straits, North Wales, using the first box girder bridge concept, built with wrought iron as the construction material. This was also the beginning of plate girder as a popular form of thousands of future steel bridges the world over. However, the Britannia bridge was badly damaged by fire in 1970 and reconstructed with steel arches replacing the tubes.

Grand Trunk bridge across the deep gorge below the Niagara Falls was built in 1855 by John A. Roebling, an outstanding engineer and widely acclaimed as the father of long span suspension bridges. He was the inventor of the strand spinning technology used in long span bridges and of forming cables from parallel wires by binding them into a compact bunch. The Grand Trunk bridge had a span of 250 m. There were two decks, one above the other with 5.5 m deep timber stiffening trusses between them. The upper deck was to carry a railway and the lower one a roadway. Four main cables of 254 mm diameter formed by binding together parallel wrought iron wires passing over four masonry towers consituted the main structural system. The deck structure was supported by vertical hangers from the cables as well as by inclined stays from the towers. This bridge is considered as the forerunner of modern suspension bridges.

Another remarkable wrought iron bridge was built in 1884—the Garabit Viaduct—over the Truyere at St. Flour, France. In this bridge, designer Gustav Eiffel's concept of a two hinged truss arch in wrought iron was the forerunner of future steel arches. Although steel was already in use elsewhere, Eiffel chose wrought iron for this bridge because in France steel was not yet considered to be reliable in quality and was also expensive. This bridge, with a 162 m parabolic arch, supports a 564 m long truss of railway, some 122 m above the river Truyere.

During the middle of 19th century, in the United States of America the early railways used composite truss bridges with wooden compression members and wrought iron tension members. Later on, use of composite truss systems with cast iron compression members and wrought iron tension members became popular.

1.2 ARRIVAL OF STEEL

It is believed that the technology of steel making was known in India in 500 BC and in China in 200 BC. However, the laborious and slow process made the end product very expensive. Consequently, steel had only limited use such as in tools and weapons. It was only in the second half of the 19th century that the technology of modern bulk production of steel was developed in England by Henry Bessemer (1856) and Charles Siemens (1867). Steel had superior qualities compared to those of cast iron or wrought iron. It was strong in tension, compression and shear. It was ductile and could be used as rivets as well. It could be rolled, cast and even drawn to form wires. However, inspite of all these advantages, it took a long time before steel could supersede iron as an alternative material for construction of bridges because of its certain drawbacks. First, steel suffered from one major disadvantage, namely the quality of different batches of steel could not be controlled and therefore its uniformity could not be guaranteed. Secondly, steel was expensive to manufacture and could not compete with iron in the market due to its higher unit price. Research and development continued in different parts of the world which resulted in improved technology of steel production and quality control. In the last quarter of the 19th century, the world price of steel dropped by about 75% and suddenly it became commercially competitive. Consequently, it was soon accepted as a viable alternative to replace iron as a construction material for bridges.

Meanwhile, significant advances were made in the field of theoretical analysis and design methods, as well as in the understanding of the behaviour of structures. These, coupled with technical skills and improved mechanical advances, heralded an era of major steel bridges towards the closing decades of the 19th century. Thus, many important bridges were constructed on both sides of the Atlantic during this period. The trend continued during the 20th century which witnessed a marked development throughout the world in diverse areas such as:

- Knowledge of the laws of mechanics
- Material science
- Industrial production
- Shop practices and facilities
- Transport system
- Erection methods
- Quality assurance

These changes contributed considerably to the parallel development of different types of steel bridges. Thus, recounting the evolution of steel bridges in strict chronological order becomes rather difficult. Therefore, for the sake of easy understanding, development of different types of bridges has been discussed in separate sections, irrespective of their sequence of development.

1.3 BEAM AND PLATE GIRDER BRIDGES

Since the early stages of development, mill-rolled beams and fabricated plate girders have always been very popular for bridges of spans upto 20 m, both for railways and highways. The reason for this popularity was primarily because of the simplicity in their design and manufacture, as well as the ease and speed of their construction at site. Initially, the deck used to be simply resting on the steel girders. Subsequently, a concrete deck was designed to act compositely with the steel girder, thereby reducing the cost. Now-a-days, Orthotropic steel deck system is often used in place of concrete. For longer bridges, economy can be achieved by making the girders continuous over a number of piers.

1.4 TRUSS BRIDGES

The concept of modern truss system was first developed by Andrea Palladio, a 16th century Italian architect. During the subsequent centuries, many truss forms were developed, but these were, in effect, variations of an arch and produced horizontal thrusts in addition to vertical reactions at the abutments. In 1820, Ithiel Town built the first 'true' truss timber bridge in America and patented it. Called the 'Town Lattice' truss, it was a true truss because it was free from arch action or any horizontal thrust. Under vertical loading this truss produced only vertical reactions at the abutments. The first patented truss to introduce iron into the timber structure was the Howe truss (1840), which was developed by William Howe. It had top and bottom chords, as also the diagonal bracings in timber and the vertical members made of wrought iron rods. This was followed (1844) by the patenting of the Pratt truss, which reversed the arrangement of the Howe truss by putting the verticals as timber members and diagonals as wrought iron members. In 1847, an American engineer called Squire Whipple published his seminal work, explaining the method of determining the distribution of axial forces in the various members of a truss, considered as a pin connected assembly. He constructed the first all-iron truss—a bow string truss—with cast iron top chord and verticals, and wrought iron bottom chord and diagonals. Subsequently, during the second half of the 19th century, steel truss bridges with different configurations came to be built, but finally, only the well known Pratt, Howe, Baltimore, Warren and K-type configurations became popular amongst bridge builders. Of these again, Warren, Pratt and K-type in pure or with variations dominated the scene during the 20th century.

1.5 CANTILEVER TRUSS BRIDGES

In the second half of the 19th century, steel, with its versatile properties, had already established itself as a useful construction material. It soon became evident that with the same quantity of material, spans with open web truss forms could be made longer if the members were built of steel instead of iron. As a result, long span cantilever truss configurations to cross wider gaps became technically feasible and economically viable.

The Fraser River bridge, Canada, was probably the first balanced cantilever steel truss bridge to be built. All the truss units including piers, links and lower chord members were fabricated from Siemes-Martin steel. It was constructed in 1886, but was dismantled 24 years later in 1910. During this period, one of the landmark cantilever truss bridges to be built was the mighty Firth of Forth railway bridge near Edinburgh, Scotland. Designed in 1881 by John Fowler and Benjamin Baker and constructed by Messrs Tancred, Arrol & Co, Scotland, this bridge was completed in 1890. At that time, it was the longest spanning bridge in the world beating Brooklyn bridge by 34 m. Carrying two railway tracks about 46 m above water level, the main bridge has two 521 m spans, each consisting of two 207 m cantilever trusses and a 107 m suspended truss section. Thus, the bridge has three huge towers from which cantilever-truss arms project to carry the two suspended spans. The main compression members comprise 3.66 m diameter tubular section of 105 m length. The depth of the trusses at piers is 107 m. About 58,000 tons of open-hearth steel and 6 million rivets were used to build the bridge.

At about the same time, the Cincinnati Newport bridge, USA, was built in 1891. This bridge, with its long cantilever spans and short suspended span, became a model, following which many railway bridges were conceived and constructed in USA. A graceful variation of the balanced cantilever configuration was the Viaur Viaduct in France. Constructed in 1902, this railway bridge consists of only two cantilever arms without any suspended span between them.

Viaur Viaduct was followed by Quebec bridge over the St. Lawrence River in Canada which was commenced in 1904. This bridge was designed to have a central span of 549 m between two giant cantilever trusses on the two main piers with a suspended span in the middle. Unfortunately, however, the bridge collapsed during erection in 1907. A new bridge was rebuilt in 1917 and became the world's longest cantilever span. A number of cantilever bridges have been built since then. Amongst these, the following deserve mention:

- East Bay bridge, San Francisco, USA, 427 m, 1936
- Howrah bridge, Calcutta, India, 457 m, 1943
- Greater New Orleans bridge, Louisiana, USA, 480 m, 1958
- Commodore Barry bridge, Pennsylvania, USA, 501 m, 1974
- Minato bridge, Osaka, Japan, 510 m, 1974

1.6 CONTINUOUS TRUSS BRIDGES

In 1909, the Queensboro bridge over the East River in New York city was built. This bridge was of a continuous truss configuration with two main spans of 360 m, a central span of 192 m and two anchor spans at two shores.

1.7 ARCH BRIDGES

Designed by J.B. Eads, and constructed in 1874, St. Louis bridge had three steel arches of lattice construction with the central arch of 158 m span and two arches of 153 m span on either side. The bridge carried a roadway on the top deck and two railway tracks on the bottom deck. The main chords of the arch were steel tubes of 450 mm diameter and were fabricated from 6 mm thick steel plates rolled in the form of a tube. St. Louis bridge was the first major steel bridge to be erected by the modern cantilever method without using any falsework from below. The arches were erected by cantilevering out from the piers towards the centre of the span. The arch units were held back by a series of tie cables made of 150 mm × 25 mm steel bars from temporary towers built on the piers.

Construction of St. Louis bridge was followed by a number of steel arch bridges around the world. The 152 m span arch bridge over the Zambezi River near the Victoria Falls in Africa was built by cantilevering from each side over the 122 m deep gorge. The bridge was built by British engineers under the leadership of Sir Ralph Freeman. In 1916, the Hill Gate bridge in New York over the East River with a span of 298 m was built. This bridge was designed by Gustav Lindenthal and was of lattice spandrel-braced two-hinged arch of high carbon steel members. It carried four railway tracks and was the longest steel arch span in the world when it was built.

Next came the 511 m span Bayonne bridge over the Kill Van Kull River in New Jersey. Designed by O. H. Amman, this bridge was built with a cheaper carbon-manganese steel in preference to nickel steel, thus ushering in the material composition of the most of the modern steel bridges. Completed in 1931, this bridge was the longest steel arch bridge until 1978, when the New River Gorge bridge at West Virginia with an arch span of 518 m was built. The site conditions permitted the bridge to be erected by means of temporary falsework, thus obviating the necessity of cantilever method of erection employed for the construction of the other major bridges.

At about the same time, was built another famous landmark bridge, namely the Sydney Harbour Bridge in Australia. Designed by Sir Ralph Freeman, and built by Dorman Long Company, England, this bridge has a Spandrel-braced two-hinge steel arch span of 509 m. The bridge was completed in 1932 and used more than 40,000 tons of nickel steel. Its design was similar to that of the Hill Gate bridge built in 1916 by Gustav Lindenthal. It carried four railway tracks and a 17 m wide roadway plus two foot paths, all suspended from the arch. Erection was done by cantilevering from two sides.

During the closing decades of the 20th century and beginning of the present century, a number of steel arch bridges were built around the world. Amongst these, the Lupu road bridge over the Huangpu River in Shanghai, China, is an important landmark. Opened to traffic in 2003, this all-welded steel bridge consists of a main span of 550 m with two side spans of 100 m each, thus becoming the longest steel arch bridge in the world.

1.8 SUSPENSION BRIDGES

The Brooklyn bridge across East River, New York, was the first steel wire suspension bridge in the world. The main cables of this bridge were made from galvanised cast steel wires spun by wire by wire method. In 1867, John Roebling was commissioned for the design and construction of this bridge of 487 m, span—nearly double the previous longest span built. Tragically, however, John Roebling died in 1869, due to an accident on site. His son, Washington Roebling undertook the responsibilty of continuing with his father's dream project. But he himself became a victim of caisson disease in 1872 and was partially paralysed. In this crisis he was actively helped by his wife Emily and she supervised the work on his behalf. She was, in fact, his principal assistant during the construction of the bridge which was completed in 1883.

The Brooklyn bridge is considered as the ancestor of all modern suspension bridges and will always be remembered as a tribute to human vision and determination. Also, the name of Roebling will always be associated with this, as well as all future suspension bridges. He introduced the method of spinning the suspension cables into the position, wire by wire, instead of manufacturing the cable first and then erecting it in the final position. The method was so simple that it has been used in many suspension bridges which followed Brooklyn during the subsequent decades.

Melan and Steinman's "deflection theory" for suspension bridges provided refinement in the analysis of suspension bridges by taking into account second order deflections of the main cable under live load. Manhattan bridge (1903) was designed by Leon Moisseiff using this recently developed theory.

George Washington bridge across Hudson River in New York was built in 1931. Designed by O.H. Ammann, this bridge had a span of 1,066 m, nearly double the previous record and had the largest span in the world for nearly a decade. Originally designed for a roadway of eight traffic lanes in the upper deck and railway track in the lower deck, this bridge was completed in 1931 without the lower deck for paucity of funds and remained as such until 1962 when a lower deck and a stiffening truss were added to carry more road traffic. The 1930s saw the contruction of some very famous bridges on the Pacific coast of the USA, viz., 6.5 km long San Francisco-Oakland Bay crossing consisting of two suspension bridges, each with one 704 m central span and two 354 m side spans with a common middle anchor pier, a tunnel through an island, a 427 m span cantilever truss bridge and approach spans. The special feature of a central anchor pier for a major multiple suspension bridge was the first of its kind in the world and has remained unique as such till date. This gigantic project was followed by the record 1,280 m span Golden Gate Bridge to connect San Francisco with Marin County to the north across the Golden Gate Strait.

The ongoing success of suspension bridges suffered a severe jolt in November 1940 when 853 m span of the Tacoma Narrows bridge across Puget Sound in Washington State, USA, collapsed only four months after it had been opened to traffic and after an hour in a wind of only 68 km/hr, which induced uncontrollable torsional oscillation resulting in the collapse of the bridge into the water. The official enquiry following the disaster attributed the failure to a lack of proper understanding and knowledge of the whole profession about aerodynamic behaviour of cable supported structures. As a logical sequence to the inquiry, design studies for all new major cable suspended bridges have generally included aerodynamic investigation as a matter of routine. Also, designs on similar lines were checked for all important bridges and a large number of them had to be strengthened to improve aerodynamic stability. This newly acquired knowledge of investigating aerodynamic stability of long span cable suspended bridges provided a much needed fillip to the new generation of bridge builders who went ahead to reconstruct the Tacoma bridge in 1950 with a wider 18.3 m deck and 10 m deep stiffening trusses. Many more suspension bridges continued to be built during the following decades. Thus, in 1957 was built the Mackinac bridge in Michigan with a 1,159 m span, which was designed by David Steinmann. Next, in 1965, the 1,298 m span Verrazano Narrows bridge across the New York Harbour entrance was built exceeding the longest span of the Golden Gate bridge by 18 m. In 1966, Severn bridge with a 988 m central span was constructed. This bridge had a revolutionary concept of an all-welded aerofoil shaped box girder suspended structure which acted both as a stiffening girder as well as support for the road deck, resulting in substantial reduction in the weight of the deck steelwork and consequently the cable sizes. One other novel feature of this bridge was that the hangers supporting the deck structure were made inclined in preference to the traditional vertical. The triangulated lattice pattern between the cable and the deck structure was meant to provide additional aerodynamic damping effect. A similar aerofoil shaped box-girder arrangement was used in Bosphorus bridge (1973) with a 1,074 m span as also in the record breaking Humber bridge (1981) in North England with a 1,410 m central span. A second crossing across Bosphorus with a span of 1,014 m was built in 1988. The 1,624 m span Stor-baelt bridge in Denmark was completed in 1998, to be subsequently exceeded by the Akaashi Kaikyo bridge, Japan, with a 1,990 m span which was opened to traffic in 2000.

1.9 BOX GIRDER BRIDGES

The pioneering concept of Robert Stephenson's box section which was used in the Britannia bridge over Menai Straits, North Wales, in 1850 was finally utilised after nearly a century as an efficient and successful popular form in many post World War II bridges. The concept was first used to build a number of concrete bridges which proved to be quite economical. Meanwhile, steel was already being used as a construction material for bridges in the form of two or more beams or plate girders to support the deck. This parallel beam/plate girder system, however, could not be used for long span crossings since girders with deeper web plates were required to take the increased shear of longer spans. These web plates were found to be susceptible to buckling failure and needed to be properly stiffened and held by cross frames at frequent intervals. Use of thicker web plates to obviate this deficiency made the design uneconomical. At this stage the concept of box girder presented a possible solution by way of joining the flanges of the girders and forming a continuous hollow section. This would not only provide easier support to the web plates at frequent intervals, but would also reduce the depth of the girder, thereby reducing the quantity of steel and making the solution economically more attractive. Shallow depth also means that high approaches are not necessary, thereby reducing the costs of the approaches. Also, lighter structure ensures all round economy—in fabrication, transportation and erection, as well as in the costs of the foundation. One other advantage of the box girder shape is that, even with relatively thin skin plates, it provides considerable torsional strength of the structure against unequal traffic loadings. While use of thin plates, in turn, makes these plates vulnerable to buckling failure, the problem can be remedied by providing longitudinal stiffeners in addition to transverse stiffeners and diaphragms. Moreover, with the advancement in the methods of structural analysis by means of electronic computers, better understanding of the behaviour of the structural form, development of sophisticated welding technology and introduction of modern weldable quality steel, it has now been realised that the steel box girder system can be designed to adequate strength and also provide a viable and really competitive alternative to concrete.

The Second World War caused heavy damage to many bridges in Europe. Post-war reconstruction activities for these bridges saw concrete, particularly prestressed concrete, occupy the centre stage of these activities. During this period concrete box girder shape was used extensively in many bridges. However, the basic concept has remained practically unchanged over several years. The advantages of steel box girders were soon recognised. It was in steel as a construction material that human ingenuity and innovative spirit finally found the opportunity to develop a variety of configurations for different circumstances. Many steel bridges using this concept were built in Europe, particularly in Germany. Research and development in the area of box girders continued in various parts of the world. Thus, the world's first box girder with aerodynamic profile was developed by Sir Gilbert Roberts of Freeman Fox and Partners, London, and used in the first Severn crossing in 1966. Apart from its functional superiority to withstand wind load, the aerodynamic profile of the box girder provides remarkable elegance to a bridge and makes it aesthetically so much pleasing.

One interesting example of the use of steel box girder is in the widening scheme of the Auckland Harbour bridge, New Zealand. The bridge with a truss configuration was built in 1959. By 1964, it was felt that the bridge needed widening to accommodate increased traffic density. To achieve this objective, it was decided to add two continuous box girders on either side of the existing bridge. Use of box girder made it possible to reduce the depth of the new structure and match the profile of the existing truss configuration of the bridge. The uncluttered look of the steel box girder further illustrates how this form can be utilised with advantage from structural as well as aesthetic points of view.

The properties of box girders were gainfully used in the three central spans of the Rio Niteroi bridge across Guanabara Bay, Brazil, built in 1974. Although most of its 8 km length was built with concrete, the 283 m wide navigational channel presented a unique challenge. This central span at this location had to be high enough to provide 61 m clearance from the navigational aspect and at the same time aviation demands from the nearby Santos Dumont airport precluded use of towers to support the bridge structure in the case of a cable suspended bridge solution. Three steel twin box girder (200-300-200 m) spans were considered to be the appropriate solution to bridge the navigational gap. These are only a few examples of the use of steel box girders to satisfy specific requirements and steel box girders continue to inspire the present generation of engineers and planners in providing solutions for many bridging problems.

The evolution of steel box girders from their humble beginning to the highly sophisticated futuristic form of today has not at all been trouble free. In fact, like suspension bridges, they too had to discover their limiting factors through disasters. In the 1970s, within a short time, four box girder bridges collapsed in different parts of the globe, drawing the attention of the world to the serious drawback in the system. The four affected bridges were in Vienna over the Danube, in Milford Haven, Wales, in Melbourne over the Lower Yarra River and the fourth in Germany over the river Rhine. "The research that followed these failures clarified the buckling behaviour of stiffened plates under complex stress pattern . . . and effects of . . . initial geometrical out-of-straightness and of welding residual stresses" [4]. Since these failures, however, in the light of the research that followed the disasters, the steel bridge design and construction standards were updated to address the problems associated with box girders and many of the long span bridges which have thereafter been built around the world have used box girder shapes as their deck system. This again is a tribute to human spirit to face a challenge and solve it. As Martin Hayden rightly said : "whether as a simple beam, a cantilever, a cable stayed deck, or forming the roadway for the longest suspension bridge, the steel box girder is undeniably the bridge for the present day." [2]

1.10 CABLE STAYED BRIDGES

Cable stayed system was first used in 1867 by John A. Roebling in the Brooklyn bridge across East River in New York in addition to the traditional suspension system. Stayed support system was also used in the Albert bridge, built in 1873 over the river Thames in London. This again was a hybrid structure, part cantilever, part suspension and part cable stayed bridge. The idea of combining suspension system with stays was again tried in 1938 by Dischinger for a proposal of a bridge to be built across the Elbe River in Hamburg, Germany. In this system, Dischinger proposed that the central part of the span was to be carried by a suspension system, while the outer parts by stays radiating from the pylon top. A similar system was proposed for some other bridges as well, but was never adopted for actual construction. However, it was the German engineers who led the field in developing the modern cable stayed bridges immediately after World War II. Thus, the first modern cable stayed bridge, namely the Stromsund bridge in Sweden, was constructed by the German engineers in 1956. The bridge had a main span of 183 m flanked by two side spans of 75 m. The bridge used a two-plane cable system with two sets of portal type pylons located at two piers. Two pairs of stays radiating from each pylon top supported the deck on either side. The deck consisted of two continuous 'I' shaped plate girders supported by these inclined stays. The poor torsional stiffness of this system allowed the forces acting on the bridge deck to be distributed transversely onto the two cable planes.

Stromsund bridge was followed by a number of cable stayed bridges to be built in Germany with significant variations in structural as well as visual arrangements. The Theodore Hauss bridge in Dusseldorf was the first cable stayed bridge to be built in Germany. The bridge had two distinct variations from the Stromsund bridge. First, the deck system was supported at the two edges of the deck by means of a number of parallel cables from four towers, an arrangement commonly know as 'harp' construction. Secondly, the towers were free standing, instead of the portal type as in the Stromsund bridge. The Severins bridge over the Rhine at Koln was the third cable stayed bridge to be built in Germany. Opened in 1960, it is an interesting variation from the two preceding bridges. It had a single 'A' shaped tower of 60 m high to support two assymetric spans of 120 m and 300 m, employing several transversely inclined radial stays emanating from the top of the tower. The stiffening girder consisted of two box girders connected by an orthotropic steel deck. In decades to come, many more cable stayed bridges were built in Germany, as also in many other parts of the world.

The knowledge and experience of the post-war aircraft industry in the field of torsional behaviour of thin walled closed sections inspired the bridge designers to perceive and foresee the vast potential of thin walled box sections to provide an answer to the various torsional problems being faced with the design of box girders for bridge decks. Introduction of advanced computer aided design in the 1960s spearheaded further studies in the fields of orthotropic steel plates and box sections. It was soon realised that the box girders could be designed with considerable torsional strength with improved capacity of lateral distribution of concentrated loads over the entire width of the bridge deck. This would allow the deck to be supported by single plane stays located at the centre of the roadway, instead of the conventional two-plane stays. The bridge across the Elbe in Hamburg was built in 1962 following this single-plane stay system. The deck for this bridge consists of a wide stiffened steel plate supported on a central rectangular box girder along the longitudinal axis of the bridge and with an outer longitudinal plate girder at each edge of the deck.

The idea of aerodynamic configuration of decks in long span suspension bridges was considered to be equally relevant to cable stayed bridges. The first application of a trapezoidal box girder in a major cable stayed bridge was in Leverkusen bridge across the Rhine (1964). Many more cable stayed bridges with different configurations of trapezoidal box girders have since been built in several parts of the world. In fact, trapezoidal shape has now become almost a standard form for box girders.

In a cable stayed system, the inclined cables support the continuous deck system at intermediate locations. These are, therefore, required to be rigid enough to provide vertical supports to the deck system in a manner similar to that provided by intermediate piers. Since the conventionally spun cables had a propensity to unwind with load and were found to stretch during service conditions, it was initially observed that the stay cables were unable to provide the required rigidity to support the deck system satisfactorily. This problem was largely resolved by intensive research in this direction which resulted in the development of 'locked coil' cables technology, whereby the coils of steel are combined in a manner which restrict the tendency to unwind with load.

Computerised analysis also opened up a new horizon in structural engineering. Engineers could now address various problems with greater confidence without taking recourse to unrealistic assumptions as pin-jointed connections even where the connections are actually riveted, bolted or welded joints. Highly indeterminate structural systems could now be solved easily by using computers. It, therefore, became increasingly easier to develop cable stayed bridges with imagination—using towers of innovative shapes and forms, decks of diverse structural systems, and stays of varying multi-cable arrangements. Thus, at one end of the spectrum we have the multi-cable Friedrich Ebert bridge across the Rhine in Bonn which was completed in 1967 with a single cable plane containing as many as 80 cables, while at the other end we have the Wye bridge (1966) and the Erskine bridge (1974), both in Britain, employing only one cable on either side of the two towers along a central vertical plane.

The Second Hooghly bridge in Calcutta, India, with a span of 457 m was planned to be the world's longest span at the time of its conception in 1971, but when completed in 1992, it trailed behind many more bridges that were completed in the intervening period. The Tatara bridge in Japan, built in 1999, has the world's longest cable stayed span of 890 m, followed by Pont de Normandie (1995) in France with a span of 856 m. Sutong bridge, now under construction at Yangtze River estuary in China, is unveiling a new era of cable stayed bridge construction. It will be a bridge with double leg pylon and double cable plane and when completed, will be a world record for its main span of 1,088 m. Apparently, the limit of such structures is not yet in evidence.

REFERENCES

[1] Bennett, D. 1999, *The Creation of Bridges*, Chartwell Books Inc, New Jersey, USA.

[2] Hayden, M. 1976, *The Book of Bridges*, Marshall Cavendish Publications Ltd., London, UK.

[3] Gimsing, N.J. 1997, *Cable Supported Bridges*, John Wiley & Sons Inc, Chichester, UK.

[4] Chatterjee, S. 1991, *The Design of Modern Steel Bridges*, BSP Professional Books, Oxford, UK.

[5] Liebenberg, A.C. 1992, *Concrete Bridges: Design and Construction*, Longman Scientific and Technical, Harlow, UK.

[6] Francis, A.J. 1989, *Introducing Structures*, Ellis Harwood Ltd., Chichester, UK.

[7] Shirley-Smith, H. 1965, *The World's Great Bridges*, Harper & Row, Publishers, New York, USA.

[8] Cottrell, A.E. 1928, *History of Clifton Suspension Bridge*, Clarks Printing Services Ltd., Bristol, UK.

Chapter 2

Learning from Failures

2.1 INTRODUCTION

The idea of learning the behaviour of structures from failures has always been a normal practice of engineers in the past. Galileo is believed to have employed case studies of failures for advancing and developing some of his pathbreaking theories. This practice was continued for centuries when builders almost always used precedents—both successes and failures—to buttress their design ideas and decisions. John Roebling studied the failures of many suspension bridges during the early 19th century and analysed them to understand the forces that must be considered to build a successful bridge. Use of mathematical theories in engineering designs is a comparatively recent development. The standard codes of practice of recent times have indeed been developed through a tradition of factor of safety against *failures*. Case studies of failures—both in the field and in the laboratories—have thus become recognised as a source of knowledge for developing structures of the future.

Of late, with the increasing technological progress in bridge building, the need for recording case histories of failures has assumed greater importance. However, in these days of litigations and arbitrations, there are many hindrances for candid discussions of engineering failures. As a result, frank analysis of errors of design and/or execution are rarely found in current technical literature.

2.2 TYPES OF FAILURES

Failure of a bridge structure can be broadly classified into two categories:

- Catastrophic and sudden failure, where the structure collapses during erection or in service. This type of failure may be life-threatening and generally attracts the attention of the public and the media.

- The second type of failure is not sudden or catastrophic, but may affect serviceability requirement. This type includes excessive deflection of the girder, premature deterioration or durability problems, vibration, etc. These failures are less attractive to public and media attention.

2.3 CASE STUDIES

Case studies of some of the landmark failures in the history of bridge engineering, their causes, as also the lessons learnt from such failures, have been described in the following paragraphs.

2.3.1 The Tay Bridge, Scotland, 1879

The first Tay rail bridge was completed in 1878. It was designed by Sir Thomas Bouch who was, as was the then practice, also entrusted with the responsibility of construction and maintenance of the bridge. Sir Thomas Bouch was, at that time, one of the most renowned living bridge engineers in Britain. He was also commissioned to design the famous Forth railway bridge.

Tay bridge carried the main line rail traffic between Edinburgh and Dundee. It was a single track railway bridge consisting of 85 wrought iron riveted lattice girder spans, of which the main 13 girders (11×75 m and 2×69 m spans) across the navigation channel were 8.2 m high through type girders. The balance 72 girders were deck type girders of smaller spans. A schematic site plan is shown in Fig. 2.1.

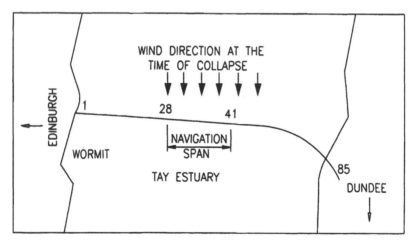

Fig. 2.1 Tay rail bridge—site plan [7]

The navigation channel girders were 26.8 m above high-water level and were supported on piers each consisting of six hollow cast iron columns, 0.38 m and 0.46 m in diameter, braced together by wrought iron bracings (Fig. 2.2). The bridge was at that time the longest in the world. It had a fairly standard design without any particularly new ideas or construction techniques to boast about.

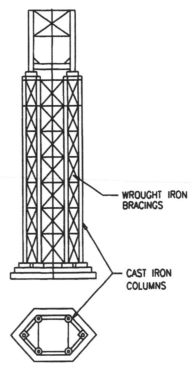

WROUGHT IRON
BRACINGS

CAST IRON
COLUMNS

Fig. 2.2 Tay rail bridge–typical iron piers supporting the navigation spans [4]

In December 1879, barely 18 months after it was officially opened to traffic, disaster struck the bridge. On the stormy night of 18 December 1879, a train of one engine and six carriages crossing the bridge from south to north was doomed when the structure of the navigation channel collapsed. There were no survivors and all 75 people aboard the train perished.

An expert committee, appointed by the Board of Trade to inquire into the causes of the disaster, interrogated several witnesses and many other persons connected with the bridge, including Sir Thomas Bouch. It came to the conclusion that the immediate cause of the disaster was the failure of the wrought iron bracing ties and the cast iron lugs on the piers to which the bracings were connected. The quality of the cast iron used in the girders and the piers suffered from bad workmanship and lack of adequate supervision. The construction of the piers was also severely criticised; they had been modified from an original design which had assumed a rock foundation, whereas solid rock did not exist. It was also revealed at the inquiry that the maintenance during the brief life of the bridge was far from satisfactory.

The other major lapse in the design was that no special provision was made for the full effects of the wind. The wind velocity along the Tay estuary at the time of the disaster was close to 100 miles per hour (161 km/hr). In those days there was no British Standards to guide a designer on the wind speed to be considered in the design. Only, the wind tables prepared by John Smeaton (about 120 years before !) was available to Bouch for calculating the wind pressure for the design of the Tay bridge. Figures of wind pressure, according to these tables were : 6 lbs per sq. ft. for 'high winds', 8 to 9 lbs per sq. ft. for 'very high winds' and 12 lbs per sq. ft. for 'a storm or tempest'. Significantly, at that time, engineers in France and USA were allowing 55 and 50 lbs per sq. ft. respectively, in their designs. Had these figures been considered in the design, the Tay bridge disaster could possibly have been avoided.

As a sequel to the Tay disaster, Sir Thomas Bouch was disgraced and was forced to resign as the designer of the Forth railway bridge. His health broke down and he died a year later.

> **Lessons learnt:**
>
> - Brittle material should not be used in bridge work. Thus, this disater rang the death knell of iron as a construction material for bridges, giving way to steel, which, by then, had firmly established itself to replace iron for future bridges. A new era was ushered in.
> - Site-specific wind load should be used in the design, including wind effects on windward and leeward girders.
> - Quality control of materials is an important aspect of any bridge work.
> - Adequate supervision during erection is imperative.
> - Communication between team members is important.
> - Post-construction maintenance is very important.

2.3.2 The Quebec Railway Bridge, Canada, 1907 and 1916

The Quebec railway bridge over the St. Lawrence River, Canada, has the inglorious distinction in engineering history as being the only bridge structure which suffered major collapse on two occasions during its erection. The first was in 1907, when the original structure collapsed during construction, mainly due to design fault, and the second one was in 1916, also during construction, when the suspended span of the re-designed bridge collapsed while being lifted into position.

The first structure was designed as a cantilever steel bridge to carry two railway tracks. It was to be the world's longest cantilever type bridge with a central span of 549 m, 27 m longer than that of the Forth railway bridge which was constructed in Scotland in 1890.

Theodore Cooper, a distinguished bridge engineer, with many years of experience in bridge building, was assigned the responsibility of designing the bridge. At that time, Cooper had only very limited staff to assist him. At 70, he was already at the end of a distinguished career and did not have enough resources to build a team to supervise such a big project. This lack of technical manpower had an adverse effect on the quality control of the various activities involved, such as, proper checking of the design, adequate supervision of shop and site activities, etc. He was physically not well also and visited the fabrication shops thrice only. He could not visit the bridge site even once !

In February 1906, when the design and detailing of the members and connections were completed, it was detected that the dead weight of the bridge was grossly under-estimated in the design, causing about 10% to 25% increase in the stresses in the members. Notwithstanding this revelation, work was allowed to proceed; the end spans were built and the south cantilever arm was completed. Cantilevering of the south suspended span was also taken up. In August 1907, when the south cantilever span was nearing midspan, it was detected that the cantilever was starting to deflect. Also, the web plates of the lower chord members adjoining the tower showed signs of buckling. Cooper was immediately informed about this development and was urged to visit the site. The situation continued to deteriorate. On 27 August, the resident engineer McClure sent a telegram to Cooper that the erection was being stopped until they heard from him. Cooper realised the seriousness of the situation and sent a cable (which was not delivered to the site) ordering work to be halted 'until after due investigation of the facts' [4]. He was probably under the impression that erection would be suspended till the matter was sorted out. Next day, the resident engineer stopped work for the day and left for New York to meet Cooper for discussions. Unfortunately, however, a contractor's superintendent, unaware of the potential danger of the situation, sent his men to work on the cantilever arm as usual on 29th August. Figure 2.3 shows the state of construction of the south half of the bridge on that day. Although, by that time deflection had increased visibly, the erection crane was moved out to the next panel for further erection of members. In the afternoon of that day, the incomplete cantilever arm of the south span collapsed and crashed in St. Lawrence River, signalling one of the worst disasters in the history of bridge construction.

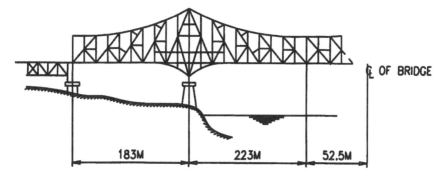

Fig. 2.3 First Quebec Bridge. State of construction of south half on 29 August 1907 [4]

According to the report of the Royal Commission which carried out the investigation of the disaster, the collapse during construction of the Quebec rail bridge was mainly 'due to the spliced joint of the compression chord being allowed to remain partly open and not riveted [1]. Also, 'it had only two rivets when it should have had eight according to the inquiry team. In fact, it had only 30% of the strength of the compression member it was connecting' [2]. Besides the above defects in the connection, the main compression chords were not designed properly; lacings connecting the four web members comprising these chords were not adequate to restrain the webs from bending. Additionally, the dead load of the bridge was grossly under-estimated causing about 10% to 25% increase in the actual stresses. The allowable unit stresses considered in the design were also too high. To sum up, errors in design, construction flaws, as also lack of communication amongst the various arms of the project team, were responsible for the failure of the bridge during construction. In any case, even if the bridge was erected successfully, it could not have been strong enough for the traffic it was intended to carry.

One other point became apparent after the Quebec disaster, namely that the empirical design rules for compression members urgently needed reviewing and updating. It was recognised that more scientific study was needed to understand the mechanism of buckling failure of long compression members which was quite different from the failure mode of tension members. The disaster also highlighted the importance of proper detailing of bridge structures including designing the details of joints and splices.

Nine years later, in 1916, erection of the redesigned second Quebec bridge—also of cantilever construction—with K trusses, was nearing completion when a second failure occured. The suspended span was being lifted from floating pontoons by jacking arrangement, when, during the third lift, the pontoons were floated clear, leaving the span suspended in the air from the lifting hangers connected to the cantilever arms. Work was going as per schedule without any hitch. What was left was only repeating the mechanical work till the span was fully lifted into position. Suddenly, one of the jacks failed with the steel truss 4.6 m in the air and the span slid off into the river.

Investigations carried out almost immediately after the accident revealed the 'initial failure occurred at the south-west corner, resulting in that corner of the span spilling off its supporting girder' [1]. The south-east corner fell next, followed by the north-east and north-west corners as the entire suspension span fell into the river.

The rebuilt Quebec bridge finally opened to traffic in 1917 and became the world's longest span beating the Forth rail bridge by 27 m. Through its construction phase, it suffered two collapses and claimed 87 human lives.

Lessons learnt:

- On completion of design work, the actual dead load of the structure should be ascertained and compared with the load assumed in the design. If necessary, the design must be re-worked.
- Study to understand the mechanism of buckling failure of long compression members and review of existing rules were needed.
- Adequate supervision during erection is necessary.
- Proper detailing of the connections in a bridge is important.
- Adequate communication amongst team members is important.

2.3.3 The Wheeling Bridge, Ohio, USA, 1854

The first suspension bridge at Wheeling over the Ohio River, West Virginia, was built by Charles Ellet in 1848 with the main span of over 305 m. It had a deck of 6 m width, with only the parapets serving as its stiffening girder. On 17 May 1854, six years after it was opened to traffic, the bridge was destroyed by wind. David Steinman recounts in his book *Bridges and Their Builders* a remarkable eye witness account of the collapse. According to the eye witness, after 'enjoying the cool breeze, and the undulating motion of the bridge, the visitors saw the whole structure heaving and dashing with tremendous force ... rose to nearly the height of the tower, then fell, twisted and writhed.... At last there seemed to be a determined twist along the entire span, with about one half (of the bridge deck) being reversed and down went the immense structure from its dizzy height to the stream below' [2]. Fortunately there was no casualty.

The Wheeling bridge was designed with normal loadings such as its own weight (dead load), loadings from bridge traffic of horses, carts, wagons, people, and cattle, as also normal wind pressure. However, the vibration pattern set up by the wind on the deck was not considered in the design. In fact, the principles of aero-dynamic stability were little understood by the profession in those days. Later, while developing his designs of suspension bridges, John Roebling, a great engineer as he was, studied the Wheeling bridge collapse and seemed to have somewhat comprehended the little-known aero-dynamic effect and in his wisdom, provided cable stays between the bridge deck and suspension cables to safeguard against aerodynamic instability.

The Wheeling bridge was later rebuilt by Charles Ellet with a stiffer bridge deck and also with additional transverse stays and was re-opened in 1860.

Lessons learnt:

- The principles of aerodynamic stability should be studied for cable supported bridges.

2.3.4 The Tacoma Narrows Bridge, USA, 1940

Nearly a century after the collapse of the Wheeling bridge, the Tacoma Narrows bridge was destroyed due to the identical reason, namely, aero-dynamic instability of the bridge deck. The engineering fraternity had, perhaps, forgotten about the lesson at Wheeling, or had, till then, grossly under-estimated the importance of aero-dynamic study of bridge decks of suspension bridges.

When it was opened to traffic on 1 July 1940, the Tacoma Narrows bridge over the beautiful Puget Sound in Washington State, was considered as one of the most graceful, slender and streamlined suspension bridges in America. This bridge was designed by Leon Moissieff who was also responsible for the design of the Manhattan suspension bridge and was one of the consultants on the Golden Gate and Oakland Bay bridges which were already in use. He was considered as one of the leading experts in suspension bridges in America at that time. Tacoma Narrows bridge, with a main span of 854 m, was the third longest suspension bridge, then constructed. It was 11.9 m wide and was stiffened along its sides by plate girders.

From the very beginning the bridge showed unusual undulations in certain windy conditions. However, the motion was limited, and was not considered unsafe. Consequently, though fondly nicknamed 'Galloping Gertie', traffic continued over the bridge. However, four months after it was opened to traffic, on 7 November and after an hour in a moderately high wind at about 68 km/hr, the span suddenly went 'into an alarming series of rolls and pitches. One side of the road dipped and the other side rose, as a ripple wave travelled the length of the span' [2]. After some time, some of the suspension hangers gave way and first a length of the deck of about 180 m fell into the river, followed by the remainder of the deck.

This spectacular bridge failure, fortunately without any casualty, was captured almost in its entirety in a movie film by an amateur photographer. This film provided invaluable data about the exact behaviour of the bridge during collapse for experts to study subsequently. Although the designer, Leon Moissieff was intitially blamed for the collapse, David Steinman exonerated him. While summing up the tragedy, he held the entire profession responsible for the disaster. 'It is simply that the profession has neglected to combine and apply in time, the knowledge of aerodynamic vibrations with its rapidly advancing knowledge and development of structural design,' he said [2]. It appears that in the euphoria of successive successes in the design of long span suspension bridges, the lessons of the failure of the Wheeling suspension bridge, which collapsed almost for the same reason a century back, was forgotten and completely overlooked by the profession.

After the collapse of Tacoma Narrows bridge, American engineers made close studies of the aerodynamic effects on a scale model of a bridge dynamically similar to the failed bridge. When tested in the wind tunnel, the model reproduced the 'gallopping' and 'twisting' oscillations very much similar to those observed in the actual structure. Studies also revealed that vortices were developed both above and below the deck, at intervals which depended on the wind speed. Figure 2.4 illustrates the phenomenon. 'The pressure differences set up by the vortices tended to amplify the vertical movement, especially if they were created at intervals which tied in with the period of vertical oscillation of the bridge itself' [4]. The wind speed itself was a critical factor. A.J. Francis [4] describes vividly the wind effect: '... at a wind speed of 62 km/hr the vertical movements were moderate (about a metre) with the two suspension cables in phase. But, when the wind speed changed to 68 km/hr the period of oscillation suddenly increased three-fold and the vertical movements leaped up to nearly 10 m. Worst still, the oscillation of the two suspension cables were now out of phase, one going up while the other went down', leading the deck to distort abnormally in the transverse direction. The open deck was not torsionally stiff enough to resist such distortion and consequently gave way leading to the disaster.

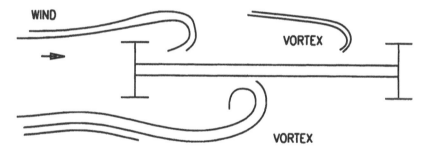

Fig. 2.4　Wind effects on Tacoma Narrows bridge [4]

The engineering fraternity in America and other countries took note of the disastrous experience of Tacoma and made it almost mandatory to include aerodynamic investigation for the design of all major cable-supported bridges built subsequently. Meanwhile, many existing suspension bridges, including the Golden Gate Bridge, were checked for aerodynamic stability and retrofitted with stiffening trusses wherever found necessary.

Lesson learnt:

• For any major bridge on flexible supports like cables, it is obligatory to examine the aerodynamic effects on the bridge structure.

2.3.5 The King's Bridge, Melbourne, Australia, 1962

Opened in 1961, King's bridge consisted of a large number of simply supported welded plate girder spans carrying a highway over the river Yarra, in Melbourne, Australia. On 10 July 1962, barely after 15 months of use, 4 girders in one of the end spans failed due to fracture after the passage of a 45-ton vehicle, although it was well under the permissible weight. Fotunately, complete collapse of the plate girders was prevented by walls underneath which had earlier been built to enclose the space under this span. As a result, no one was injured.

The bridge consisted of a deck system of reinforced concrete slab supported on four welded steel plate girders of 30 m span. Typical details of the plate girders are shown in Fig. 2.5. Each plate girder was built up by welding two flange plates and a web plate. An additional flange plate was welded to the underside of each bottom flange in the central part of the span to cater for increased bending moment at this location. The steel used was of medium-high strength quality with somewhat lower ductility.

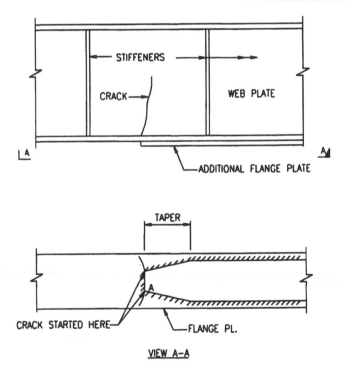

Fig. 2.5 King's bridge, Melbourne [4]

It was revealed during investigations that, in the girders which failed, the cracks in most cases started in the main bottom flanges close to the transverse welds at the ends of the additional flange plates (location 'A' in Fig. 2.5). Investigations also revealed that some of these cracks had been there even before the girders were painted, i.e., in the workshop itself! The cracks apparently progressed upwards through the flange plate towards the web plate—in many cases half-way up the web. In several cases the flanges were completely separated. In some cases the entire girders were severed. It was only because of the presence of the walls underneath, that a major disaster was avoided. As it came to be known during investigation, many of the other girders also suffered similar distress. It was only providential grace that the failure did not take place at a busier area over which the bridge crossed, avoiding a catastrophy involving heavy casualty.

Investigations also revealed that the 80 mm long transverse welds at the ends (location 'A' in Fig. 2.5) were welded last, after the longitudinal welding had been completed. There was, thus, a complete restraint against contraction when the transverse weld 'A' was deposited. Consequently, this weld was unable to shrink freely, resulting in high residual tensile stress which made the area prone to transverse cracking. Apparently this aspect was overlooked during detailing and also while finalising the welding sequences at the workshop.

The other aspect which contributed to the failure, was the lack of understanding about the nature and behaviour of the high strength steel used in the bridge. Presence of manganese and chromium and somewhat higher percentage of carbon increases the tendency of 'notch-brittleness' in high strength steel and also makes the metal adjacent to the weld somewhat hard and brittle on cooling and consequently, prone to crack development. Necessity of special care in welding, particularly pre-heating, use of special electrodes, etc., were not realised by the participating agencies.

Lessons learnt:

- Proper care should be taken in the design and detailing of welded steel bridges to avoid cracking.
- Special care, such as pre-heating, use of special electrodes, etc., should be considered for welding high strength steel material which may be prone to brittle fracture.
- Sequence of welding must be carefully examined before taking up fabrication.

> • Adequate supervision and inspection at every stage is an important
>   aspect in fabrication of bridge steelwork.

2.3.6 The Westgate Bridge, Melbourne, Australia, 1970

On 15 October, 1970 at 11.50 am, the 112 m long west cantilever section of the main approach span of the Westgate bridge, under construction over the Yarra River in Melbourne, suddenly crashed, killing 35 people and injuring many more. This happened only four months after a similar box girder span buckled and collapsed in Milford Haven, Wales, on 2 June 1970. There were two more box girder failures around the same time, one in Vienna over the Danube and the other in Koblenz over the Rhine. These and several other box girder failures drew the attention of the engineering fraternity the world over, which led to a thorough re-examination of the design principles of thin plate box structures.

The main approach spans of the Westgate bridge were welded steel girders of tropezoidal box section (Fig 2.6). These tropezoidal boxes were formed by thin (maximum 22 mm thick) plates. The top and bottom flanges and the inclined outer webs were framed up with the two inner vertical webs and stiffened by stiffener plates and cross diaphragms at intervals. Transverse stiffeners were placed at 3.2 m apart. The box girder was to act compositely with the concrete deck when completed. There were seven cantilever box units, each of 16 m length to make up the 112 m length of the span. To make the construction easier, each box unit was pre-fabricated in two halves. Each half span (comprising of seven half boxes connected by bolts) was to be lifted on the piers in two halves and spliced along the centre line by means of HSFG (high strength friction grip) bolts. In the span that failed, because of the unsymmetrical cross section of each half, a mis-alignment occurred. The central vertical deflection of the two halves differed by over 110 mm. The free inner edges of the top flanges were inadequately stiffened and were incapable to cater for the compression forces during erection. These also showed signs of distress and suffered local buckling. 'In an effort to get the buckle out, some of the transverse bolts joining these boxes were removed. This so weakened that side of the span that its weight was partially transferred to the other half span, which could not sustain the additional load and the complete structure collapsed by buckling at mid-span.' [4].

Westgate bridge and other box girder bridge collapses that took place around that period underscored, yet again, the importance of good

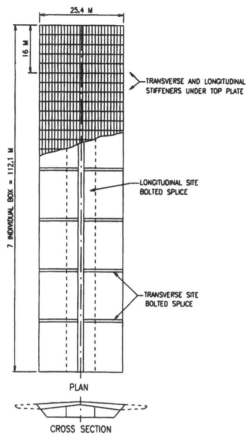

Fig. 2.6 Westgate bridge [4]

communication between the designers and the construction personnel at site. Lack of adequate standards on design and construction of thin plated box sections was also highlighted by these tragedies. A British committee, headed by Dr. Alexander Merrison, made comprehensive investigations into these box girder failures and made wide ranging recommendations. The prevailing standards and codes related to thin-walled box girders were reviewed and adjusted particularly in respect of buckling strength of thin plates.

Lessons learnt:

- Study to understand the behaviour of thin-walled box girders and review of existing rules were needed

2.4 CONCLUSION

In the preceding section, case studies of some landmark bridge failures since the late 19th century and the lessons learnt therefrom have been discussed. There is no doubt that such case studies have helped designers of subsequent generations to avoid similar flaws in their designs. Thus, it is imperative for bridge engineers of today to know the history of failures, lest they unknowingly repeat the mistakes of the past. It was true in Galileo's time, it is true now, and will also be true in the future. Therefore, it is necessary that every major failure is dispassionately investigated and analysed, and the results widely publicised for the benefit of the profession.

REFERENCES

[1] Hamond, R. 1956, *Engineering Structural Failures*, Odhams Press Ltd., London, UK.

[2] Bennett, D. 1999, *The Creation of Bridges*, Chartwell Books Inc, New Jersey, USA.

[3] Dowling, P.J., Knowles, P.R., and Owens, G.W. (eds.) 1988, *Structural Steel Design*, The Steel Construction Institute, London, UK.

[4] Francis, A.J. 1989, *Introducing Structures*, Ellis Horwood Ltd., Chichester UK.

[5] Carper, K.L. 1996, *Construction pathology in the United States*, Structural Engineering International, February, 1996, International Association for Bridge and Structural Engineering (IABSE), Zurich, Switzerland.

[6] Petroski, H. 1995, *Case histories and the study of structural failures*, Structural Engineering International, November, 1995, International Association for Bridge and Structural Engineering (IABSE), Zurich, Switzerland.

[7] Martin, T. and Macleod, I.A. 1995, *The Tay rail bridge disaster*, Civil Engineering, May, 1995, Proceedings of the Insitution of Civil Engineers, London, UK.

[8] Ghoshal, A. and Ghosh, U.K. 2002, *Learning from failures*, Paper presented in the International Seminar on Steel and Composite Bridges, organised by Indian Institution of Bridge Engineers, Mumbai, India, April, 2002.

Chapter 3

Design Philosophies and Safety

3.1 INTRODUCTION

One of the prime considerations for a structural designer is to ensure that the structure is safe. To achieve this, it is imperative that the strength of every element comprising the structure must be *at least* equal to or *preferably* greater than the force expected on it over the life of the structure. Consequently, in most cases, some reserve strength (over and above the required strength) is included in the structure. Traditionally, this reserve strength is known as the factor of safety and is expressed as a ratio of the strength of a structure or member to the force expected on it. In the case of steel, which has a well-defined yield point, the factor of safety could be related to deformation (yield strength) or to failure (ultimate strength).

3.2 UNCERTAINTIES

In spite of the designer's best efforts, it is not easy at the design stage to assess with certainty the degree of safety that the proposed structure would possess in actual practice. This is because of the presence of many uncertainties which govern the factor of safety. As a consequence, absolute safety of the designed structure can hardly be guaranteed—only a low risk of failure can be assured. Some of these major uncertainties are discussed in the following paragraphs.

3.2.1 Material Strength

Structural materials vary in quality and strength. Even steel which is manufactured in controlled conditions shows large variations in strength from the average value. In fact, no two specimens of the same quality of steel will have identical strength. Figure 3.1, for example, shows the histogram of the results of 73 tensile tests on structural silicon-steel of supposedly the same strength used for the towers of the Golden Gate bridge in San Francisco, USA. The ordinates in the figure represent the frequencies in

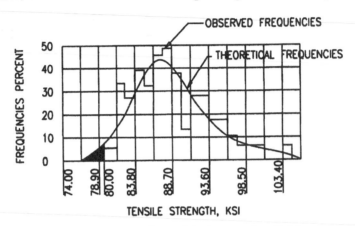

Fig. 3.1 Tensile test results of steel in towers for Golden Gate bridge (Reproduced from: A.M. Freudenthal, The Safety of Structure, Trans. ASCE, Vol. 112, p. 125, 1947, appearing in [5])

respect of the different test results of tensile strengths. These test results are shown as abscissas. It will be noted that the range of strength is extremely wide—from 80.3 ksi (553.8 N/mm^2) to about 104.5 ksi (720.5 N/mm^2). This particular set of results has been represented in the figure by a smooth curve which closely follows these results. This probability curve which fits the experimental data is termed 'theoretical frequency distribution curve'. One of the properties of a probability curve is that the total area under this curve is proportional to the total frequency. Thus, the area lying under the curve and between any two ordinates, indicates the probablity that the tensile strength will lie between the corresponding abscissas. Other physical properties such as yield point, modulus of elasticity and weight show similar distribution of values about a mean. When a set of results is disposed symmetrically about a mean value, and takes a bell-shaped form, the curve is known as normal distribution curve (Fig. 3.2).

3.2.2 Loads

Loads in a bridge are broadly of two kinds, viz., dead loads and live loads. The former include not only the load from the structure itself, but also the 'fixed' loads which the structure has to support as a part of permanent fixtures, such as railway tracks, ballasts, footways, handrails, etc. These

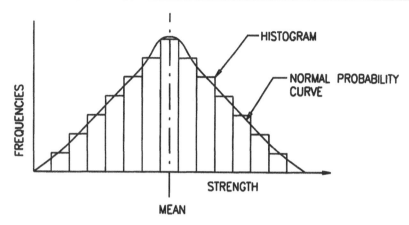

Fig. 3.2 Normal distribution curve

loads can be assessed with reasonable accuracy. The second kind of loads, which are called live loads, include vehicle loads (in road bridges), axle loads (in railway bridges), crowd loads and similar loads. A bridge has also to resist other occasional loads such as wind pressure, seismic forces, etc. In case of steel bridges, which have a comparatively slender and lighter structure than concrete bridges, the effect of live loads becomes more significant. Generally, design codes and national standards specify the live loads to be considered for the design of bridges. However, it cannot be guaranteed that the theoretical live loads will never be exceeded during the design life of the bridge. Thus, a gusty wind causing pressure much above the design pressure, excessive snow loads or an overloaded lorry not envisaged in the design codes, cannot be altogether ruled out. Usually, most of the liveloads follow statistical principles. As for example, Fig. 3.3 shows

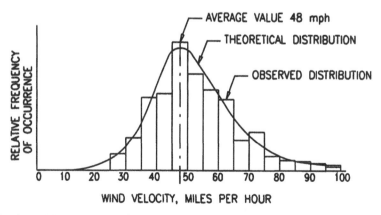

Fig. 3.3 Maximum wind speeds in New York over period 1884–1950 [1]

records of maximum wind speed measured over 5-minute periods in each month between 1884 and 1950 in New York at a height of 140 m. From this diagram, the probability of heavy winds can be determined mathematically. Similarly, if enough data are available, frequency diagrams for other loads such as snow loads, lorry loads, can also be plotted and the probability of excess loadings can be reasonably determined.

3.2.3 Analysis and Behaviour of the Structure

It was in the middle of the 18th century that scientific approach of analysis was first applied in the design of structures. Previously, design rules were largely dependent on traditions and experiments on models. By the second half of the 19th century, developments of various methods of theoretical analysis of structures made it possible for engineers to determine the internal forces of most structures with varying degrees of precision. Things have moved forward considerably since then. Nowadays, with the use of computers, structures of almost any configuration can be analysed with considerable accuracy. In spite of these advances, however, uncertainty about the strength of a completed structure cannot be altogether eliminated. This is primarily because the members of the structure are not likely to have exactly the same properties which were considered in the analysis. Also, the degrees of rigidity of joints assumed in the analysis may not have been achieved in actual practice.

3.2.4 Type of Structure

The risk of failure is not necessarily uniform for all components of a bridge structure. When a structure has, within itself, multiple load carrying mechanisms, in case of failure of one mechanism, the load is likely to be carried by another. On the other hand, if a structure has only a single load carrying mechanism, failure of a single element may cause collapse of the structure. As for example, in the case of a two-cable suspension bridge, if one of the hangers fail, the load would automatically re-distribute itself amongst the other hangers and the structure will still not collapse. However, failure of main cables on one side would almost certainly result in the collapse of the deck system. Therefore, the type of structure adopted for a steel bridge may contribute to the uncertainties which govern the factor of safety.

3.2.5 Corrosion and Fatigue

Atmospheric corrosion, stress corrosion and fatigue are the three primary causes of distress in steel bridges during their operational life. These conditions may bring about severe loss of strength of members at certain locations, leading to uncertainty about the overall strength of the bridge structure.

3.3 RELATION BETWEEN LOAD AND STRENGTH

Figure 3.4a shows a combined distribution curve for load and strength in respect of a bridge structure. The frequencies are represented as ordinates against loads and strengths which are shown as abscissas. Mean values of loads and strengths are represented as L and S. The upper tail of the distribution curve for load overlaps the lower tail of the distribution curve for strength. This overlap indicates that there is a probability that at some time, the structure may be subjected to an exceptionally heavy load beyond its strength. In other words, there is a risk of collapse. One method to avoid this situation is to impart more strength into the structure and shift the distribution curve for strength sufficiently away to avoid the overlap.

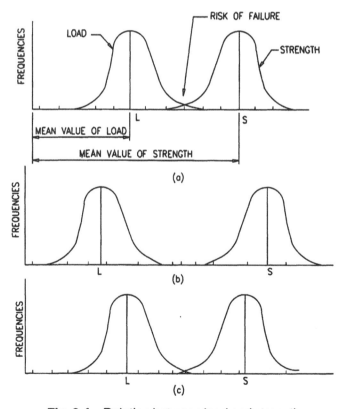

Fig. 3.4 Relation between load and strength

Figure 3.4b shows this scenerio. But this solution is likely to entail prohibitive additional cost. An alternative compromise solution is to shift the distribution curve of strength from the distribution curve of load to an extent, such that the area of overlap is small. In that event, the risk of failure, though not completely absent, is tolerable and may be accepted by the society. Figure 3.4c illustrates such a solution. In such a situation, although there exists a possibility that the structure may fail, the designers can still improve significantly on the statistical failure rate by applying their knowledge and understanding of the subject. Thankfully, it is revealed from experience that the actual failure rate is quite small and much lower than the statistically projected rate. In general, failures result from gross errors or negligence, rather than from random or systematic errors which form the basis of statistical analysis.

3.4   DESIGN METHODS

Basically, there are two methods for design of steel bridges:

- Allowable stress design
- Limit state design

Salient features of these two methods are discussed in the following paragraphs.

3.4.1   Allowable Stress Design

In this method (also known as permissible/working stress design) structural members are designed so that the unit stress caused by the design loads does not exceed a pre-defined allowable stress. The allowable stress is defined as a limiting stress (generally taken as the yield stress) divided by a factor of safety. This factor (of safety) caters for the acceptable risk of failure *vis-a-vis* the uncertainties discussed earlier in this chapter.

Structual analysis is done to obtain the actual stresses in all the members or elements for a number of specified combinations of loadings. These stresses are then checked against the pre-defined allowable stresses. These allowable stresses generally depend on load combinations. For example, allowable stresses for combinations of wind or earthquake forces with dead load and live load forces are usually larger than those for dead load plus live load forces only. This is to take into account the lower probability of wind or earthquake forces to occur concurrently with dead and live load forces. The allowable stresses corresponding to different combinations of loadings are generally specified in the national codes.

In the allowable stress design method, apart from structural failure, serviceability requirements such as deflection, vibration, etc., also need to be considered and properly checked. Guidelines for these conditions are also specified in the codes.

The main attraction of this method is its simplicity based on linear elastic theories. The entire structure is designed to fall well within the elastic range of the material of the components. Steel has a fairly well defined yield point, upto which the stress-to-strain ratio is relatively proportional and there is no permanent deformation. (In the elastic range, strain will return to zero if the load is removed.) Within such parameters, material of the components is considered to behave in a linear fashion and working stresses are calculated from linear elastic theories. Thus, stresses arising from different loads (such as, dead load, live load, wind load, etc.) can be added together.

Inspite of its inherent simplicity, however, the method has its share of irrationality also. First, in this method, only one fixed factor of safety is considered. Thus, the degree of uncertainties concerning the loads in terms of frequency or magnitude is ignored. In effect, the same factor of safety is applied for dead loads (which can be assessed with a fair amount of accuracy) and live loads (which vary quite widely and whose occurrence cannot be prediced with precision). This, indeed, is an unrealistic consideration.

Secondly, the factor of safety is applied on the yield stress determined on a test piece in tension. But in a steel structure all members are not necessarily subjected to only tension; some may be subjected to compression or to bending or even to combined axial load and flexure. Failure modes for tension, compression and flexure are not uniform. In case of tension, a member can be stressed upto its ultimate strength, while for compression and bending, the critical limit is buckling. Thus, the reserve strength in the three modes are not same, although the factor of safety is considered same.

Also, allowable stress method does not take into account the behaviour of a steel structure at failure. Steel has the ability to tolerate high theoretical elastic stresses by yielding locally and re-distributing the loads. Therefore, as some of the components of a structure reach the yield point, re-distribution of forces takes place amongst the other less critical components. This phenomenon is not considered in the fixed factor of safety concept in the allowable stress method.

Subsequent study and researches to overcome these shortcomings led to the development of plastic theory which, in turn, led to an alternative philosophy based on nominal loads and stresses used in the calculations in terms of statistical concepts. This philosophy, commonly known as 'limit state design' forms the basis of new generation codes.

3.4.2   Limit State Design

This approach was initially developed in Europe and has since been adopted in many countries of the world, albeit, with certain amounts of variations. It utilises the concept of plastic range for the design of structural members and incorporates load factors to cater for the uncertainties of loading conditions.

In this approach, limit state is defined as 'a condition which represents the limit of structural usefulness'. Therefore, the designer has to ensure that the structure has an *acceptable probability* against reaching a limit state throughout its life. It, therefore, implies that the probability of failure cannot be ruled out altogether and that no structure is absolutely safe.

Limit states may be grouped under two general categories:

Ultimate limit state

In this limit state, assessment is made of the total collapse condition which endanger the safety of people and/or require major or total reconstruction. This category includes strength (i.e., safety against yield, buckling, collapse, etc.), stability against overturning, fatigue fracture, and brittle fracture.

Serviceability limit state

This limit state represents the limit of acceptable performance in service and includes deflection, vibration and durability (e.g., corrosion).

It is obvious that ultimate limit states are more important between the two categories and necessarily attract larger safety factors to reduce the risk of occurrence to an acceptable level. Thus, by using limit state approach, a designer is able to set a higher probabilty against reaching the ultimate limit state than the other less important serviceability limit state.

In the limit state approach, partial safety factor format is generally used. In this format, safety factors for each condition are applied separately to the loads, and to the strength of the member, and then the results are compared to ensure that the factored load or load combination does not exceed the factored strength (capacity) of the member. It is, therefore, possible to recognise the differing probabilities in the prediction of load and member strength. As for example, safety factors for dead loads can be kept low, as these can be assessed with reasonable accuracy, compared to those for live loads whose accuracy cannot be guaranteed. Values of these safety factors and their application procedures are explicitly stated in codes of practice.

3.5 SPECIFICATIONS AND CODES OF PRACTICE

As has been discussed earlier, most structural failures can be attributed to gross errors during design or construction. Use of incorrect assumptions as the basis of design, conceptual misunderstanding of the structural behaviour are examples of such errors. Also, lack of clarity in the drawings and specifications, coupled with serious errors during construction may contribute to failure. In order to counter such situations, many countries produce guidance documents in the form of standard specifications or codes of practice. These documents, however, can provide guidance in a general manner only. In order to interpret the rules correctly, the designer has to have a thorough understanding of the behaviour of the material and the system he has adopted as well as the design uncertainties discussed earlier in the chapter. As for example, safety of steel bridges depends heavily on good detailing, particularly to guard against the effects of fatigue and corrosion. Standard codes may not provide complete guidelines in these areas and the designer has to use his own ingenuity and experience to complement the recommendations of the codes.

Most of the industrialised countries have their own specifications and codes. In the USA these documents are ASTM or AASHTO standard specifications, while in UK and Germany these are BS and DIN. In India these are IS standards issued by Bureau of Indian Standards, or the codes issued by the Indian Railways or the Indian Roads Congress. In Europe, Eurocodes are being newly introduced to replace the separate standard codes of member countries. It is not intended that this book would focus on any particular document; rather, the material presented would conform generally to modern engineering practice.

REFERENCES

[1] Francis, A.J. 1989, *Introducing Structures*, Ellis Harwood Ltd., Chichester, UK.

[2] Seward, D. 1994, *Understanding Structures*, The Macmillan Press Ltd., London, UK.

[3] *Commentary on BS 5400: Part 3: 1982, Code of Practice for the Design of Steel Bridges*, The Steel Construction Institute, Ascot, UK.

[4] Kulicki, J.M. 1999, *Design philosophies for highway bridges*. In: Bridge Engineering Handbook, Wai-Fah C. and Lian, D. (eds.) CRC Press, Boca Raton, USA.

[5] Gaylord, E.H. (Jr.), Gaylord C.N. and Stallmeyer, J.E. 1992, *Design of Steel Structures*, McGraw Hill Inc, New York, USA.

Chapter 4

Characteristics of Steel

4.1 INTRODUCTION

Steel is an alloy of iron with carbon and other elements, some of them being unavoidable impurities (sulphur and phosphorus), while others are added deliberately (copper, nickel, chromium, molybdenum, vanadium, etc.). The manufacturer certifies the chemical composition and physical properties of the steel produced. The designer generally considers these properties and selects the appropriate steel to be used in the bridge structure. However, it is possible that due to various reasons the steel may fail at a stress below the nominal yield value certified by the manufacturer. The designer should be aware of these reasons in order to produce an economical and durable design. Some of the primary reasons for such premature failures and the methods for avoiding them form the subject matter of the present chapter.

Distress during service life of steel bridges may be attributed to the following three main reasons:

- atmospheric corrosion
- stress corrosion
- fatigue

Although these three reasons are strictly separate mechanisms, they are somewhat inter-related, in a sense that any one of these may trigger one or both of the other two mechanisms to contribute to the distress. Thus, in many cases, all these three mechanisms may be contributory factors for the distress. Apart from these mechanisms, other aspects discussed in the following sections are: brittleness, lamination, residual stresses, weldability and weathering steels.

4.2 ATMOSPHERIC CORROSION

Atmospheric corrosion is defined as the deterioration of steel due to its reaction with environment. Essentially, corrosion is an electro-chemical process of flow of electricity and chemical changes in steel, occurring in stages. The process can be understood in terms of a battery, i.e., anode and cathode, connected in the presence of an electrolyte (a conducting medium). The process first takes place locally, by formation of anodes (steel substrata), and cathodes (surface mill scales) similar to battery cells. In the next stage, when the surface is moist, the cells start functioning in the presence of an electrolyte media produced from the moisture film and the acidic substance present in the atmosphere. In combination with oxygen from air, hydrated ferric oxide or red rust is formed:

$$4\,Fe \quad + \quad 3\,O_2 \quad + \quad 2\,H_2O \quad = \quad 2\,Fe_2O_3H_2O$$

$$\text{(iron/steel)} \qquad \text{(oxygen)} \qquad \text{(water)} \qquad \text{(rust)}$$

It is to be particularly noted that corrosion occurs only in the presence of water as well as oxygen. In the absence of either, corrosion cannot occur.

After a period of time, accumulation of the corrosion products (red rust) in the first area causes the corrosion process to be somewhat stifled. This results in the formation of a new anodic zone, thereby spreading the corrosion area. In cases where the original anodic area is not stifled, the corrosion process continues deep into the steel, resulting in pitting or 'pock marks'.

The immediate or direct effect of corrosion is loss of cross sectional area of the member. This increases the stress in the member, and indirectly makes the member vulnerable to stress corrosion and fatigue cracking.

It is necessary for the designer to examine the environment of the location of the proposed bridge and recommend adequate protection (painting) system. It has been reported that many steel bridges have suffered distress in the past, because of premature arrival of corrosion due to selection of an inadequate painting system. The subject of corrosion protection system has been covered in considerable detail in Chapter 26.

4.3 STRESS CORROSION

It has been observed that in an aggressive environment, members subjected to sustained tensile stress, either due to overloading or due to detailing adopted, are prone to corrosion. As the cross sectional area of such a highly stressed member is reduced due to corrosion, the resultant increase in stress may initiate cracks. Thankfully, stress corrosion cracks are not common in conventional steel girder bridges. These are found mostly in specific areas, such as eye bars and pin connections in suspension and cable-stayed bridges, where details are such as to cause high concentration of stress.

4.4 FATIGUE

Bridges, particlarly railway bridges with greater live load incidence, are subjected to high fluctuation of stresses in the members. This fluctuation of stresses reduces the ultimate strength of steel considerably, with failure occurring at stress levels much lower than the yield stress. This phenomenon of localised permanent structural change due to fluctuating stresses that may initiate and propagate cracks in the member is termed fatigue. Thus, a member may be able to withstand a single application of the design load, but may fail if the same load is repeated for a large number of times, say 10,000,000 cycles. This reduction of strength is dependent on two factors, namely, number of load repetitions (cycles) and the range of stress.

Figure 4.1 shows the typical forms of cyclic stress loadings. Figure 4.1a shows the stress pattern for a member subjected to variation in tensile stresses while Fig. 4.1b shows the stress pattern for a member subjected to reversal of stresses (tension and compression).

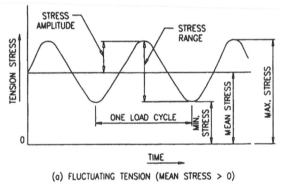

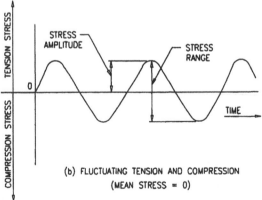

Fig. 4.1 Typical cyclic stress loadings

The fatigue data are generally presented in a curve called the S-N curve, illustrated typically in Fig. 4.2. Here the total cyclic stress range (S) is plotted against the number of cycles (N). S-N curves are based on test data obtained experimentally by conducting tests on a series of identical test specimens subjected to cycles of constant amplitude and measuring the number of cycles required for failure. The results are plotted, using logarithmic scales for both axes.

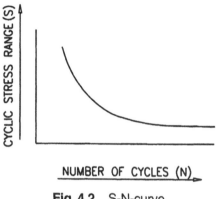

Fig. 4.2 S-N-curve

The principal design specifications and national codes for bridges include provisions for fatigue design strengths corresponding to different combinations of stress range and cyclic loadings, based on the results of laboratory tests. These should be followed for design of members and joints to cater for fatigue effects.

To avoid fatigue failure, the factor, which perhaps, is of greatest importance, is the detailing of members and joints. Fatigue failure occurs at the tension zone of members and is generally initiated by stress concentration at locations such as sharp changes in cross sections, sharp corners and notches. Sharp notches produce significant stress concentrations and result in extremely low fatigue resistance. Also, details of certain items such as cover plates, splices and stiffeners in welded flexural members (beams and girders) need careful consideration during drawing. Consequently, appropriate care should be taken in designing members and developing structural details of bridge elements subjected to cyclic loading.

In welded joints, the fatigue strength of steel tends to be reduced due to change in the structure and the properties of the steel used. As a result, welded components in bridges are more prone to fatigue cracks than riveted or bolted ones. Also, a crack developed in one component of a riveted or bolted connection normally stops at the hole and does not travel beyond to the next component. In cases of a welded joint, however, a crack developed at the weld tends to progress and may affect both the connecting components, thereby damaging the entire member.

Other factors affecting fatigue resistance of a member are residual stresses in steel and work hardening due to cold rolling.

In order to reduce the harmful effects of fatigue, the following aspects need particular attention of the designer :

1. Details which are likely to produce severe stress concentrations should be avoided.
2. Gradual changes in sections (avoiding re-entrant notch like corners) should be adopted.
3. Eccentricities in connections should be avoided or kept to the minimum.
4. Details with 'stress-raisers' should be avoided.
5. Butt welds are preferable to fillet welds.
6. Weld details that introduce high localised constraints should be avoided?
7. Double-sided fillet welds are preferable to single-sided fillet welds.
8. Fillet welds across the direction of stress should be avoided.
9. Intersections of longitudinal and transverse welds should be avoided.
10. Continuous welds should be recommended in preference to intermittent welds.
11. Proper inspection during fabrication and erection should be done. This will ensure proper riveting, tightening of bolts and depositing of welds.
12. Welding procedure should be carefully formulated.
13. Suitable non-destructive testing (NDT) should be adopted.

4.5 BRITTLENESS

Brittleness is characterised by a fracture that propagates rapidly at a relatively low tensile stress and at low temperature, usually occurring suddenly with little or no prior plastic deformation. On the other hand, ductile fracture in a member is preceded by a considerable plastic deformation until the load exceeds the value corresponding to its yield stress. Notch ductility is a property of steel which indicates its resistance to brittle fracture.

Collapse of King's bridge, Melbourne, in 1962 due to failures in the welded tension flanges is one of the most well-known examples of the lack of understanding amongst the engineering community about the implications of brittle fracture failures in steel girders. Causes of this failure have been discussed in detail in Chapter 2. It is thus important that the designer is familiar with this phenomenon, in order to eliminate the recurrence of such catastrophic failures in new bridge structures.

The key factors that influence brittleness in steel are :

- *Metallurgical features:* Depending on their chemical composition, heat or mechanical treatment, some steel may be more brittle than others.
- *Temperature of steel in service:* Structural steel undergoes a ductile-to-brittle transition as the temperature falls. Therefore, fracture may occur at even low stresses when the ambient temperature drops below freezing point. Geographical location of a bridge thus assumes particular significance from this point of view.
- *Service conditions*: Brittleness may be increased due to certain distribution pattern of force field, such as at locations of stress concentrations due to abrupt change of sections, notches, cracks, etc. Thick plates, wide plates and deep webs normally carry higher risk of brittle fracture due to their complex stress pattern caused by higher degree of restraint. Other conditions which induce brittleness include: cold working during fabrication and rapid rate of loading.

Internal stresses due to welding contraction also enhance the risk of brittle fracture. In particular, fillet welds laid across tension flanges of girders and intermittent welding induce brittleness of the steel in the weld region. Weld defects such as under-cutting, slag inclusion, porosity or cracks, enhance the risk of brittle fracture. These aspects are covered in more detail in the chapters dealing with welding.

4.5.1 Avoiding Brittle Fracture

In order to avoid brittle fracture, the following aspects need special attention of the designer:

Quality of steel

The material should have adequate strength and notch toughness properties, i.e., it should have the capacity to be ductile at the service temperature.

There are many types of tests to assess notch toughness of a material. Some of these are:

- KIc test
- CTOD test
- Charpy V-notch test

The first two are large scale laboratory tests on full plate thickness samples. The third, i.e., the Charpy V-notch test is a small scale laboratory test on a small machined sample. It is popular for its reasonable cost and quick results. In this test, a specimen of a small square bar with a machined notch across the centre of one side is hit on the face behind the notch by a striker mounted on the end of a swinging calibrated pendulum. This is done by first lifting the striker to a specified height and then releasing it to hit and break the specimen, rising to a height on the other side (Fig. 4.3). A pointer moved by the pendulum over a scale indicates the energy absorbed in breaking the specimen. This value is termed, impact value, and is utilised to assess the notch toughness of the steel. The test is generally carried out at different temperatures such as room temperature, 0, –10, –20 degrees Centigrade, etc.

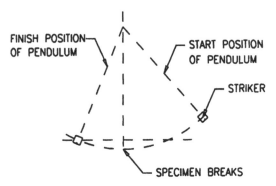

Fig. 4.3 Schematic diagram showing charpy V-notch test

Design of details

Design of details plays an important role in reducing the risk of brittle fracture. As has been discussed in the preceding paragraph, brittle fracture is initiated almost always at points of stress concentration. It is, therefore, imperative that notches and sudden geometric changes such as re-entrant angles are eliminated in any detail and large radii are used at changes of sections.

Stresses in the structure should be made to flow as smoothly as possible, without obstruction. Therefore, the details of the structure and joints should be so developed as to reduce the chances of sudden changes in the stress flow pattern.

Fabrication

In order to avoid the risk of brittle fracture in a bridges structure, certain aspects of fabrication work need to be carefully noted:

- Manual flame cutting of plates leaves the edges uneven and may cause stress concentration. These should be trimmed by grinding or machining. However, edges of machine-flame-cut plates generally provide a smooth finish and would not need further trimming. In any case, care should be taken not to leave any harmful irregularities in the edges.
- Punching of holes is a cold work process and can cause embrittlement and cracking. This should be avoided. The degree of embrittlement depends on the type of steel and plate thickness. Also, punching is likely to produce minute radial cracks at the edge of the hole. This can initiate brittle fracture when the member is stressed. To eliminate the risk of brittle fracture, holes may be punched to a reduced diameter and then reamed to the required diameter.
- Shearing of plates is also a cold work process making the edges susceptible to brittle fracture. Such plates, also need to be trimmed, usually by grinding or machining to a smooth finish.
- Notches, changes in sections, etc., should be made with as large radii as possible.
- Cold working during fabrication process (e.g., bending), particularly in higher strength steels, is likely to decrease notch toughness and must be avoided.
- Intermittent welding should be avoided. Every time the arc strikes (particularly when the welding is not done), the parent metal is subjected to a risk of embrittlement.
- Small fillet welds on relatively heavy members are susceptible to cracking and, therefore, should be avoided.
- Pre-heating of the components to be welded reduces the chance of embrittlement of steel.
- Adequate inspection during fabrication work supported by non-destructive testing would greatly enhance the quality of fabrication *vis-a-vis* brittle fracture.

4.6 LAMINATION IN ROLLED STEEL PRODUCTS

Theoretically, structural steel is assumed to be homogeneous, i.e., the steel material is uniform over its volume. On the basis of this assumption, steel is considered to be an isotropic material, i.e., a material that possesses the same elastic and strength properties in all axes. In actual practice, however, during the rolling process, planes of weakness or planer defects, typically running parallel to the surface, are created. These defects are termed laminations and generally originate from the ingot or slab from which the plate or section is rolled. As the plates and sections are rolled, the non-metallic matter (such as steel making slag or other foreign bodies which were entrapped in the ingots or slabs), produces weak layers parallel to the direction of rolling. As a consequence, the plane parallel to the direction of rolling tends to be weaker than the other two planes. Thus, the load carrying capacity in the transverse direction (i.e., through thickness direction), is significantly weakened at the location of lamination. In cases where the lamination is only local and in a limited area, it may not affect the load bearing capacity where stresses are wholly parallel to the main axes of the member. However, large laminations are likely to impede the load bearing capacity of the member, particularly in locations close to welded joints, where weld contraction produces high stresses in the transverse direction. A related phenomenon is *lamellar tearing* which is generally associated with welding on thicker sections subject to stresses in the transverse direction. This phenomenon has been discussed in greater detail in Chapter 16.

Lamination can be detected by non-destructive testing (NDT) such as ultrasonic testing. This is a simple test and can be carried out at any fabrication shop by a portable testing machine (see Chapter 20).

4.7 RESIDUAL STRESSES

Residual stresses are those stresses that remain in structural elements after rolling or welding. These are mainly caused by uneven cooling after rolling or welding. For example, in a rolled wide flange beam, the flange tips start cooling first, followed by the zone at the centre of the flange. The delayed cooling of the interior parts induces compressive stress along the flange tips. The centre of the flange cools slowly and develops tensile residual stresses as the adjacent cooler parts tend to prevent its shrinkage. For equilibrium, the tensile residual stresses at the centre of the flange are balanced by the compressive stresses elsewhere in the section. Figure 4.4a illustrates a typical residual stress pattern in a wide flange rolled beam. In case of rolled plates, the plate edges tend to cool first, followed by the central portion. Thus the plate edges show residual compressive stresse. Figure 4.4b shows the stress pattern in a rolled plate. In welding and flame cutting, the metal is subjected to a very high temperature along a localised strip. The cooler remaining portion of the metal resists the shrinkage (due to cooling) in the strip, inducing high tensile residual stress along the strip. A typical residual stress pattern for a plate with flame cut edges is shown in Fig. 4.4c. When a beam is fabricated from flame cut plates, the stress pattern in the edges before the welding is reflected in the final stress pattern as shown in Fig. 4.4d.

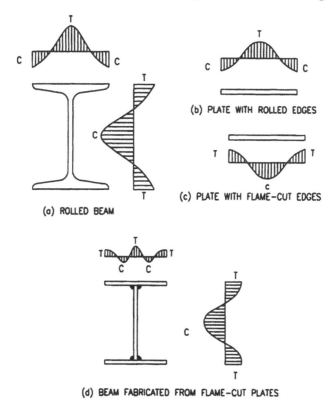

(a) ROLLED BEAM

(b) PLATE WITH ROLLED EDGES

(c) PLATE WITH FLAME-CUT EDGES

(d) BEAM FABRICATED FROM FLAME-CUT PLATES

Fig. 4.4 Typical residual stress distributions

As the compressive and tensile residual stresses in any cross section balance each other, these stresses, in effect, do not lead to any resultant axial force or bending moment on any cross section. However, the presence of residual stresses may cause premature yielding due to an applied load in a localised portion, where residual stress is of the same nature (tension or compression) as the stress due to the applied load. The effect may be significant in compression members where the residual compressive stress may decrease the allowable buckling load from the theoretical value. However, the current codal provisions generally take care of this effect.

4.8 WELDABILITY

Weldability is the distinctive feature of steel which allows it to be welded without occurrence of cracking and other defects which make the resultant joint unfit for structural purpose.

Carbon and manganese are primarily the strengthening elements in steel. However, increased content of these elements makes the steel harder, thereby impairing the weldability of the steel. In order to attain better physical properties in steel, while keeping the carbon level low, other alloying elements are generally added during the process of steel making. Thus, elements like chromium, molybdenum and vanadium are added to increase the strength, while copper and nickel are added to improve the corrosion resistance. These elements also adversely affect weldability of the resultant steel. The relative influence of chemical contents on the weldability of a particular steel is guided by the value of 'Carbon Equivalent' (CE). Traditionally the following formula of the International Institute of Welding (IIW) has been used to assess the weldability of steel:

$$CE = C + (Mn/6) + (Cr + Mo + V)/5 + (Ni + Cu)/15$$

where the chemical symbols represent the percentage of the respective elements in the steel. It may be noted that the amount of each alloying element is factored according to its contribution towards hardening of the steel. Several other variants of this formula are available and are recommended by different authorities. However, the formula given above is widely followed in various countries.

In structural steel, the value of carbon equivalent ranges between 0.35 and 0.53. Higher yield stresses can only be obtained by increasing the percentage content of various alloying elements, resulting in further increase of CE. As a general rule, a steel is considered weldable when CE is less than 0.4. With increased value of CE, special measures in welding, such as use of low-hydrogen electrodes and pre-heating of the components to be joined, become important.

The level of hardness in steel is measured by Vicker's Pyramid Number (VPN). In this system, an indenter is forced into the surface of the steel and the size of the impression is compared with a pre-set standard. VPN for steel to be used for fabrication should ideally be in the range of 190 to 200. This should be taken into account in the design stage while specifying the material to be used.

4.9 WEATHERING STEEL

Weathering steels were developed in America in the 1930s and were very popular for use in the fabrication of coke and ore wagons. They were first used for structural purposes in the 1960s. Quite a number of structures, in particular bridges, have since been constructed using these steels. The best known weathering steels are in the brand names of 'Cor-ten', produced by United States Steel Corporation and 'Mayri R', produced by Bethlehem Steel Corporation in the USA. These steels have now been licensed for manufacture in many countries in the world.

4.9.1 Performance

Weathering steels are high strength low alloy weldable structural steels in which the resistance to atmospheric corrosion has been improved by addition of small amounts (generally upto 3%) of alloying elements such as chromium, copper, nickel , phosphorus, etc. The rate of corrosion in these steels is lower than that in plain carbon steel and when used in many climatic conditions, these steels form a relatively stable layer of adherent hydrated iron oxide coating on the surface. This coating resists further ingress of moisture, thereby retarding the corrosion rate, until it reaches a low terminal level. Figure 4.5 shows a typical comparison of corrosion rates for mild steel and weathering steel. Conventional coatings of corrosion resistant paint system are not usually necessary in these steels as the steel protects itself from corrosion due to the presence of the oxide coating. The colour of the coating varies from red-orange to purple-brown or almost black, depending on the conditions of the exposure and provides an aesthetically pleasing colour in most environments.

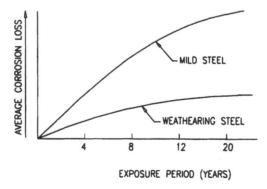

Fig. 4.5 Comparative corrosion rates of mild steel and weathering steel in atmosphere (Reproduced from: J.C. Hudson and J.F. Stanners, Journal of Iron and Steel Institute, 271–89, 1955, appearing in [2])

The type of the coating and the time required for it to form depend on the following:

- The alloy content in the steel
- The corrosivity of the environment
- The frequency with which the surface is wetted by dew or rainfall and dried by wind and sun.

The environment in which a weathering steel bridge is located plays an important role in how the material will perform. In marine environments and where sulphur is present in the atmosphere, formation of the oxide coating is somewhat hindered. It should also be noted that wetting and drying of the surface are required to produce the protective coating. Therefore, under a continually wet or damp sheltered condition, or where the environment is too dry to have the necessary wet/dry cycle, weathering steels do not perform in the same way as in a freely exposed condition. Consequently the corrosion rate goes up. In such conditions, weathering steels may need application of coatings of corrosion resistant painting system. Thus, use of weathering steel in bridge structures should be made only after careful deliberation about the environmental condition.

4.9.2 Precautions

In unpainted weathering steel bridges, the following precautions are required to be taken during design, detailing and construction stages:

- Expansion joints in reinforced concrete decks are the commom locations where water leakage causes damage to the steelwork underneath. A trough under the joint may be useful in diverting the water away from the bridge steelwork. As a safety measure, superstructure steelwork adjoining the expansion joint may be coated with anti-corrosive paint.
- Details which act as traps for debris or water should be avoided. Provision of weepholes to drain out such water should be considered.
- Wherever possible, box girders should be hermetically sealed to prevent ingress of moisture. Alternatively, weepholes or similar arrangements for ventilation and circulation of air should be provided; otherwise, the interior surfaces should be painted.
- Concrete pier caps and abutment walls should be adequately protected to minimise staining from the iron oxide coating during the first two-three years, prior to formation of the protective coating.

- Steel should be blast cleaned to remove the mill scale in order that a uniform oxide coating is formed on its surface. Care should be taken in handling the bare steel sections and sub-assemblies during fabrication and erection to avoid scratches, dents etc. Slags after welding should be removed carefully. Staining by concrete, mortar, oil etc., must be meticulously avoided.

4.9.3   Welding

Weathering steels can be used in a wide range of applications which involve welding by the usual methods such as manual metal arc, submerged arc, gas shielded welding, etc. Because of higher level of alloy additions, the CE values of weathering steels are high compared to most high yield steels. However, experiences in the USA and the UK, where these steels have been in use for a considerable period of time, show that these steels can be easily welded if appropriate precautions applicable to steel of similar level of CE and strength are taken.

In bridges using unpainted weathering steel, it is important that the weld beads exposed to public view should have the colour similar to that of the steel components. To cater for this requirement, welding consumables (electrodes for manual welding and wire-flux combinations for submerged arc welding) having a composition which closely matches that of weathering steel, have been developed. These are, however, comparatively expensive. As a measure for reducing the cost, in multiple pass welds, only the final passes (on the surface) may be done using the more expensive special consumables, while the rest of the passes (underneath) done with normal consumables. This alternative system, however, requires stringent supervision and may not be economical from the fabricator's point of view.

4.9.4   Bolting

In a bridge with unpainted weathering steel, it is necessary to use bolts, nuts and washers having similar electro-chemical potential to that of the weathering steel. Special grade of nut and bolt materials compatible to different grades of weathering steels have been developed. Similarly, matching washer materials have also been developed. Connections using such bolts, nuts and washers have been in use successfully in the USA and the UK. Galvanised or electroplated steel connections are not suitable for use in weathering steel structures, since insulating washers are required to counter the electro-chemical action within the joint. Otherwise, in time, the joint will be left with fasteners which are liable to be corroded earlier than the weathering steel.

REFERENCES

[1] Llewellyn, D.T. 1992, *Steels: Metallurgy and Applications*, Butterworth-Heinemann Ltd., Oxford, UK.

[2] Chandler, K.A. and Bayliss, D.A. 1985, *Corrosion Protection of Steel Structures*, Elsevier Applied Science Publishers, London, UK.

[3] Owens, G.W., Knowles, P.R. and Dowling, P.J. (eds.) 1994, *Steel Designers' Manual* (fifth edition), Blackwell Scientific Publications Ltd., Oxford, UK.

[4] Gaylord, E.H. (Jr.) and Gaylord, C.N. (eds.) 1990, *Structural Engineering Handbook*, McGraw-Hill Publishing Co, New York, USA.

[5] Brockenbrough, R.L. and Merrit, F.S. (eds.) 1994, *Structural Steel Designers' Handbook* (second edition), McGraw-Hill Inc., New York, USA.

[6] Tardoff, D. 1985, *Steel Bridges*, British Constructional Steelwork Association Ltd., London, UK.

Chapter 5

Loads

5.1   INTRODUCTION

At the outset, the bridge designer has to make an assessment of the loads and forces likely to come on to the structure during its design life. In doing so, he is generally helped by different codes specifying these loads and forces. In cases of special structural systems, however, these specifications may not be sufficient and the designer may have to delve into specialised literature in order to collect additional information and loading data.

From the point of safety and economy, it is very important that the initial assessment of these loads are made as accurately as possible. Assumption of loads larger than those actually required would lead to unnecessary heavy structure, entailing increased project cost. On the other hand, assumption of lower loads would lead to unsafe design.

A bridge structure is designed to carry broadly two types of loads, viz., permanent loads and temporary loads.

Permanent loads are those loads which act on a bridge throughout its life. These comprise the self-weight of the structure and the superimposed dead loads which are permanently carried by the bridge structure. The term 'dead load' is normally used synonymously with permanent loads to include both the self weight of the structure and the superimposed dead loads.

Temporary loads are those loads which act on the bridge for only a short period of time. These include:

- Live load
- Impact
- Longitudinal force
- Centrifugal force
- Wind load

- Earthquake load
- Erection load

An overview of these loads and forces and their application on the bridge structure are presented in the following paragraphs.

5.2 DEAD LOAD

As discussed in the preceding section, dead load of a bridge includes the self-weight of the structure as well as the superimposed dead loads. Thus, dead load comprises the loads from the following:

- Bridge structure including deck, walkways, kerbs, parapets and crash barriers
- Wearing surface and potential future overlays in case of road bridges
- Ballast, tracks, sleepers and fixtures in case of railway bridges
- Cables, fixtures, etc., for electrified traction system of railway bridges
- Signs, signals and other bridge furniture
- Formwork which becomes part of the structure
- Services and utilities, such as cables, water lines, gas lines, etc.
- Other elements deemed permanent by the owner or designer

As regards unit weights of materials, these are generally available in standard specifications. However, unit weights of materials commonly used in bridges are given in Table 5.1. Designers may also use actual dead weights of materials specified for the structure.

It is a good practice to indicate the computations of dead loads in the design notes as permanent records. These may be useful for analysis purpose during possible rehabilitation design in the future.

It is a common practice to assume that the self weight of the bridge structure is uniformly distributed over the entire length of the span. For parallel chord simply supported truss bridges, this assumption is fairly correct, since the chords are heavy at the middle and lighter near the ends, while for web members the situation is reversed. However, for other types of structures, particularly for cantilever or arch type bridges, the self weight is not normally uniform over the entire span. In such cases, the various panel dead loads are first assumed, and the design is computed on that basis. The designed sections are then checked against the assumed weight. If these are not within a reasonable limit, the process is repeated. It is always preferable that the assumed dead load is too high rather than too low.

Table 5.1 Unit weights of materials commonly used in bridges

Material	Unit weight	
	(kg/cubic metre)	*(kN/cubic metre)*
Steel or cast steel	7,850	77.00
Aluminium	2,750	26.97
Cast iron	7,200	70.62
Wrought iron	7,700	75.53
Reinforced concrete	2,400	23.54
Plain concrete	2,300	22.56
Bricks	2,100 – 2,400	20.60 – 23.54
Stone masonry	2,200 – 2,950	21.58 – 28.94
Timber	480 – 1,200	4.71 – 11.77
Asphalt	2,300	22.56
Tarmacadam	2,400	23.54
Compacted sand, earth or gravel	1,950	19.13
Loose sand, earth or gravel	1,600	15.69

In general, bridges, particularly road bridges, are subjected to greater dead load than buildings. Furthermore, dead load effects are significantly more in longer spans than in shorter ones. Thus, in bridge girders, error in the estimate of the dead load is bound to affect the calculated stresses appreciably. It is, therefore, imperative that after the design is completed, the dead load assumed in the design of the bridge is compared with that of the actual design. In case the actual dead load exceeds the one initially assumed, the analysis and design should be re-done, or suitable corrections incorporated in the design. There have been cases of failure of bridges in the past due to wrong assessment of dead load in the design. As for example, in the first Quebec railway bridge, across St. Lawrence River in Canada, which collapsed in 1907 during erection, it was found that the dead weight of the bridge was grossly under-estimated in the design, causing about 10% to 25% increase in the stresses of the members (see Chapter 2). Although this discrepancy was not the direct cause of the failure, its indirect contribution cannot be overlooked. In any case, it would have certainly affected the theoretical carrying capacity of the bridge, had it been possible to complete the bridge as per the original design.

Estimate of the dead load of a bridge can be made from experience or by reference to other similar existing structures. Some organisations use charts and empirical formulae for assessing the self-weight of different types of bridges, which have been developed over the years from their previous experience. Care should, however, be taken while using such charts and formulae, since the specifications of the bridge under design may differ in respect of arrangement of load distribution, live loads, impact factor, etc., from those of the earlier bridges which form the basis of the reference charts and formulae.

5.3   LIVE LOAD

Live loads on a bridge can be grouped into three classes, namely:

- Roadway
- Railway
- Footway

Theoretically, a bridge should be designed for the actual live loads which it is expected to carry. In this context, a few issues need consideration.

5.3.1   Kind of Traffic to Allow for

This aspect generally depends on the location of the bridge and an appreciation of the future developments in the area. Furthermore, a decision has to be made by the authorities whether footways are to be added to the proposed highway or railway bridge and, also whether a combined railway and roadway bridge is desired.

5.3.2   Allowance for Future Developments

This aspect presents a somewhat uncertain problem, since the highway and railway loads are being constantly upgraded. The early bridge engineers normally understressed their materials, so that even with the increase in the load of the traffic, the material was not overstressed, being merely used more efficiently by utilising the reserve strength. This accounts for the reason of some of the early bridges being still in service. In modern design, on the other hand, the stresses are, in some cases, approaching the yield, thereby eliminating the allowance for future increase in the maximum loads. The present engineers, therefore, have to ensure that the design loads fairly represent the actual loads expected during the life of the bridge. In developed countries this task is perhaps easier than in underdeveloped ones, where it is very tempting to allow for light present-day loads in order that the budget of the project is kept within limits.

5.3.3 Co-existent Traffic on the Whole Width and Length of the Bridge

This brings us to the third problem, viz., how much traffic should be considered at the same time on a bridge. For example, in a bridge consisting of, say, two main girders with cross girders placed at intervals to support loads from the stringers, each cross girder has to carry the heaviest axle loads that pass over the bridge. But it is most unlikely that the entire bridge will be crowded with a closely packed train of such heaviest axle loads. Therefore, it has become a common practice to use 'standard design loadings' which comprise a combination of heavy concentrated load on each traffic lane for the design of members like deck, stringers and cross girders, together with an equivalent uniformly distributed load over the length of the bridge as global load for the design of the main girders. These design loadings vary from country to country and are only simplified versions of the actual vehicle loads. They do not, necessarily, follow even the pattern of the actual vehicle loads. Thus, the specified intensity of the traffic is only a guess.

5.4 IMPACT

Heavy vehicles, such as trucks and locomotives, cause dynamic effects while travelling on a bridge at certain speeds. The effects may be local hammer effects as well as global vibrations. These dynamic effects are commonly termed 'impact'. Unlike earthquake effects which produce forces in longitudinal, transverse and vertical directions, vehicular impacts induce forces mainly in the vertical direction. Various situations contribute to this effect. These include:

- Application of the load suddenly
- High speed of the moving vehicle
- Deficient suspension system of the moving vehicle
- Mass of the moving vehicle relative to the mass of the bridge
- Span of the supporting structure (impact increases as the span decreases)
- Irregularities in the railway track or the surface roughness of the roadway
- Lurching action of locomotives in the case of railway bridges, which causes periodical shifting of the load from one wheel of the axle to another

- Unbalanced or eccentric weights of the counterweighted locomotive driving wheels. These weights balance the reciprocating and rotating parts of the locomotive mechanism, and produce unbalanced forces and 'hammer effect'. These pulsating loads tend to set the bridge structure into vibration. This vibration is the most dominant dynamic effect in railway bridge design.

The effect of impact is expressed as a percentage of the live load on a structure, and is termed 'impact factor' or 'impact coefficient'. The impact load is determined as a product of impact factor and live load. Thus, if the impact factor is 30%, the live load is multiplied by 1.30 in the calculation of the forces to include impact effect.

The expression of the impact factor specified by most national codes is normally empirical in nature, and does not attempt to account for all the variable parameters contributing to this effect. However, the factor allows for a conservative idealisation of the problem.

5.5 LONGITUDINAL FORCE

Longitudinal force on a bridge results from one or more of the following situations:

- Braking effect from the application of the brakes to the wheels of a vehicle while on a bridge
- Tractive effort or acceleration of a vehicle while on a bridge
- Frictional resistance offered to the movement of free bearings

For spans supported on sliding (elastomeric) bearings at both ends, the longitudinal force is to be equally divided between the two ends of the span. For spans having fixed bearings at one end and free bearings at the other, the entire longitudinal force is considered to be acting through the end with fixed bearings. The value and the application mode for longitudinal force are generally provided in national codes. In railway bridges, where long welded rails are used, some codes recommend that the effect of such continuous welded rails (CWR) should also be included while considering longitudinal forces.

5.6 CENTRIFUGAL FORCE

When the traffic lane or track over a bridge is situated on a curve, the effect of the centrifugal force of the moving load must be calculated. The value and the application mode of the centrifugal force are determined in accordance with the recommendation provided in the governing code.

5.7 WIND LOAD

It has always been common knowledge that blowing of wind produces pressure on any surface it strikes. But until the second half of the 19th century, hardly any concern seems to have been expressed by the bridge designers for the effect of wind. It was the collapse of Tay railway bridge in Great Britain in a storm in 1879 that brought the subject of wind force in sharp focus in the minds of the bridge engineers (see Chapter 2). There has been extensive research on the subject by a host of researchers ever since, which is still continuing.

Assessment of forces exerted by wind on an object in its path is a complex problem in aerodynamics. However, for designing conventional medium span bridges, these forces are generally approximated as a static horizontal load uniformly distributed over the exposed region of the bridge and the traffic on it. The exposed region of the bridge is the aggregate surface areas of all components of the superstructure in the direction of the wind. Thus, for wind acting perpendicular to the longitudinal axis of the bridge, the surfaces as seen in the elevation of the windward side of the bridge plus an allowance for exposed leeward surfaces, are to be considered. The wind should be in such a direction that the resultant stresses in the members under consideration are the maximum. The loading due to wind is dependent on many factors and varies from region to region. These are specified in different national codes which are to be followed by the designer.

For bridges with low natural frequency, the dynamic effect of wind and the oscillation caused by it is very important. Thus, the dynamic response of a long span bridge becomes a very important consideration. The collapse of Tacoma Narrows bridge (USA) in 1940, after an hour of a moderately high wind of about 68 km/hr (44 m/hr), generated interest in the research on the problems of aerodynamic instability of long span bridges amongst the engineering community in America and other countries (see Chapter 2). For analysis of these structures, special literature, over and above the related codes, may have to be studied by the designer.

5.8 EARTHQUAKE LOAD

The forces caused by earthquakes are extremely destructive as well as unpredictable. Although earthquakes may occur in any parts of the world, severe shocks are generally confined to fairly well-defined regions, such as the Pacific Ocean including Japan, the Pacific coast of America, particularly California, and other stretches in South Asia and Mediterranean region. Even in these zones, certain areas may not have experienced a major earthquake for hundreds of years, and then, suddenly, be devastated by it.

Earthquakes set in motion three types of waves, of which the one that travels over the earth's surface seems to be most destructive. These waves cause the ground beneath the structure to move rapidly to and fro and in turn, induce accelerations in three perpendicular directions at the base of the structure. Of these, the horizontal accelerations in the direction along the length of the bridge and perpendicular to the axis of the bridge are the most critical. These are generally considered in the analysis of the bridge structure. The vertical acceleration is considered for the design of the bearings against uplift.

The response of a bridge structure to an earthquake is a very complex phenomenon. It depends on a number of factors, viz.,

- Physical properties of materials and the dead weight of the structure
- Intensity of the shock and the distance of the bridge from the epicentre of the earthquake
- Natural period of vibration of the structure
- Type and condition of the soil

The response takes the form of an equivalent *static* earthquake loading expressed as the product of seismic coefficient (depending on the above factors) and the weight of the structure. This loading is applied at the centre of gravity of the structure to calculate the forces in the different bridge elements. The seismic coefficient is computed on the basis of recommendations provided by the national codes.

5.9 ERECTION LOAD

During erection, some of the members of a bridge may be subjected to additional forces induced by construction equipment or loads temporarily placed on the bridge or due to erection methodology. In some cases, erection stresses may even be opposite in nature to the design stresses for service condition. These situations should be taken into account and necessary temporary bracings or strengthening of members or support structures are to be provided for.

In cases where additional forces are introduced due to a method of construction preferred by the contractor, the contractor normally provides for all temporary strengthening of the members or support structures. In such cases, the erection scheme should be examined by the designer also.

5.10   OTHER LOADS

A bridge structure may be subjected to other loads and forces also, caused by a number of situations. These include:

- Accidental load due to skidding of vehicle or derailment, resulting in collision with parapet or bridge member
- Blast load due to explosion or terrorist attack
- Impact from ship passing underneath
- Creep and shrinkage of concrete deck
- Snow load on deck and girder members
- Settlement of supports
- Temperature effects

These aspects need to be considered and data be collected from appropriate sources which may include national standards.

REFERENCES

[1] Francis, A.J. 1989, *Introducing Structures*, Ellis Harwood Ltd., Chichester, UK.

[2] Gaylord, E.R. (Jr.), Gaylord, C.N. and Stallmeyer, J.E. 1992, *Design of Steel Structures*, McGraw Hill Inc, New York, USA.

[3] Bresler, B. and Lin, T.Y. 1960, *Design of Steel Structures*, John Wiley & Sons, Inc, New York, USA.

Chapter 6

Primary Structural Systems

6.1 INTRODUCTION

Bridges are constructed mostly for carrying highway, railway and pedestrian traffic across natural gaps such as rivers, deep gorges, etc., or artificial gaps, such as railway tracks, roadway underneath, etc. Bridges are generally characterised in three span ranges:

- Short span: upto 30 m
- Medium span: between 30 m and 150 m
- Long span: beyond 150 m

6.2 TYPES OF BRIDGES

Bridges can be of *deck type*, *semi-through type* or *through type*. A *deck bridge* carries the traffic on the top of the structure. This type of bridge offers greater comfort to the public by way of uninterrupted view of the surrounding landscape and should be adopted wherever possible. This type also has a structural advantage as the top chord or top flange, which are in compression, are restrained laterally by the deck, thereby enhancing the strength of the component against buckling. Also, since the deck system is situated above the girders, the girders can be placed closer together, thereby reducing the lengths of the lateral members, and resulting in saving in material. In case of a multi-span deck type bridge, the piers need not be as high as in a through type bridge. This would result in considerable saving in the cost of substructure. Deck type bridges are normally used where the clearance between the underside of the bridge and the waterway or roadway beneath is sufficient. However, very often the clearance requirements under the bridge restrict the construction depth and consequently deck type construction cannot be allowed. In such a situation, the bridge structure has to be raised and the deck placed at a lower level, and either a semi-through or a through type bridge is to be adopted. In a *semi-through bridge* the deck is placed somewhere below the top of the bridge. Since overhead bracing cannot be provided due to lack of headroom, this arrangement may present some problems in the design. However, this arrangement often provides a correct solution for a set of given conditions. And finally, the third type is *through bridge* where the deck is situated at or near the lower chord of a truss bridge. This type of bridge, where overhead bracings are provided, usually results in maximum economy for medium span bridges and is almost unavoidable for longer spans. However, no matter what type of bridge is considered, the designer has to arrive at the best solution by selecting the most appropriate structural system. The various deck and structural systems available to the designer are discussed in the following sections.

6.3 DECK SYSTEMS

In recent years, two major advancements in the design of bridge decks have been developed, viz., composite construction and orthotropic steel deck systems. In the former method, the concrete deck slab is connected to the steel beams with shear connectors and is made to act compositely with the beams both in longitudinal as well as transverse directions. In the orthotropic steel deck system, the steel deck plates are stiffened by steel ribs welded to the bottom of the deck plate. Wearing surface of bituminous asphalt pavement (or costlier, but more durable epoxy asphalt system) is laid over the top of the deck plate. The steel deck plate is designed to serve as a flange of the integrated girder system. In both the systems, the deck participates as the top flange of the main beams of the structure. The consequent reduction in the structural depth of the beams contributes to the economy in steel requirement as also in the construction of approaches. An added advantage is that the slender lines of the bridge using these systems offer considerable aesthetic appeal.

6.4 PRIMARY STRUCTURAL SYSTEMS

Primary structural systems of steel bridges can be classified into four groups in terms of their configurations. These groups and their constituent systems are:

(a) *Beam and plate girder bridges* : These comprise:
- Single span, simply supported beam bridges
- Multiple-span, simply supported plate girder bridges
- Continuous plate girder bridges
- Rigid or portal frame bridges

- Box girder bridges

(b) *Truss bridges:* These comprise:
  - Simply supported truss bridges
  - Continuous truss bridges
  - Cantilever truss bridges

(c) *Arch bridges:* These comprise:
  - True arch
  - Tied arch
  - Part tied arch

(d) *Cable-suspended bridges:* These comprise:
  - Cable stayed bridges
  - Suspension bridges

Table 6.1 shows the the normal economic span ranges for these systems. It should be noted that the figures indicated in the table relate to the bridges in general terms; there are, however, instances of bridges with spans exceeding those shown in the table.

Table 6.1 Normal span range of bridge structural systems

Bridge structural system	Normal economic span in metres
Simply supported beam and plate girder	Upto 30
Continuous plate girder (uniform depth)	Upto 120
Continuous plate girder (haunched or curved soffit)	50 to 200
Rigid or portal frame	40 to 200
Continuous box girder	100 to 250
Simply supported truss	30 to 150
Continuous truss	150 to 400
Truss in cantilever construction	250 to 550
Arch bridge	200 to 500
Cable stayed bridge	250 to 600
Suspension bridge	350 to 1,300

The main characteristics of these structural systems are briefly discussed in the following paragraphs.

6.5 SIMPLY SUPPORTED BEAM AND PLATE GIRDER BRIDGES

These bridges are generally economical for short spans (upto 30 m). The majority of bridges fall under this system, because for many crossings of small rivers, railways or secondary highways, a single span upto 30 m is sufficient. These bridges come under three categories:

6.5.1 Rolled Beams

These are formed by hot rolling process in steel mills. The most common type of rolled beams used in bridges is Universal Beam or Wide Flange Beam with parallel flanges.

6.5.2 Rolled Beams with Cover Plates

In these beams, additional plates are welded to flanges in order to enhance the flexural strength of the beams. This obviates the necessity to use a larger size rolled beam, or plate girder, thereby effecting economy to the system. However, with this arrangement, there is a potential danger of fatigue induced cracks due to stress concentrations at the ends of the added plates. This aspect is discussed in detail in Chapter 9.

6.5.3 Plate Girders

These girders have I type cross section—similar to rolled beams—but are fabricated from steel plate elements by using angles connected together with rivets or bolts, or by directly welding the web plate to the two flanges to form an I section. With this system, greater economy can be achieved as the designer can specify almost the exact size of the plates to achieve the required capacity. Further economy of steel can be achieved by varying the depth or haunching the girder. However, added fabrication costs may offset a part of the advantage; hence overall cost effect needs to be considered while opting for curved or haunched plate girders.

For beam or plate girder bridges, the following three arrangements are commonly used:

(a) The roadway deck is supported directly on the top flanges of a series of longitudinal beams extending from abutment to abutment, and placed at comparatively close centres. In this arrangement the deck spans transversely between longitudinal beams. It is expedient to provide cross bracings at ends and at intermediate locations. With this arrangement, the torsional rigidity of the structure will be enhanced and the entire deck system will act as a grid, thereby permitting better transverse distribution of concentrated loads to the longitudinal girders. A typical beam road bridge of this arrangement is shown in Fig. 6.1.

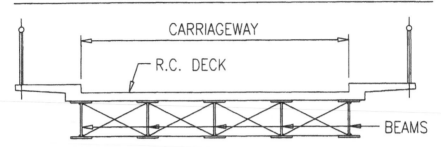

Fig. 6.1 Road bridge with longitudinal beams only

For railway bridges, one pair of plate girders is normally used to support each track. Thus, for a double track bridge, it is common to have two single track bridges placed side by side on common abutments or piers. For stability reasons, spacing of the girders is generally kept a little more than rail spacing. The girders are braced at top flange level with lateral bracings. Cross bracings or diaphragms in the vertical plane are provided at each end and at intermediate locations. A typical single track railway plate girder bridge is shown in Fig. 6.2. Railway plate girder bridges are generally limited to 20 m spans while highway bridges may be economical upto 30 m spans.

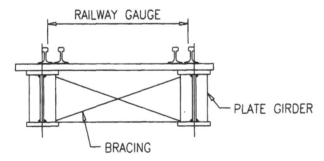

Fig. 6.2 Railway plate girder bridge

(b) In the second arrangement, the deck is supported on the top flanges of a series of cross beams placed at comparatively close centres which, in turn, are supported by a smaller number (sometimes only two numbers) of longitudinal girders spanning between the abutments. A typical road bridge of this arrangement is shown in Fig. 6.3.

(c) A variation of the preceding arrangement is to have only two main girders with cross beams spanning between the main girders, and stringers spanning longitudinally between the cross beams. In such an

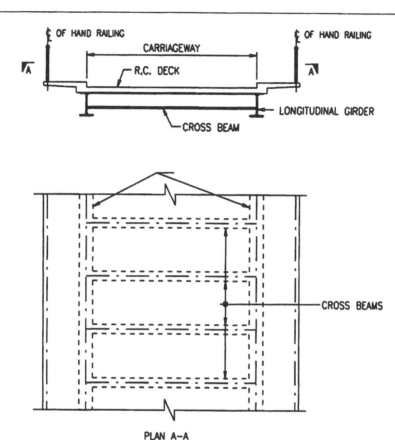

Fig. 6.3 Road bridge with cross beams and longitudinal girders

arrangement, the RC deck slab may be supported by the main girders, the cross beams as well as the stringers. Figure 6.4 shows such an arrangement. Spacings of the stringers can be suitably selected to keep the concrete deck thickness to the minimum permitted. Spacings of the cross girders can also be suitably chosen to keep the stringer size economical, and also to reduce or eliminate the number of intermediate cross frames or diaphragms.

6.5.4 Semi-through Plate Girder Bridges

Where a limited clearance below the structure is available, a semi-through type girder may have to be used. In this arrangement, the cross beams are framed into the girders just above their lower flanges, with the roadway deck supported on the top flanges of the cross beams (Fig. 6.5). In case of railways,

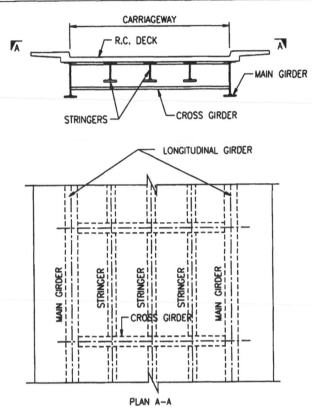

Fig. 6.4 Road bridge with stringers, cross girders and main girders

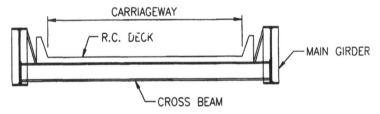

Fig. 6.5 Semi-through road bridge with main girders and cross beams

a pair of stringers may be placed near the top level of the cross beams to carry the track system. In a ballasted track system, ballast may be laid on concrete or steel plate decking supported by the cross beams. In older bridges, ballast is laid on trough plates, placed transversely over the lower flanges of the two main longitudinal girders.

6.5.5 Multiple Span, Simply Supported Plate Girder Bridges

The single span bridges discussed in the previous section may not be economical beyond 30 m span range, for which intermediate piers will be necessary for supporting multiple plate girders (Fig. 6.6a). Thus, to span a gap of 50 m, two 25 m long girders supported by a centrally placed pier may be used. Similarly, for larger gaps, three or more girders supported on two or more piers may be used. However, locations of such intermediate piers and consequently the span lengths of girders depend on a number of factors:

- Pier positions are determined by obstacles underneath the bridge alignment such as, rivers, railway tracks, roadway lanes, buried services, etc.
- The minimum clearance below or the maximum deck level above the bridge determine the maximum construction depth for the bridge. This, in turn, limits the span length.
- Unsatisfactory soil conditions, such as presence of boulders in the bed level, may call for expensive pier foundations. In such cases, reducing the number of piers and using longer spans may be more economical.
- Depending on the location of the site, construction of tall piers becomes significantly expensive if the height increases beyond a certain limit. In such cases, economy may favour longer spans.
- In rivers with a history of adverse flow conditions, such as heavy current, floods, tidal effects, etc., construction costs of piers may dictate choice of longer spans.
- In navigable waterways the locations of navigation channels usually dictate the span lengths.

In locations where differential settlements of foundations are matters of concern, requirement of longer spans may be met by providing cantilever girders with a central suspended span (Fig. 6.6b).

6.6 CONTINUOUS PLATE GIRDER BRIDGES

In cases where the differential settlements of piers and abutments are not significant, a continuous multiple span plate girder may be used with concrete slab acting compositely with the beam, instead of the multiple single span system (Fig. 6.6c). With this system, span moments will get reduced considerably resulting in possible reduction in the depth of the girder. This reduction in the depth of the girder will not only lead to saving

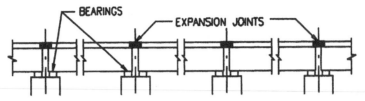

(a) MULTIPLE SPAN SIMPLY SUPPORTED GIRDERS

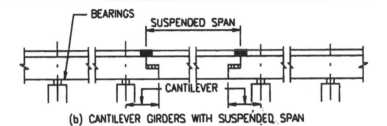

(b) CANTILEVER GIRDERS WITH SUSPENDED SPAN

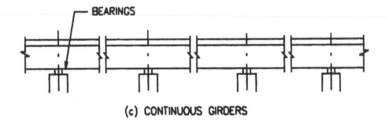

(c) CONTINUOUS GIRDERS

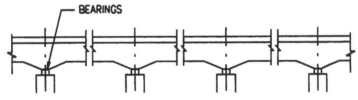

(d) CONTINUOUS GIRDERS WITH HAUNCHES

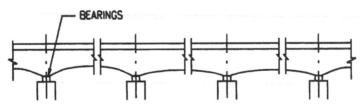

(e) CONTINUOUS CURVED SOFFIT GIRDERS

Fig. 6.6 Plate girder bridges

in the costs of the deck and the sub-structure, but also other costs. Other advantages of continuous plate girder bridges are discussed in Chapter 9.

It is common to increase the span of the girder to around 100 m by increasing the capacity of the girders at different locations by varying the thicknesses of the flanges and/or webs while keeping the overall line of the bridge horizontal. Alternatively, the girders may be made deeper at piers with haunches or smooth curve in the soffit (Figs. 6.6d and e).

For spans in the lower range, there is stiff competition from pre-stressed concrete girders. However, when several spans are to be built, a steel alternative may become effectively competitive.

6.7 RIGID OR PORTAL FRAME BRIDGES

Rigid or portal frame bridges have steel supporting legs as an integral part of the horizontal girder; thus both the superstructure and the substructure are made of steel. Figure 6.7 shows a few forms of this type of bridge. Arrangement (a) is the simplest form of a rigid frame and is suitable for single span openings. Arrangement (b) shows a three span form with inclined legs that offers an aesthetically attractive and economical solution by way of reducing the span length, thereby reducing the depth of the girder as well. Such rigid frame bridges generally enjoy the advantages of continuous bridges discussed earlier. However, the foundations are required to resist significant horizontal thrust and may entail complicated details, causing disadvantage from economical point of view.

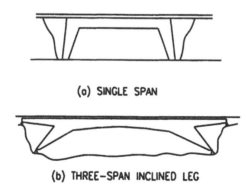

(o) SINGLE SPAN

(b) THREE-SPAN INCLINED LEG

Fig. 6.7 Rigid or portal frame bridges

6.8 BOX GIRDER BRIDGES

The girder and slab type of construction can be used beyond 100 m and upto about 250 m by using fabricated steel box sections as main beams. Some

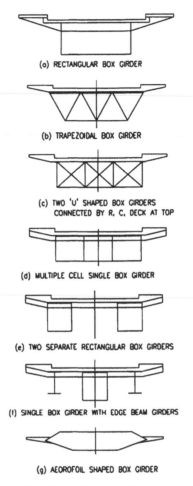

(a) RECTANGULAR BOX GIRDER

(b) TRAPEZOIDAL BOX GIRDER

(c) TWO 'U' SHAPED BOX GIRDERS
CONNECTED BY R. C. DECK AT TOP

(d) MULTIPLE CELL SINGLE BOX GIRDER

(e) TWO SEPARATE RECTANGULAR BOX GIRDERS

(f) SINGLE BOX GIRDER WITH EDGE BEAM GIRDERS

(g) AEOROFOIL SHAPED BOX GIRDER

Fig. 6.8 Typical forms of box girders

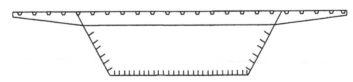

Fig. 6.9 Cross-section of steel box girder with orthotropic steel plates

typical forms of box girders are shown in Fig. 6.8. The geometry of the box girders allows more steel to be concentrated in the flanges without using very thick plates, thus avoiding the metallurgical problem associated with the welding of thick plates. The geometry also provides the box girders with exceptional torsional stiffness, resulting in better transverse load distribution.

This characteristic enables the span to be extended upto about 250 m without significant instability problems. For the same reason, box girders are suitable for use in curved bridges. The depth of the structure can be comparatively shallower leading to economy in the approach embankment and aesthetically pleasing lines.

For spans greater than 100 m, the dead weight of the concrete deck becomes disproportionately high *vis-a-vis* the total loading and, therefore, proves uneconomical. In such cases, the concrete deck may be replaced by orthotropic steel deck over which the wearing coarse may be directly placed. A typical cross section of a box girder with orthotropic steel plate is shown in Fig. 6.9. An alternative solution is to use light weight concrete to act compositely with the steel girder, for which only narrow top flanges are required (see Fig. 6.10).

6.9 SIMPLY SUPPORTED TRUSS BRIDGES

Open web girder bridges, also commonly known as truss bridges, are in effect 'open web beams' and can be used in place of beams in almost any type of bridge. In crossings over 30 m, where continuous plate girder bridges are not feasible because of possible differential settlements of piers and abutments, an expedient solution would be the traditional simply supported truss bridges. Like plate girders, these have been in use for a long time. These are very popular in the medium span range of 30 m to 150 m and are used with ease for railways and highways. The types of bridge trusses used generally are shown in Fig. 6.11.

The characteristics of some of these trusses are briefly discussed in the following paragraphs:

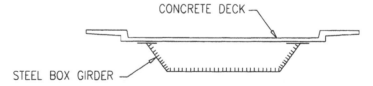

Fig. 6.10   Steel box girder with concrete deck

6.9.1 Warren Trusses

Warren trusses with parallel chords are the most commonly used form of truss (Figs. 6.11a, b, c and i). In this type, alternate diagonals are in compression. They are often used with verticals to reduce the panel size (Figs. 6.11b,

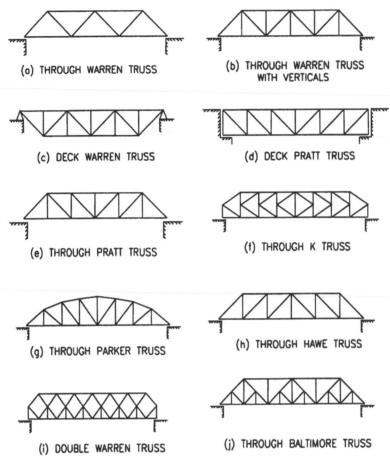

Fig. 6.11 Types of bridge trusses

and i). However, Warren trusses without vericals, present aesthetically pleasing appearance and have become popular recently in Japan and the USA. In this type, intermediate sway bracings are eliminated and truss type portals are replaced by beam portals, resulting in 'open' appearance. Additionally, detailing, as well as construction work become considerably simplified.

6.9.2 Pratt Trusses

Pratt trusses are also of parallel chord configuration and have diagonals sloping downwards towards the centre of the span (Figs. 6.11d and e). Thus, the shorter web members, i.e., the verticals are in compression, rather than the longer web members, a situation which provides considerable advantage from design point of view.

6.9.3 Parker Trusses

These are similar to Pratt trusses, but have variable depth (Fig. 6.11g), with top chord on a curved alignment. These trusses are used when economical depth required at mid span exceeds the depth required for clearance at the end portals. In such cases, the magnitude of the forces in the chords can be nearly equalised as the curve closely follows the bending moment diagram. With this arrangement, the inclined top chords carry part of the shear, thereby reducing the forces in the diagonals. Theoretically, a parabolic profile of the top chord would result in maximum economy in the weight of the chord steel. However, this profile entails the diagonals to have a variety of slopes which is likely to offset this economy. Therefore, as a compromise, the truss profile is normally made to match with reasonable diagonal slopes, i.e., between 40 and 60 degrees with the horizontal. Compared to parallel chord trusses, these trusses attract a marginal increase in the cost of fabrication, but for medium and long spans, the increase in the cost is likely to be more than offset by the saving in material.

6.9.4 K-trusses

These trusses (Fig. 6.11f) provide satisfactory solutions for spans with increased length where deep trusses with short panels are required for economy. These trusses have two diagonals in each panel to intersect at midheight of a vertical which look like the alphabet 'K'. With this arrangement, the inclination of the diagonals is kept within the desirable limits and the required depth of the truss for economy can be achieved. Also, the short panels would keep down the cost of the floor system.

Truss bridges can also have a two level deck system, one carrying the railway and the other carrying roadway; or one carrying the 'up' traffic and the other carrying the 'down' traffic.

6.10 CONTINUOUS TRUSS BRIDGES

Truss bridges may be made continuous over a number of piers (Fig. 6.12). However, since the stresses in the members are quite sensitive to the settlement of supports, this type of bridge is suitable only in cases where the differential settlements of abutments and piers are not significant. Needless to add, with this system, larger spans are possible compared to simply supported trusses. Continuous truss bridges have been in use in span ranges of 150 m to 400 m.

A continuous truss is comparatively more rigid and is a statically indeterminate structure.

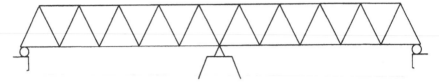

Fig. 6.12   Continuous truss bridge

6.11   CANTILEVER TRUSS BRIDGES

For extending the reach of the truss bridges in the span range of 250 m to 550 m, continuous trusses supporting suspended spans have been effectively and intelligently used all over the world. Essentially, these bridges comprise of two projecting arms emanating from support piers and forming a span, either with their tips touching, or by supporting a suspended span. The overturning moments of the cantilever arms are generally balanced by suitable counterweights on the arms extended behind the piers. For longer spans, the dead load of the bridge itself helps in the balancing act. The main advantage of this type of bridge is that unequal settlements of foundations do not cause any additional stress in the members. One other main advantage of cantilever truss bridges is that it can be erected without using any falsework, as the bridge can be erected member by member from the support piers by means of cranes placed on the bridge itself. This method is generally known as cantilever method of erection. A typical cantilever bridge is shown in Fig. 6.13.

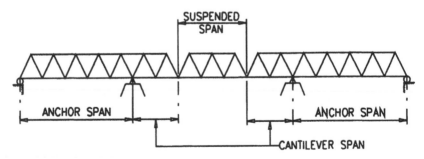

Fig. 6.13   Cantilever truss bridge

Steel cantilever bridges became quite popular for long span railway bridges because of their greater rigidity compared to the suspension bridges.

However, during the last decades of the twentieth century, with the development of cable stayed bridges, cantilever bridges became an increasingly less attractive solution for long span railway bridges.

6.12 ARCH BRIDGES

Basic principles of arch form have been successfully used in bridge construction for many centuries. In many locations they fit the contours of the country and are aesthetically very pleasing. In steel construction also they have proved economical in a span range of 200 m to 500 m. They are well adapted in high crossings and gorges. Like cantilever bridges, they can also be constructed without falsework. Versatility of steel has been utilised to create arches of different types. Some basic configurations are shown in Fig. 6.14.

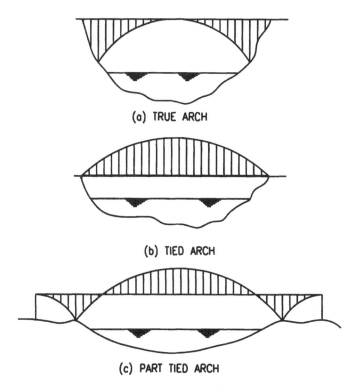

(a) TRUE ARCH

(b) TIED ARCH

(c) PART TIED ARCH

Fig. 6.14 Typical arch bridges

In arch bridges, most of the forces are transferred from the deck to the arch by compression. The structural elements of the arch ribs also act in compression and develop large outward horizontal thrusts at the spring points. Thus, arches are ideally suited for rocky gorges which furnish efficient natural abutments to receive such heavy thrusts without any movement. In cases where the ground conditions cannot guarantee movement-free foundations, it is expedient to introduce hinges at the spring points and at the crown. Alternatively, a tied arch may be used. In this type, a tie girder is placed longitudinally at the springing level. This tie is used to balance the horizontal thrust, leaving only the vertical forces to be resisted by the abutments. The tie is suspended from the arch rib by means of hangers and carries the deck system of the roadway. Tied arch has also been used gainfully where construction depth does not permit the arch to be located below the road deck

Arch bridges may be designed as curved beams or trusses, and structurally they can be:

- 3-hinged : usually hinged at the abutments and at crown
- 2-hinged : usually hinged at abutments only
- 1-hinged : seldom used in practice
- Fixed : rigidly built in at abutments and continuous throughout the rib.

Provided that the abutments can withstand the moments resulting from full fixity, the fixed arch will prove most economical. However, because of ease of construction, 2-hinged arches are commonly used in bridge work. It is obvious that 3-hinged arches are least economical and do not warrant adoption in major bridge work.

Hangers of through arch bridges are generally vertical. However, truss like diagonal hangers have also been used. This system has an advantage that it reduces the bending moments in the arch rib and makes the structure very economical.

Plated ribs can be used economically for short span arches. For medium and long spans, ribs of trussed form or box section are suitable

6.13 CABLE STAYED BRIDGES

Cable stayed bridges have lately become a preferred choice in span ranges of 250 m to 600 m, because of their economic viability and pleasing appearance. They have thus become effective alternatives to steel arch or cantilever type bridges, filling up the gap between the girder bridges of moderate spans and long span suspension bridges. Although their origin can be traced back to more than 400 years, modern cable stayed bridges were developed by German engineers in the post-war years in order to economise on quantity of steel for reconstruction of their war devastated bridges.

A cable stayed bridge essentially consists of a main girder system supported on abutments, and in addition, elastically supported along the span by inclined cables, emanating from one or more towers or pylons. A typical cable stayed bridge is shown in Fig. 6.15. The cables form the tension elements of a truss system while the deck structure acts as the compression element. Thus, the structure is self-anchoring and, therefore, unlike suspension bridges, does not require a particularly good foundation condition for anchoring the cables. However, the deck needs to be designed for the axial forces from the horizontal components of the cable forces. The damping effects of the inclined cables make cable stayed bridges less prone to wind-excited oscillations as are observed in suspension bridges.

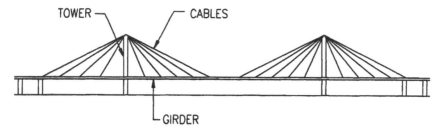

Fig. 6.15   Typical cable stayed bridge

With the development of box girders having considerable torsional strength, single plane stay system for cable stayed bridges has become possible. Recent developments in the area of computer aided design for the analysis of steel structures have enabled engineers to address non-standard problems with greater confidence. It is, therefore, becoming easier to develop cable stayed bridges with considerably greater imagination.

6.14 SUSPENSION BRIDGES

Suspension bridges offer an economical solution to the longest spans and are, perhaps, the only solution for spans in excess of 1,000 m. They are also regarded as competitive for spans as short as 100 m, depending on specific site constraints. An example of such a situation is presence of large boulders in river beds and high velocity of water flow, at the foothill regions of mountain ranges, where pier foundations are difficult to locate and become prohibitively expensive. Suspension bridges can offer ideal solutions to such problems.

Suspension bridges comprise towers, high tensile steel cables, anchorages, suspenders, deck and stiffening trusses. Figure 6.16 illustrates the main components of a common suspension bridge. Loads from the stiffened deck are transmitted via the comparatively closely spaced suspenders to the main cables which are slung over tall towers within the overall length of the bridge and anchored to anchorages on the ground at each end of the bridge. Depending on the ground conditions the anchorages may be of gravity type, i.e., relying purely on mass for their stability or of the rock anchorage type, i.e., plugged by concrete in the adjoining rock and relying on the strength of the rock and its own mass. In some cases, the main cables are attached to the stiffening trusses, which makes the structure self anchored, i.e., without requiring any external anchorages.

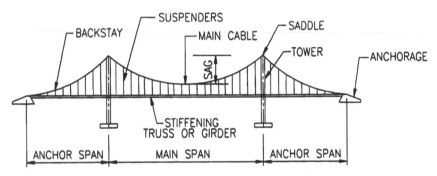

Fig. 6.16 Typical suspension bridge

Most suspension bridges are stiffened at the deck level by stiffening trusses or girders for equalising deflections due to concentrated live loads and distributing these loads to main cables via the suspenders. Athough the main cables are flexible by themselves, the significant high forces induced in them make these cables stiff enough against the vertical loadings.

Use of high strength cables in tension as the basic structural elements makes the suspension bridges to be of relatively low self-weight, hence comparatively economical for longer spans. This characteristic, in turn, also makes these bridges rather flexible, hence susceptible to vibrations. It is, therefore, imperative that an aerodynamic study be made for all major bridges of this type of construction.

REFERENCES

[1] Gaylord, E.H. (Jr.), Gaylord, C.N. and Stallmeyer, J.E. 1992, *Design of Steel Structures*, McGraw Hill Book Co. Inc., New York, USA.

[2] Dowling, P.J., Knowles, P.R. and Owens, G.W. (eds.) 1988, *Structural Steel Design*, The Steel Construction Institute, Ascot, UK.

[3] Owens, G.W., Knowles, P.R. and Dowling, P.J. (eds.). 1994, *Steel Designers' Manual*, (fifth edition) Blackwell Scientific Publications, Oxford, UK.

Chapter 7

Selection of
Structural System

7.1 INTRODUCTION

Selection of an appropriate structural system and establishment of its essential features are crucial to the success of any bridge project. This activity is normally carried out at the initial 'conceptual design stage', when a number of options of structural systems are studied and evaluated. It is necessary, therefore, to allocate adequate resources at this stage itself, in order that the most appropriate system is selected. This chapter discusses some of the commonly used criteria for evaluating different options and the techniques which can be employed for such evaluation for ultimate selection of the most appropriate solution. The basic strategy is to compare a number of options against a set of evaluation criteria.

7.2 EVALUATION CRITERIA

Evaluation critera may be divided into two distinct categories, viz., 'hard criteria' and 'soft criteria'. Hard criteria are those which cannot be compromised under any circumstance. Safety, serviceability, functional requirements, clearances, allowable depth of the structure, special preferences of the client, etc., come under this category. Soft criteria, on the other hand, are those which are open for compromise. These may include costs, aesthetics, ease of construction, durability, etc. Since hard criteria must be satisfied, evaluation for comparison is generally carried out on the basis of soft criteria only. Some of the commonly used criteria for evaluating different structural systems are:

(a) Costs
(b) Aesthetics

(c) Speed and ease of construction

(d) Environmental considerations

(e) Ease of maintenance

7.2.1 Costs

Total construction cost of a bridge comprises of the costs of superstructure, sub-structure and foundation. Primary factors that determine these costs are lengths of spans, number of piers and types of foundations. Topography of the site, navigational clearance and the sub-soil condition largely influence the locations of the piers and abutments. These, in turn, determine the number of piers, type of foundation and length of spans. Other factors that contribute to the costs are methods of fabrication and erection. Method of fabrication depends mainly on the facilities available with the fabrication contractor. If he has modern equipment, proper tools and tackles, and adequately expreienced personnel, he is likely to produce satisfactory results within the estimated cost. Thus, in order to produce the required quality and quantity of fabrication, it is necessary that the fabricator should have the infrastructure by way of men, machinery and money. Otherwise the project is likely to get delayed and cost more. Also, the method of erection can affect both the design and the costs. The designer, therefore, needs to consider these aspects also during preparation of the design and for arriving at a rational cost estimate.

The traditional practice for the financial evaluation of a bridge project is to base the decision on the initial costs of different systems with little regard to future maintenance costs. It is, however, increasingly being felt that the evaluation should be done on a suitable basis which considers the long term effects of different options. *Life Cycle Costing* (LCC) can fulfil this objective. The philosophy behind LCC is that, not only the initial costs, but also all future maintenance costs to be incurred during the subsequent life span of the competing structures are to be considered for evaluating the total cost. Thus, in an aggressive environment, a steel bridge using weathering steel as the construction material may initially cost more than an alternative solution using ordinary steel, but would not require painting during its life time and, therefore, may prove to be more economical over a period of time if analysed with LCC.

In this system, the future costs are brought together into a single value, often by applying the traditional 'present value' concept on a base date corresponding to the initial costs. In the simplest form, present value can be calculated from the following formula:

$$\text{Present Value (PV)} = \frac{C}{(1+r)^t}$$

where C is the future costs at current prices, r is the discount rate expressed as a decimal, and t is the time in years when the future cost is incurred. As an example, if the discount rate is 12% and a sum of $ 10,000 is required for maintenance after 5 years, the present value of the sum will be:

$$\text{PV} = \frac{10,000}{(1.12)^5} \text{ or about } \$ 5,675$$

The present values of all maintenance costs to be incurred at intervals of say 5 years upto a specific period (design life of the structure) can be calculated in a similar manner and added to the initial costs to give the net present value (NPV) of a particular system under consideration. Similar values of other systems may also be calculated and the results compared for financial evaluation.

Life cycle costing involves four parameters, viz., initial costs, future costs, predicted life of the structure and the discount rate. While initial costs can be reasonably estimated, there are some misgivings about the accuracy in forecasting the other parameters. Although these perceived difficulties are real to varying degrees, there is no doubt that the method can be utilised as a very useful qualitative tool for decision making. It addresses the maintenance costs in proper perspective and offers an informed and rational way of evaluation of alternative solutions. It also brings to focus that the future maintenance costs are required to be given due importance in the design process itself and that the designers should be aware of the future consequences of their present actions. However, to make it more effective, the designer needs reliable inputs, such as database of maintenance cost and service lives of different solutions as well as appropriate discount rates.

7.2.2 Aesthetics

Normally a bridge structure—situated in the rural or the urban area—is visually prominent and often dominates the surrounding lanscape. Effectively, it becomes a focal point to hundreds of thousands of commuters travelling daily by road or by rail. Therefore, in the selection of the structural system, aesthetic considerations should play an important, if not a deciding, role. The French call their bridges 'objects of art', which expresses so well what a bridge really should be.

Aesthetics is a matter of feeling and taste and relate more on the subjective (judgmental) rather than objective (logical) way of thinking. Engineers, on the other hand, by their education and training, develop a predominantly logical approach to design and often tend to become insensitive to the judgmental approach required for meaningful evaluation of the aesthetic aspects of their work. This, however, is not to say that all engineers are insensitive to aesthetics. There have been many outstanding designs of bridges by eminent engineers that adorn the face of the earth.

A few basic parameters for aesthetic structures are discussed in the following paragraphs:

Harmony with environment

A bridge structure should be in harmony with its environment and enhance its quality, rather than offend it as an intruder. It is best attained in a bridge structure by merging its lines with those of the surrounding landscape or cityscape.

Proportions

A bridge is likely to be viewed from various angles, from various distances and at different times of the day and night. Consequently, good and harmonious proportions in 'three-dimensional space' which carry a sense of balance are important for it to look elegant. Good proportions are also necessary in the relative sizes of the various parts that comprise a bridge. Thus, in a continuous bridge, the overall length of the superstructure should have good proportion between the height of the supporting piers and the distances between them. There should also be good proportions between the depth and the span of a girder bridge, between the sag and the span of a suspension bridge, and between the height and the span of an arch bridge. Thus, not only is it sufficient for a bridge to be only functionally adequate, structurally safe and economically competitive, but it needs to be well-proportioned as well.

Order

Disorder cannot lead to beauty. Therefore, order should be imposed in bridge structures. Symmetry is an age old proven element of order and should be used wherever possible. Selection of one uniform system throughout for a bridge is a good example of order. Mixing different systems in the same bridge, such as an arch with suspension bridge, may create problems of disorder. Similarly, too many directions in the same span may create confusion. Likewise, different parts of the structure should be connected by a smooth flow of lines and too many numbers of projecting parts or edges should be avoided. However, heavy overdose of order may result in monotony and dullness. Occasional breaking of the order by introducing other design elements could, in some cases, enhance aesthetic beauty.

Colour

Selection of colour in a bridge is very important from aesthetic point of view. Some colours such as grey-green, patina-green, blue blend the structure with the environment. Some colours, on the other hand, emphasise the structure itself, such as the red used in the Golden Gate bridge in San Francisco. In all cases, the colour should be appealing and attractive to the viewer.

Appearance of strength

Advancement in technology and analysis of structures has encouraged a very definite preference for slender structures in the design of bridges. There should, however, be a desire not to select too slender members as the structure may appear to be too weak to the observers. The structure, in fact, should impart feelings of confidence and reliability to the observer. In this connection, it must be understood that a bridge is primarily built for the user and it is the user who should feel safe. Therefore, the structure should not only be strong and stable, but should also appear to be so. Often, this also becomes an aesthetic requirement.

7.2.3 Speed and Ease of Construction

Speed and ease of construction are important considerations for selection of any structural system. Duration of a bridge project has an indirect effect on the overall cost. Consequently, a structural system with prefabricated components, which can be erected quickly and without any serious problem, would certainly have a definite edge over a system with a slow and complicated method of erection.

7.2.4 Environmental Considerations

Environmental considerations play an important role in the selection of the structural system. Studies need to be carried out to identify potential effects that the structure may have on the existing environment. Environmental considerations may preclude the use of some types of structures and may allow some others. As for example, hydraulic situations in a particular site may dictate selection of longer spans or shaping the foundations and piers in a manner to minimise their blocking effect on the flow of water.

Wherever *in-situ* concrete is used, there is a risk of adverse impact on the environment. On the contrary, a steel option is likely to have much less adverse impact on environment. Furthermore, a steel bridge is likely to be lighter in weight compared to a concrete alternative. This difference in weight will reduce the size of the substructure and foundation with a consequent reduction of its adverse effect on the environment.

7.2.5 Ease of Maintenance

A maintenance friendly structural system is considered superior to a system where maintenance is difficult. Therefore, ease of maintenance is an important criterion in the selection process of the system.

7.3 EVALUATION METHODS

Evaluation of different structural systems can be done by using a number of methods. Two of these methods, viz., (a) *Controlled Convergence*, and (b) *Scoring* are discussed in this section [3]. Both these techniques rely primarily on quantification as well as judgmental assessment.

7.3.1 Controlled Convergence Method

In this method, one of the structural systems is considered as a datum and all other systems are evaluated in respect of this datum. Each system is given a relative value against each criterion:

(a) ' + ' for more satisfactory than the datum
(b) ' – ' for less satisfactory than the datum
(c) ' S ' for same value as the datum.

This method is illustrated by an example. In this example there are six structural systems which are to be evaluated for final selection. Table 7.1 shows the estimated costs in millions of US dollars in respect of each system. While the construction costs indicate investments against each system, the periodic current costs in respect of future maintenance work over a specified

Table 7.1 Costs for different structural systems (million US dollars)

Particulars	Structural systems					
	1	2	3	4	5	6
Construction cost	7.25	7.45	7.15	7.90	7.95	7.98
Maintenance costs (PV)	0.50	0.95	0.95	0.90	0.95	1.00

period have been calculated to show their respective present values. It may be noted from this table that the investment for system 3 is the lowest, followed by system 1. However, the present value of the future maintenance costs in respect of system 1 is significantly lower than that of system 3.

Now, turning our attention to Table 7.2, we see that system 3, which has been arbitrarily chosen as the Datum, has been marked 'D' against each criterion. Also, relative values ('+', '−', or 'S') are assigned against each criterion in respect of each system.

Table 7.2 First evaluation by controlled convergence method

Criteria	Structural systems					
	1	2	3	4	5	6
1. Life cycle costs :						
a. Construction cost	−	−	D	−	−	−
b. Maintenance cost	+	S	D	−	S	−
2. Aesthetics :						
a. Harmony	−	+	D	+	+	−
b. Proportion	+	S	D	S	S	S
c. Order	−	+	D	S	S	S
d. Colour	S	S	D	S	S	S
e. Appearance of strength	S	S	D	S	S	S
3. Speed and ease of construction	+	−	D	−	−	−
4. Environmental considerations	+	−	D	−	+	+
5. Ease of maintenance	+	−	D	−	−	−
Summary:						
Total 'S'	2	4	0	4	5	4
Total '+'	5	2	0	1	2	1
Total '−'	3	4	0	5	3	5

In the Controlled Convergence method, if a particular structural system stands out as superior to the system chosen as Datum, then evaluation should be repeated with the superior system as the Datum. Also, if a particular criterion has only 'S' in respect of all the systems, then this criterion need not be included in the subsequent matrix, as this will have no effect on the evaluation process. Similarly, a system that has not been given any '+' and has at least one '−', can be removed from the matrix. Ideally, in a matrix there should be no '+' anywhere, in which case the 'Datum' is the clear winner. However, this situation is hardly achieved. Thus, in cases where a few '+' marks appear in a matrix, either *'convergence'* or *'trade-off'*, or both the principles can be used to arrive at a decision.

Convergence

The basic principle of convergence is to try to improve the values of the competing systems by addressing their negative sides. In other words, the design should be modified so that the negative signs are made positive.

Trade-off

In this excercise, the relative advantages and disadvantages are to be considered in respect of each criterion *vis-a-vis* the different structural systems and a compromise decision is to be arrived at.

Applying the foregoing in the illustrative example shown in Table 7.2, system 1 is clearly superior to the Datum, i.e., system 3, and therefore, a fresh matrix considering system 1 as the Datum is prepared. This has been shown in Table 7.3. It may be noted that the criteria '2d' and '2e' have been deleted in this table as these had only 'S' against every system in Table 7.2.

Table 7.3   Second evaluation by controlled convergence method

Criteria	Structural systems					
	1	2	3	4	5	6
1. Life cycle costs:						
a. Construction cost	D	–	+	–	–	–
b. Maintenance cost	D	–	–	–	–	–
2. Aesthetics:						
a. Harmony	D	+	+	+	+	–
b. Proportion	D	–	–	–	–	–
c. Order	D	+	+	+	+	+
3. Speed and ease of construction	D	–	–	–	–	–
4. Environmental considerations	D	–	–	–	–	–
5. Ease of maintenance	D	–	–	–	–	–
Summary:						
Total 'S'	0	0	0	0	0	1
Total '+'	0	2	3	2	2	0
Total '–'	0	6	5	6	6	7

Considering the revised matrix shown in Table 7.3, it is noted that system 1 leads system 3 on overall totals, but is deficient in two critera, viz., Construction Costs (1a) and Aesthetics (2a and 2c). Referring to Table 7.1, it will be noted that although the cost of construction of system 1 is marginally higher than that of system 3, the total Life Cycle Cost of system 1 is considerably less than that of system 3. This aspect alone should tilt the balance in favour of system 1.

7.3.2 Scoring Method

In this method, instead of using '+', '−' and 'S' symbols, grading is done by assigning scores (by points) to the systems in respect of each criterion. A common approach is to use a range from 1 to 10 in the ascending order,—the least satisfactory system gets 1 point, while the most satisfactory system gets 10. In order to highlight bad options, sometimes a range (say 2) spread equally on both sides of '0' (i.e, −2, −1, 0, +1, +2) is considered. For highlighting the bad options, another method is to use a 4-point scoring system of 1, 2, 5, 14, where 1 is given to the best system, while 14 to the most unsatisfactory system. Thus the bad systems automatically stand out and can be rejected.

The relative importance of each criterion can be taken into account by applying *weights* to them with scores in each row of the matrix being multiplied by the 'weight'. The sum total of the weighted scores for all the systems (columns) can then be compared for arriving at the preferred system.

As an illustrative example, the six structural systems considered in the Controlled Convergence method are now evaluated by the Scoring method, adopting a scoring range of −2 to +2. Weights (in a scale of 100) have also been applied against each criteria. Table 7.4 shows the evaluation by the

Table 7.4   Evaluation by scoring method

Criteria	Weight	Structural systems					
		1	2	3	4	5	6
1. Life cycle costs:							
a. Construction cost	30	+ 1	0	+ 2	−1	−1	−1
b. Maintenance cost	30	+2	−1	−1	−1	−1	−1
2. Aesthetics:							
a. Harmony	3	0	+2	+1	+2	+2	0
b. Proportion	3	+2	+1	+1	+1	+1	+1
c. Order	3	0	+2	+1	+1	+1	+1
3. Speed and ease							
of construction	7	+2	0	+1	−2	−2	−2
4. Environmental							
considerations	12	+2	−1	0	−1	+1	+1
5. Ease of maintenance	12	+2	0	+1	−1	−1	−2
Total	100	+158	−27	+58	−86	−62	−80

scoring method. It may be noted that the Structural System No. 1 emerges as winner in this method also.

REFERENCES

[1] Troitsky, M.S. 1999, *Conceptual bridge design*. In: Bridge Engineering Handbook, Wai-Fah, C. and Lian, D. (eds.), CRC Press, Boca Raton USA.

[2] Leonhardt, F. 1999, *Aesthetics—Basics*. In: Bridge Engineering Handbook, Wai-Fah, C. and Lian, D. (eds.), CRC Press, Boca Raton USA.

[3] McLeod, I.A. and Hartvig, S.C. 1999, *Issues and strategies for evaluation of structural design options*. In: Structural Engineering International, August, 1999, Journal of the International Association for Bridge and Structural Engineering, Zurich, Switzerland.

Chapter 8

Deck Systems

8.1 INTRODUCTION

In modern bridges cast *in situ* concrete decks acting compositely with the steel floor girders are commonly used in steel bridges. In cases where busy traffic underneath may cause problems for erection of temporary stagings required for supporting the shutterings, pre-cast concrete decks are often used. Alternatively, self supporting stay-in-place steel forms may be used to eliminate the temporary stagings.

Where dead load is required to be reduced, steel grids (gratings) may be used as decking. This system, apart from reducing the capital cost by way of lighter supporting structures and main girders, lowers the cost of removal of snow during winter and also eliminates the cost of providing drainage system for rain water, as neither snow nor rain water can remain on the grid floor. Sometimes a light concrete layer is applied to the steel grid, in which case, however, provision for removal of snow and rain water will have to be made. Grid flooring system is often used for movable and temporary bridges.

The other type of bridge deck used in modern bridges is orthotropic steel plate deck system. Both steel-concrete composite deck and orthotropic steel plate system are widely used in modern steel bridges. The salient features of these two systems are discussed in the following sections.

8.2 STEEL-CONCRETE COMPOSITE DECK

8.2.1 General

In a non-composite girder, a concrete slab simply rests on the top of a steel girder. A phenomenon known as slippage occurs at the interface. Both the elements, i.e., the slab and the girder deflect independently when the load is placed on the slab. They behave as independent beams with two distinct neutral axes. The geometry of each component defines the neutral axis and moment of inertia for each component. Thus, the tops of both the slab and the girder are in compression and the bottoms of the slab and the girder are in tension. This results in a slip which is the tensile strain at the bottom of the slab plus the compressive strain at the top of the girder. The resulting effect is that the slab would be extending out over the ends of the girder. The phenomenon is illustrated in Fig. 8.1a. The likely strain diagram is shown in the right. If the girder and the slab are integrated, the two elements will resist the load as a single unit. This integration is accomplished through the incorporation of shear connectors between the slab and the girder. Such a girder is termed a 'composite girder'. The phenomenon is illustrated in Fig. 8.1b. In this arrangement, the neutral axis is located somewhere in the middle of the section between the top of the slab and the bottom of the girder. With proper integration, this 'slab-girder' combination will act as a unit with the top of the slab in compression and the bottom of the girder in tension and no slippage in between.

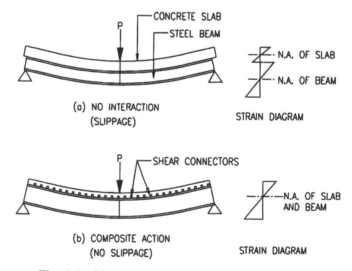

Fig. 8.1   Non-composite and composite sections

8.2.2   Shear Connectors

The primary function of shear connectors is to ensure integrated behaviour of the RC slab and the steel girder. This is achieved by designing the shear connectors to satisfy two conditions:

(a) To transmit the longitudinal shear along the contact surface without slip. For this purpose several shear connectors are fixed along the length of the steel girder. These connectors must be capable of transferring the horizontal shear at the interface.

(b) Shear connectors must also prevent vertical separation of the *in-situ* slab from the steel girder at the contact surface. Apart from the direction of the load, the other situations which may cause vertical separation of the slab from the girder are torsion or triaxial stresses in the vicinity of the connections.

Shear connectors are metal elements extending vertically from the top of the top flange of the supporting steel girder and embedded into the slab concrete. These are generally classified into three categories:

Rigid type shear connectors

Figure 8.2 shows some of the typical rigid type shear connectors. These consist of short lengths of tees, channels or angles welded on to the top flange of the steel girder. As can be seen from the figure, these connectors are detailed to be bend-proof with little inherent power of deformation. In general, these connectors also incorporate some device to anchor down the concrete slab to prevent it from being separated from the steel girder in the direction normal to the contact surface. This is often done by introducing longitudinal reinforcement bars through holes provided in the rigid connectors, as shown in the figure.

Resistance of rigid type connectors is derived from the bearing pressure of concrete, distributed evenly over the surface because of the stiffness of the

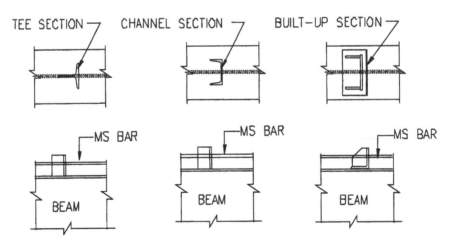

Fig. 8.2 Rigid type shear connectors

connectors. Failure of this type of connectors is generally associated with crushing of concrete.

Flexible type shear connectors

Figure 8.3 shows some common flexible type shear connectors. Unlike the rigid type connectors, flexible type connectors are not stiffened and are allowed to deform. These are commonly studs, channels, angles or tees welded to the steel girders in the manner shown in the figure. As the head of the stud and the horizontal upper flange of the channel prevent vertical separation of the slab from the girder, no special device is required for these connectors to prevent the separation. However, in the case of the angle or the tee connectors, some special device is required for this purpose. The figure shows a round bar bent in the form of inverted 'U' welded to the vertical leg, which serves the purpose.

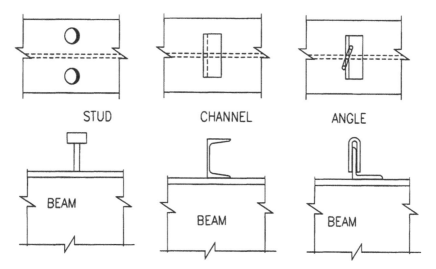

Fig. 8.3   Flexible type shear connectors

Resistance of flexible connectors is derived essentially through the bending of the connectors. The upper portion of the connectors is relatively flexible compared to the rigid bottom portion which is welded to the girder. Tests show that yield stress is reached in the channel connectors before the crushing of the concrete. Thus, the yield stress of the flexible connectors determines the strength of the connectors for a particular characteristic strength of concrete used in the slab.

Bond or anchorage type connectors

Figure 8.4 illustrates some typical bond or anchorage type connectors. These normally consist of the following:

- Inclined bars with one end welded to the top flange of the steel girder and the other end suitably bent and embedded into the concrete.
- Round bars in the form of helical stirrups welded in the bottom to the top surface of the top flange and embedded into the concrete.

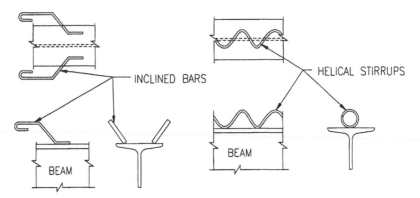

Fig. 8.4   Bond or anchorage type connectors

Resistance of these connectors is derived from their bond and/or anchorage action with concrete slab. These connectors are seldom used nowadays.

8.2.3   Advantages of Composite Construction

A composite girder has a stiffer cross section and the integrated slab can resist considerable lateral deformation. This type of girder has a higher load capacity than a non-composite girder made up of the same girder and slab. This results in saving in material consumption and economical use of steel for longer spans. In addition, a composite girder has a marked advantage over a non-composite girder whose design is governed by deflection limitations. It also makes possible significant reduction in the depth of the steel girder. This reduces the overall depth of construction leading to lower embankment for bridges, resulting in saving in the cost of embankment. Figure 8.5 illustrates the point [4].

Furthermore, toughness of a composite structure is much in excess of that of its non-composite counterpart.

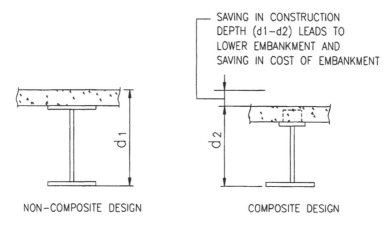

Fig. 8.5 Saving in cost of embankment

8.2.4 Construction Methods

There are two methods of construction for composite bridges, viz., with or without temporary props (or supports).

In a propped construction, the deck slab is cast with support from the steel girders and the girders are supported on temporary props until the slab concrete has hardened and attained the required strength. In such a construction, the composite section carries all loads including the dead load of the structure (i.e., the concrete slab and the steel girders).

On the other hand, in an un-propped construction, steel beams support their own weight, as well as the weights of the forms and slab during casting and curing. Thus, the composite section resists only the loads which are applied after the slab has hardened. While propped construction requires smaller steel sections than that required for un-propped construction, this apparent economy is often offset by the cost of temporary props and supports, particularly where special provision has to be made for un-interrupted flow of traffic underneath. Thus, the economic impact often dictates the construction method to be adopted for a particular composite steel bridge.

8.2.5 Effective Flange Width

A typical form of composite deck consists of a series of parallel inter-connected 'T' shaped girders, each made up of a steel girder and a portion of the concrete deck slab. The portion of the concrete slab which acts as the top flange of the T-shaped girder is termed the effective flange width. In a normal deck system, the interior steel girders have concrete slabs on both sides, while the exterior girders have slabs on one side only, as shown in Fig. 8.6. For the purpose of design, the effective flange width recommended by the governing code is to be followed. The effective flange width represents the portion of the deck which, together with the steel girder resists the loads, and is used for computing section properties of the composite section.

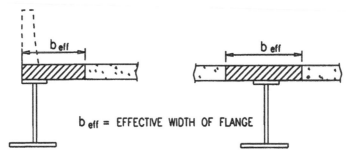

Fig. 8.6 Effective width of flange for slab for exterior and interior girders

8.2.6 The Transformed Section

For calculating the composite section properties, the composite slab-girder is transformed into a modified cross section, where the concrete slab becomes an equivalent area of steel (Fig. 8.7). Keeping the thickness of the concrete slab unaltered, the transformed width of the concrete slab is calculated by dividing the effective width of the slab by the modular ratio:

$$b_{tr} = b_{eff}/K \cdot m$$

where b_{tr} = Transformed width of the slab
b_{eff} = Effective width of the slab

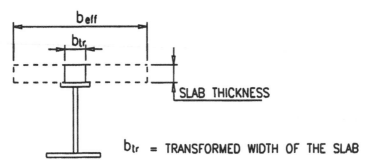

Fig. 8.7 The transformed section

$$m = \text{Modular ratio} = E_s/E_c$$

E_s = Modulus of elasticity for steel

E_c = Modulus of elasticity for concrete

K = Multiplier accounting for creep, as recommended by the governing code

The transformed width of the slab is used for computing the properties of the composite section.

8.2.7 Composite Girder Design Procedure

Design procedure of a composite girder essentially comprises the following steps:

1. A section is assumed for the steel girder
2. The concrete is transformed into an equivalent area of steel
3. The properties of the transformed section are computed
4. Capacity of the transformed section is checked against the load effect as per governing code
5. Shear connectors are designed as per governing code
6. Other design details are finalised as in a non-composite design.

8.3 ORTHOTROPIC STEEL DECK

8.3.1 General

The outstanding structural properties of orthotropic steel plate deck system have made it one of the most important elements in modern steel bridge construction.

The word orthotropy is derived from the expressions orthogonal for *ortho* and anisotropy for *tropy*, meaning that an orthotropic deck has dissimilar elastic properties in the two mutually perpendicular directions.

Orthotropic steel deck system is a comparatively modern concept which first came into use for long span bridges in the 1950s, replacing the earlier 'battle deck' floor system introduced in the 1930s for reducing the dead weight of road bridges. A typical battle deck floor system is shown in Fig. 8.8. In this system, steel plates were supported on longitudinal I beam stringers placed on (or framed into) transverse cross girders. The deck plate would participate in the stresses of the individual stringers as a part of their top flanges, but did not participate in the cross girder stresses, nor did it contribute to the strength of the main girders. Tests conducted on battle deck

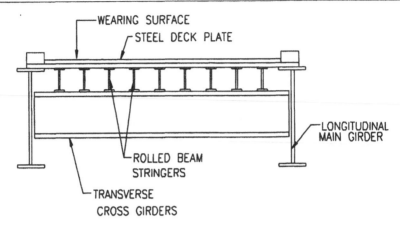

Fig. 8.8 Typical "battle deck" floor system

floors, however, showed that the strength of such steel deck plate loaded by a wheel had much higher strength reserve than predicted by ordinary flexural theory. This was recognised in the semi-empirical formulae recommending a 40% increase on the allowable stress by American Institute of Steel Construction (AISC) [5].

Figure 8.9 shows a typical arrangement of orthotropic steel deck bridge. Here the steel deck plate serves as the top flange of the longitudinal stiffening ribs, the transverse cross girders and the longitudinal main girders. The stiffened deck plate behaves as an integral part of the supporting members which together form the primary member of the bridge, having three separate sectional properties. These are: (1) bending resistance along the longitudinal axis of the bridge, (2) bending resistance in the direction transverse to the longitudinal axis of the bridge, and (3) torsional resistance about the longitudinal axis of the bridge. Because of the flexibility of the deck with the longitudinal ribs acting as beams on elastic supports, any concentrated load placed on the deck plate is efficiently distributed over a wide area to several adjacent cross girders. A light weight and thin wearing surface is normally placed on the deck plate, thereby completely eliminating the heavy concrete floor. Overall saving of weight and efficient use of materials make orthotropic steel deck system immensely attractive.

8.3.2 Structural Behaviour

In a conventionally designed bridge, the individual structural components, viz., the deck, the stringers, the cross girders and the main girders are designed to perform separate, clearly defined functions. Thus, the deck

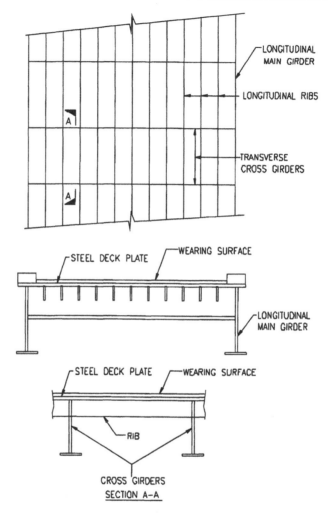

Fig. 8.9 Typical arrangement of orthotropic steel deck bridge

directly supports the wheel loads and transmits these to the stringers, which react on the transverse cross girders. These cross girders, in turn, transmit the loads to the main girders. Thus, the individual members act independently and do not contribute to the strength or rigidity of the other members. The transverse stability of the bridge is normally provided by top and bottom lateral bracing systems.

In orthotropic steel deck bridge, on the other hand, the functions of the structural components are closely inter-related. The deck, the stringers and the cross girders are integrated as one structural element with the steel deck plate acting as a common top flange. This stiffened deck (along with the longitudinal ribs) becomes a part of the main girders as their top flange. The deck also provides adequate transverse rigidity to the main girders. Therefore, a separate lateral bracing system in the deck level is no longer required. Also, the closely spaced grid structure with the steel plate deck has a good load distributing capacity of concentrated wheel loads to adjoining elements. Thus, the safety against failure in this system is considerably greater than that of a conventional bridge floor. A local load causes an elastic and, eventual plastic stress re-distribution to the adjoining elements. This eliminates immediate failure of the overloaded element. If the load is further increased beyond critical limits, the eventual failure would be a local one, and the entire structure would not be affected.

8.3.3   Design Considerations

The stresses in any element of a loaded orthotropic steel bridge deck is due to the combined effects of the various functions that the deck is called upon to perform. For convenience of design, these stresses are assumed to result from bending of four types of members.

Member Type 1 is defined as the deck plate supported by welded ribs as shown in Fig. 8.10a. The deck plate acts locally as a continuous member, and directly supports the loads placed between the ribs and transmits the reactions to the ribs.

Member Type 2 consists of the deck plate and the longitudinal ribs spanning between cross girders and is normally continuous for at least two spans (Fig. 8.10b).

Member Type 3 comprises the stiffened deck plate and the transverse cross girders spanning between the two main girders (Fig. 8.10c).

Member Type 4 consists of the main girders and the stiffened deck plate spanning between the supports and is normally continuous for at least two spans (Fig. 8.10d). For the purpose of computation of stresses, the effective cross sectional area of the deck plate along with the cross sectional area of the longitudinal ribs is considered as the top flange of the main girders.

Thus, the deck plate acts as the common top flange for all the three members, viz., the ribs, the cross girders and the main longitudinal girders, thereby making the system to behave in an integrated manner. Discussions on detail design is not within the scope of the present text. For further detailed treatment on the subject, relevant literature and codes should be consulted.

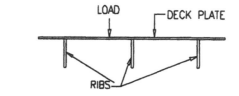

(a) MEMBER TYPE-1 : DECK PLATE
(SUPPORTED BY RIBS)

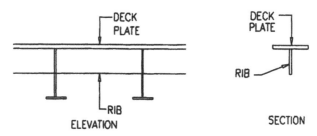

(b) MEMBER TYPE 2 : DECK PLATE AND RIB
(SUPPORTED BY CROSS GIRDERS)

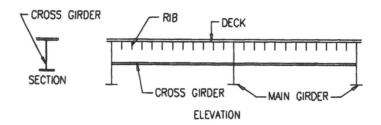

(c) MEMBER TYPE 3 : STIFFENED DECK PLATE AND CROSS GIRDERS
(SUPPORTED BY MAIN GIRDERS)

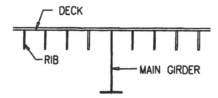

(d) MEMBER TYPE 4 : MAIN GIRDER AND STIFFENED DECK PLATE
(SUPPORTED BY ABUTMENTS / PIERS)

Fig. 8.10 Four types of members in analysis of orthotropic plates

8.3.4 Components

Longitudinal ribs

In general, two basic types of longitudinal ribs are used, viz., open type and closed type. Figure 8.11 shows some typical rib configurations. The ribs are generally spaced at about 300 mm centres.

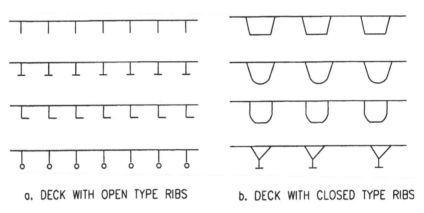

Fig. 8.11 Typical rib configurations

Open type ribs comprise flat bars, inverted T sections, angles, bulb sections, welded to the deck plate. Closed ribs generally comprise trapezoidal, rounded, triangular, and combined shapes. Bent or rolled pieces of steel are welded to the deck plate to form closed ribs. A single angle (rolled or folded plate), rotated by 45 degrees and both the legs welded to the plate may form a triangular closed rib. Additional inverted T section welded at the tip of the triangular closed rib increases the flexural rigidity of the triangular rib considerably. Closed rib deck system is essentially a series of miniature box girders placed side by side with considerable torsional rigidity and can contribute substantially to distribute the load transversely. On the other hand, torsional rigidity of the open type ribs is very small and consequently these ribs have relatively small capacity of wheel load distribution in the transverse direction. They also require closer cross girder spacing. These factors make a steel deck with open type ribs less attractive economically, being heavier than a steel deck with closed type ribs. Also, the quanity of welding required for an open-rib system is significantly more than that in a closed-rib system. The advantage of the open-rib system is, however, its relatively simple fabrication procedure, requiring only a normal degree of precision, and ease in site splicing. The bottom of the open-rib system is also easily accessible for inspection and maintenance. On the other hand, closed-rib system has only half the rib surface to protect from corrosion, which reduces the mainenace cost considerably.

Generally, there are two details for connecting the longitudinal ribs of trapezoidal cross section to the deck plate: the edges of the ribs can either be bevelled to match the slope of the webs (Fig. 8.12a) or cut square (Fig. 8.12b).

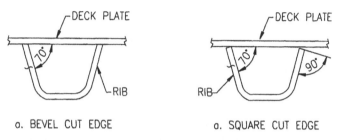

Fig. 8.12 Edge details of ribs

The ribs can be formed to the required trapezoidal shape by means of brake press. Alternatively, the trapezoidal ribs can be obtained by rolling process. Normally, ribs are formed from steel plates of required width. Sometimes, special mill rolled sections of the shape shown in Fig. 8.13 are used for this purpose. In such plates, the central portion of the plate width which will form the bottom flange is rolled thicker and the two outer portions which will become webs are rolled thinner, thereby providing more material where required and further reducing the weight of the bridge. Since the trapezoidal ribs can be nested one above the other, these can be stored and transported economically [6].

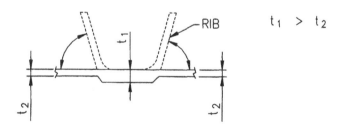

Fig. 8.13 Specially rolled plates for trapezoidal ribs [6]

Transverse cross girders

Transverse cross girders may comprise of inverted T sections (rolled T sections, or T sections cut from rolled I sections, or built up T sections) using the deck plate as the top flange. Alternatively, rolled shapes (I beams, channels, etc.) or full depth diaphragm plates (e.g., in box girders) welded to deck plate can also be used. Their spacings may be upto about 4.5 m, depending upon design conditions, economy (material cost vis-a-vis fabrication cost) and other factors.

The longitudinal ribs are normally connected to the transverse cross girders by welding. These may also be made continuous through slots in the webs of the cross girders for at least two spans. This detail is preferable, particularly in the portions of the deck subjected to tension.

Site splices of longitudinal ribs and panels

Typically, site splices of longitudinal ribs can be detailed in two ways:

1. The ribs are made discontinued at the transverse cross girders and joined preferably by single bevel butt welds to the webs of the cross girders. The bending stresses in the ribs are transferred through the web of each cross girder by means of welds. With this detail, however, there is a risk of lamellar separation of the cross girder web because of transverse force applied through it. Also, the detail requires two welds per rib at each cross girder. Figure 8.14 illustrates typical splice details for ribs discontinuous at cross girders.

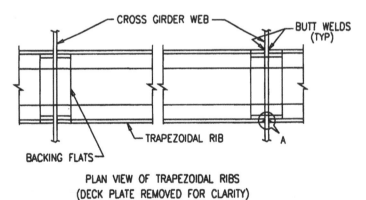

PLAN VIEW OF TRAPEZOIDAL RIBS
(DECK PLATE REMOVED FOR CLARITY)

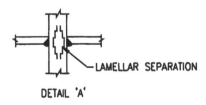

DETAIL 'A'

Fig. 8.14 Typical splice details for ribs discontinuous at cross girders [6]

2. The ribs are made to run continuous through the cross girder webs which are locally slotted to suit the profile of the ribs (Fig. 8.15). The splices in the ribs may be located at every 12 m to 18 m, depending on the length available, and away from the cross girder locations. With this detail the risk of lamellar tearing in the webs of the cross girders is eliminated. Also the quantity of welding is reduced considerably.

It is common practice to pre-fabricate the deck in panels of suitable sizes in the shops, transport these panels to the site, erect and connect these

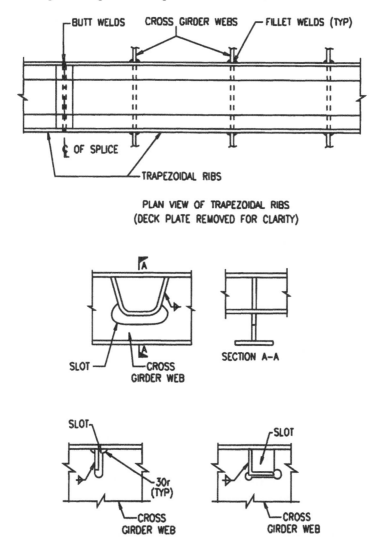

Fig. 8.15 Typical splice details for ribs continuous at cross girders [6]

by welding or bolting at site. This arrangement offers maximum use of automatic down hand welding and efficient fabrication methods, including use of jigs for the production of the deck sections. The size of the pre-fabricated panels will depend on the transportable size and/or the facilities available at site.

Bolted site connections are easy to perform and do not require particularly expert workmen at site. However, the protruding splice plate and bolt heads make the wearing surface at splice locations thinner. Also, loss of cross sectional area due to bolt holes in the tension region reduces the strength of the member to some extent. These problems are not present in welded site connections, but they need considerable skill to avoid deformation of the plates due to transverse and longitudinal shrinkage of welds.

Wearing surface

In order to achieve the objective of reducing the self weight of the structure, it becomes imperative that, in a road bridge, an orthotropic steel plate deck should have a light, and yet durable and skid-resistant wearing surface. This wearing surface should satisfy the following requirements:

1. The surfacing should be light-weight and thin. It should, however, be thick enough to adeqately cover the protrusions and the irregularities of the deck plate.
2. It is essential that the surfacing is durable and possesses a long, trouble-free service life. In a busy bridge, replacement of the surfacing becomes costly, not only to the owner, but also to the user as it disrupts the traffic flow.
3. The covering should be skid-resistant throughout its life. Vehicle tyres normally wear and polish the aggregates on the top surface, thereby reducing their resistance against skid. Hard, durable and polish resistant aggregates should, therefore, be selected. Some of the aggregates can be scattered over the surface to improve skid-resistance.
4. It must help corrosion protection of the steel deck by being impervious to water and de-icing chemicals, resistant to cracking, and well-bonded to the top of the deck plate. In effect, it should act as an integral part of the total orthotropic deck system.

Wearing surface for orthotropic deck is generally a bituminous mix such as asphalt (or polymer resin) concrete and asphalt (or polymer resin) mastic. Both types require a bond coat to be applied first on the steel deck plate to hold the surfacing in place against acceleration or braking of vehicles, as also to serve as a protective layer against corrosion. The thickness of the wearing surface ranges between 40 mm to 60 mm laid in two courses. For materials and methods of application, reference should be made to appropriate specifications. However, some materials used in the wearing surface are proprietory and patented, in which case manufacturers' specifications are to be followed.

A stiff and well bonded wearing surface of about 50 mm thickness may contribute in the distribution of the wheel load of vehicles and increase the rigidity of the steel deck plate. In addition, it may also provide some reduction in the fatigue stresses in the steel deck, ribs and welds.

8.3.5   Corrosion Protection

In orthotropic deck system, two items which are inaccessible after erection, deserve particular attention, viz., inside of closed ribs and the top surface of the deck plate. The surfaces which are accessible can be painted for corrosion protection.

As regards the closed ribs, normally, these can be made airtight during fabrication. Since atmospheric corrosion cannot occur without the presence of air, once the inside of the closed ribs is made airtight, the necessity to make any provision for corrosion protection becomes minimal. However, when closed ribs are field spliced with bolts, hand holes required for bolting make the inside vulnerable to corrosion. In such cases, airtightness is provided by welding diaphragm plates inside the rib section on either side of the splice.

The bituminous-mix wearing coat over the steel deck plate does not provide any satisfactory protection against corrosion. Therefore, the bond coat which is first applied on the top surface of the steel deck serves as the protective layer against corrosion.

8.3.6   Advantages

Orthotropic steel deck system has many advantages, of which the following are most important:

Savings in weight of steel in superstructure

For spans exceeding about 50 m, the weight per unit area of structural steel in orthotropic steel deck bridges is normally less than that of similar steel bridges of conventional design. With the increase of span, this saving increases significantly.

Studies on some European bridges of conventional design, which were destroyed in World War II and replaced by new structures using orthotropic steel plate decks, showed savings in weight per unit area of the deck to be in the order ranging between 25% and 44%. The corresponding saving in total weight of the superstructure ranged between 52% and 62% [5].

In the case of long span suspension bridges, 60% to 70% of stresses in the cables and towers are contributed by the dead load of the superstructure. Thus, reducing the weight by using orthotropic steel deck system would make the structure substantially lighter, entailing reduction in the cost of material [8].

Orthotropic deck system has been used for replacing the concrete decks of many existing bridges, thereby reducing the dead load and consequently increasing the live load carrying capacity of these bridges. Examples are: George Washington bridge, New York, Throgs Neck bridge approach, New York, Benjamin Franklin bridge, Philadelphia [10].

Mass is an important factor for designing bridges in the earthquake prone areas. The lower the mass, the lower will be the seismic forces. For such areas, orthotropic plate deck system can be gainfully used for reducing the effects of seismic forces for new as well as existing bridges. The reinforced concrete deck of the Golden Gate bridge in San Francisco built in 1937, was replaced by an orthotropic deck system in 1985, thereby reducing the seismic forces in the towers and other bridge components, besides considerably increasing the live load carrying capacity of the bridge [10].

Savings in substructure

Saving in the dead weight of the superstructure steelwork has direct effect on the substructure design. The savings in the cost of the substructure may be due to reduction in footing sizes, or in the number of piles. Also, the number of supports (piers) can be reduced by increasing the span lengths, since longer superstructure spans can be feasible with orthotropic plate deck design. Studies indicate that the savings in the order of 5% to 15% can be achieved in the foundation cost, if orthotropic plate deck system is used [5].

Ease of erection

Another advantage is the ease of erection of orthotropic steel deck bridges. The deck can be pre-fabricated in the workshops in large modular units of suitable sizes. These can then be transported to site, and easily erected even in adverse weather conditions.

The weight of the main girders using orthotropic steel deck is comparatively low. This provides significant advantage in the cantilever method of erection.

The other advantage at construction site is elimination of concreting of the deck after completion of the steel framework. This reduces the erection period considerably.

Saving due to reduction of the depth of the structure

Orthotropic steel plate deck system offers a very thin deck structure, thereby reducing the construction depth considerably as compared to the conventional reinforced concrete deck structure. For this reason, inspite of its higher cost, orthotropic steel deck systems have been in use for even small span railway bridges in Germany [8]. The savings in the construction of approach embankments, particularly in the urban environment, may exceed the cost of a small span bridge. In such cases, an orthotropic steel deck system may result even in lower total cost.

8.3.7 Applications

Orthotropic steel plate decks have been used in many bridges—from small bridges to some of the major bridges of the world. The range includes plate girder bridges, truss bridges, arch bridges, cable stayed bridges, suspension bridges and movable bridges. These are briefly described in the following paragraphs.

Plate girder and box girder bridges

Initially, most applications of orthotropic steel plate decks were in plate girder bridges. Figure 8.9 shows a typical plate girder bridge with orthotropic plate deck system. As already discussed, the integrated deck system reduces the dead weight of the superstructure considerably, making it possible for the span of the plate girder bridge to be increased beyond the conventional limits. Orthotropic deck system can be gainfully utilised in various plate girder arrangements, such as two girder bridges, multiple girder bridges, as also bridges with two pairs of main girders.

This type of deck system has been frequently used in box girder bridges. These bridges can have a single cell box with vertical webs, a multi-cell box with multiple vertical webs, or a system with inclined webs. Wide box girders possess considerable stiffness, because of their top or bottom flanges being stiffened by longitudinal ribs and intermediate diaphragms. An alternative system for a wide bridge is to have a single cell central box girder with inclined struts from the bottom flange of the box girder supporting a cantilevered deck on two sides. Figure 8.16 shows a few box girder arrangements. There are many examples of existing box girder bridges with orthotropic deck system.

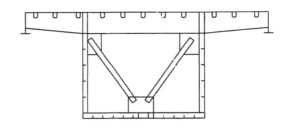

SINGLE CELL STEEL BOX GIRDER

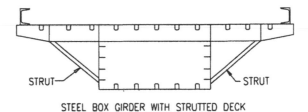

STEEL BOX GIRDER WITH STRUTTED DECK

Fig. 8.16 Typical box girder arrangements

Truss bridges

In truss bridges the orthotropic deck is designed to participate in resisting the chord stresses. An example of a truss bridge with orthotropic steel deck is the Fulda River bridge at Bergshaunsen in Germany. Built in 1962, this seven span deck type bridge has two trusses spaced at 2.3 m apart. The deck plate and the longitudinal ribs, made up with triangular box with inverted T sections at the tip, participate in the top chord stresses. The deck system also provides lateral restraint to the top chords against buckling. The German Federal Railways also use a standard orthotropic deck system in steel truss superstructure for their railway bridges. Standardisation of the deck adds to the savings in the cost of construction of such bridges.

Arch bridges

A steel arch superstructure can utilise orthotropic plate deck system to its advantage. In a stiffened tied arch bridge, the plate deck not only contributes to the flexural rigidity of the stiffening girders, but also participates in resisting the tensile stresses in the girders acting as ties. Port Mann bridge in Vancouver, Canada, is an example of such a bridge. The steel plate deck can also be used in a normal stiffened arch system to contribute to the flexural rigidity of the stiffening girders and make these girders shallower. Apart from saving in material costs, this arrangement renders a slender appearance to the structure, making it aesthetically very pleasing.

Cable stayed bridges

Structural steel box girders incorporating orthotropic steel deck system have been used in many cable stayed bridges in the past. Because of the triangular disposition of the load carrying members (viz., the stays, the deck, and the pylons), the orthotropic deck structure is subjected to compressive stresses, in addition to flexural stresses. Structural steel box girders with orthotropic steel deck respond favourably to such a combination of stresses.

Suspension bridges

Orthotropic steel deck system in box girder superstructure has been used extensively in many suspension bridges around the world. Although use of this system reduces the dead weight of the superstructure, which is likely to affect the stability adversely, this effect is more than offset by the increase of the flexural and torsional rigidity of the box girder superstructure incorporating orthotropic steel deck system. Also, as discussed earlier in this chapter, many existing suspension bridges in the USA have been retrofitted by replacing the original concrete decks by orthotropic steel plate decks to accommodate higher traffic loading and to extend the useful or fatigue life of the bridges. The world's longest concept suspension bridge, proposed to span the straits of Messina, between Sicily and Italy, also incorporates three wing-shaped box girders with orthotropic steel plate system.

Movable bridges

Orthotropic steel deck system has many advantages for use in movable bridges. Of these, saving of dead weight is by far the most important consideration. In addition, this deck system provides considerable rigidity inspite of shallow construction depth to the bridge, as also a solid riding surface. The swing bridge across a 42 m wide navigation channel near Naestved in Denmark (1997), the bascule bridge in Krakeroy in Norway (1957), the Danziger vertical lift bridge in New Orleans in the USA are some examples of use of the orthotropic steel deck system in movable bridges.

REFERENCES

[1] Duan, L., Saleh, Y. and Altman, S. 1999, *Steel-concrete composite I-girder bridges*. In: Bridge Engineering Handbook, Wai-Fah, C. and Lian, D. (eds.) CRC Press, Boca Raton, USA.

[2] Tonias, D.E. 1995, *Bridge Engineering*, McGraw-Hill Inc, New York, USA.

[3] Viest, I.M., Fountain, R.S. and Singleton, R.C. 1958, *Composite Construction in Steel and Concrete*, McGraw-Hill Inc, New York, USA.

[4] Pritchard, B. 1992, *Bridge Design for Economy and Durability*, Thomas Telford Services Ltd., London, UK.

[5] Wolchuk, R. 1963, *Design Manual for Orthotropic Steel Plate Deck Bridges*, American Institue of Steel Construction, New York, USA.

[6] Blodget, O.W. 2002, *Design of Welded Structures*, The James F Lincoln Arc Welding Foundation, Cleveland, USA.

[7] Xanthakos, P.P. 1994, *Theory and Design of Bridges*, John Wiley & Sons, Inc, New York, USA.

[8] Mangus, A.R. and Sun, S. 1999, *Orthotropic deck bridges*. In: Bridge Engineering Handbook, Wai-Fah, C. and Lian, D. (eds.) CRC Press, Boca Raton, USA.

[9] Stainsby, D. 1994, *Floors and orthotropic decks*. In: Steel Designers Manual, Owens, G.W. and Knowles, P.R. (eds.), Blackwell Scientific Publications, Oxford, UK.

[10] Ghosh, U.K. 2000, *Repair and Rehabilitation of Steel Bridges*, A.A. Balkema, Rotterdam, Netherlands.

Chapter 9

Beam and
Plate Girder Bridges

9.1 INTRODUCTION

Beam and plate girder bridges are used both as simply supported bridges and continuous plate girder bridges. In this chapter the various design aspects of these bridges are discussed.

9.2 ROLLED BEAMS

Rolled I section steel beams are the most economical and convenient type of construction for short-span bridges. To form a simple bridge, rolled beams are placed at regular intervals parallel to the direction of the traffic, between abutments or piers, and the concrete deck is cast on the top flanges of the beams. These longitudinal beams are called stringers. Not much fabrication is required for such a bridge; fixing only a pair of vertical stiffeners and drilling two pairs of holes for bearing plates at each end support should generally suffice. The concrete deck, placed over the top flanges, generally provides lateral support against buckling of the top flanges. Diaphragms are often placed at the ends and at a few intermediate locations to provide lateral supports during erection (before the deck is cast) and help in distributing the loads. In order to keep the fabrication cost to the minimum, diaphragms and their connections should be made simple. Figure 9.1 shows a beam bridge arrangement with diaphragms made of light channel sections bolted to the beams.

9.3 ROLLED BEAMS WITH COVER PLATES

When a rolled beam is used, the flexural strength of the composite section can be increased by welding a cover plate to the bottom flange of the beam. In

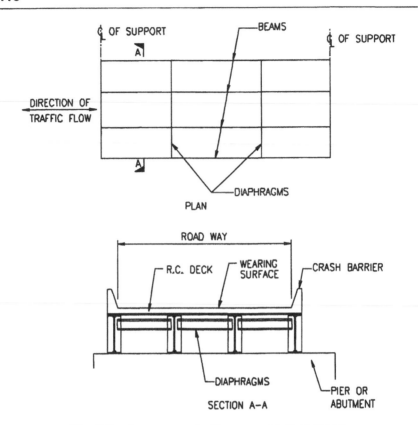

Fig. 9.1 Arrangement of beams and diaphragms

a simply supported span, economy can be achieved by using a smaller rolled section than required for maximum moment and welding a cover plate to the bottom flange in the location of the maximum moment. Figure 9.2 shows the graphical representation of a partial length cover plate. To facilitate fillet welding, the cover plate should be either suitably wider or narrower than the width of the flange of the rolled beam. Figure 9.3 illustrates the point.

The welded cover plate should be limited to only one in any single flange. The fillet welds connecting the plate to the beam should be designed to cater for the horizontal shear between the plate and the flange. While use of a cover plate in a simply supported span is quite common, use of cover plated beam in a continuous span is generally discouraged and fabricated plate girder is preferred instead. Also, the maximum thickness of the cover plate is generally limited to one and a half times the thickness of the flange to which it is attached. Thickness and width of a cover plate may be varied by butt welding plates of different thicknesses or widths with smooth transition.

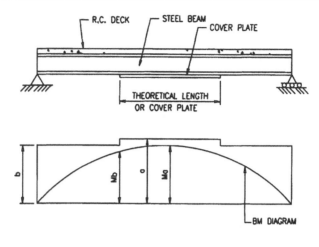

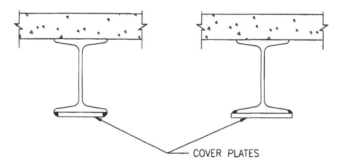

o = CAPACITY OF COMPOSITE BEAM WITH COVER PLATE
b = CAPACITY OF COMPOSITE BEAM WITHOUT COVER PLATE
Mo = BENDING MOMENT AT MID SPAN
Mb = BENDING MOMENT AT CUT OFF POINT

Fig. 9.2 Graphical representation of a partial length cover plate for a composite beam

Fig. 9.3 Typical cover plates

Figs. 9.4 and 9.5 illustrate how a smooth transition can be made by reducing the thickness or width of the larger plate to correspond to that of the smaller plate. Transition can also be made by varying the surface contour of the butt weld itself. Required radiographic tests are to be done on these plates prior to attachment to the flange.

At the ends of the cover plate, the abrupt change in the flange cross section causes severe stress concentration in the fillet weld, making the zone to be prone to fatigue related cracking. Cracks may initiate at the toe of the weld and propagate through the flange. Cracks can even lead to total collapse of

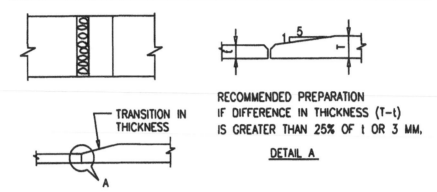

TRANSITION IN THICKNESS

RECOMMENDED PREPARATION IF DIFFERENCE IN THICKNESS (T−t) IS GREATER THAN 25% OF t OR 3 MM,

DETAIL A

Fig. 9.4 Method of transverse butt welding of plates of unequal thickness [11]

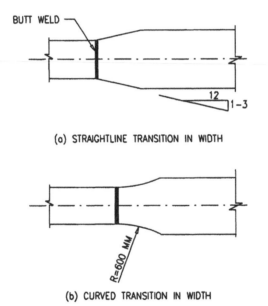

(a) STRAIGHTLINE TRANSITION IN WIDTH

(b) CURVED TRANSITION IN WIDTH

Fig. 9.5 Suggested detail of transition in width of flange plate [6]

the beam. To avoid the possibility of such cracks, many details have been suggested for the ends of cover plates. In this context two important conditions need to be considered. First, the tensile force which is uniformly distributed across the width of the cover plate is to be transferred simply and directly into the flange without any appreciable stress concentration. Secondly, the change in the cross sectional area of the cover plate at the curtailment region should be gradual, in order that the change in the developed bending stress is also gradual, so that the fatigue strength of the beam at this location does not get reduced. In the background of these conditions, some of the details suggested by different authorities are now discussed:

1. Elimination of the welds across the ends of the cover plates (Fig. 9.6a). This solution, however, may not be preferable from durability point of view, as a continuous weld is needed to seal the ends to prevent ingress of moisture underneath the plate, whch may otherwise lead to rust formation.

2. Bolting at the ends of the cover plate instead of using transverse welds (Fig. 9.6b). This detail, however, is not aesthetically pleasing and may not be a popular solution.

WELD ⌐NO WELD ACROSS END END OF FILLET WELD— ⌐BOLTS

(a) (b)

Fig. 9.6 Alternative details at end of cover plate

3. As discussed earlier, the first condition for consideration, viz., transferring the force directly into the flange without causing appreciable stress concentration may be achieved by providing a large transverse fillet weld across the ends of the cover plate. A large transverse fillet weld—in contrast to a small fillet weld—would help to transfer the force more uniformly through the surface between the fillet weld and the flange into the end of the cover plate, and improve the fatigue strength of the joint. See Fig. 9.7a and 9.7b.

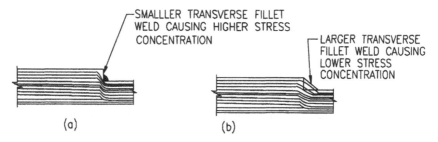

⌐SMALLLER TRANSVERSE FILLET WELD CAUSING HIGHER STRESS CONCENTRATION

⌐LARGER TRANSVERSE FILLET WELD CAUSING LOWER STRESS CONCENTRATION

(a) (b)

Fig. 9.7 Effect of large transverse fillet weld on stress concentration [6]

4. The second condition, viz., gradual change in the cross sectional area of the cover plate at the curtailment region favours a taper in the width of the cover plate at the ends. With a thick cover plate, the plate can be tapered also in the thickness to improve the gradual transition of the sectional area. Figure 9.8 shows suggested proportions of the taper in the width as well as in the thickness. Termination of the cover plate with tapered detail, however, needs some further consideration. Presence of the web in this region, makes the zone rather rigid, with little chance of localised yielding to prevent build up of possible high stress concentration. A large fillet weld, as suggested in the previous paragraph (Fig. 9.7b), would help to transfer the force more uniformly.

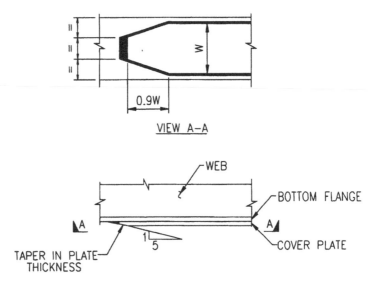

Fig. 9.8 Suggested proportion for tapering of cover plate [11]

5. Some texts suggest that both the cover plate and the fillet welds at the ends should be ground to a taper of 1 in 3 to avoid fatigue related cracks [10].

6. In order to avoid the possibility of fatigue crack due to transverse weld at the ends of the cover plate, some authorities recommend that the cover plate should extend beyond the theoretical cut-off point at each end by a distance not less than any of the following [4]:

   (a) Two times the nominal cover plate width for cover plates not welded across their ends.

(b) One and a half times the nominal cover plate width for plates which are welded across their ends.

(c) The cover plate should be extended beyond the theoretical curtailing point to within 1.5 m from the end of the girder.

It is thus not easy to come to a decision in adopting a particular detail for the cover plate ends, and the designer has to cosider various aspects and balance the pros and cons of a specific detail he wishes to adopt.

For road bridges, spacings of stringers depend on the specified minimum thickness of the concrete deck slab. The maximum spacing of the stringers should be fixed in such a way that the minimum thickness of the slab can be used. If the spacing is more, the required thickness of the slab will also be more, thereby increasing the dead load and requiring heavier stringers, and in turn, cutting into the savings from use of fewer stringers. Also, the loads on the outer stringers are likely to be different from those on the inner ones.

9.4   WELDED PLATE GIRDERS

9.4.1   General

The main advantage of plate girders is their ability to customise the sizes of the primary components to the specific bending moment and shear dictated by the design. This customisation of the cross section leads to an economy in the material consumption of the steel superstructure. A point of caution needs to be raised in this context. A girder with too many variations in the cross section configuration may represent a girder with least metal, but may not be the most economical girder. This is because excessive variation in plate sizes increases the fabrication cost, which the designer must guard against while finalising the component sizes of the plate girder.

In a constant depth simply supported plate girder, the depth of the web is generally determined by the maximum bending moment, while its thickness is dependent on maximum shear. Also, the minimum thickness of the web is governed by the depth to thickness ratio, which is generally specified in the code. Thus, as the depth of the web increases, its minimum acceptable thickness also increases. However, the required area of the flange decreases with the increase of the depth of the web. Also, with webs fitted with longitudinal stiffeners, the required thickness is generally lower. It therefore follows that, in order to arrive at an economical solution, a number of alternative arrangements are required to be tried out and compared. The various aspects to be considered while selecting the components of a plate girder are discussed in the following paragraphs.

9.4.2 Flanges

In a simply supported composite plate girder, the usual practice is to keep the width and the thickness of the top flange uniform. The strength of the composite section can be varied by varying the area of the bottom flange. This can be achieved by varying either the thickness or the width of the bottom flange. The thickness of the flange can be varied in two alternative ways:

- By butt welding lengths of different thicknesses in the transverse direction.
- By fillet welding a cover plate on to the bottom flange plate.

Since butt welding is a costly process, transverse butt welded flange plate for a short span is likely to form a high proportion of total welding cost and may not prove to be an economical solution. Therefore, in such cases, the second alternative, viz., use of a cover plate may prove to be economical.

Where plates of unequal thicknesses are butt welded to form a single length of flange plate, the thicker plate is usually chamfered to the thickness of the thinner plate with a slope of not greater than 1 in 5 (Fig. 9.4). In case the difference in thickness of the plates is less than 25% of the thickness of the thinner plate, or not more than 3 mm, the slope can be made up in the weld itself.

Transition in the width of the bottom flange can be done in the manner shown in Fig. 9.5. However, this practice is not very popular nowadays because of high fabrication cost.

A number of factors control the selection of the type of butt weld to be adopted. These include:

- Site or shop welding
- Availability of suitable equipment for edge preparation
- Availability of manipulators for easy turning over of the work
- Size of electrodes available

A double 'V' preparation is most commonly used for shop welding flange plates where manipulators are available for rotating the work for welding in down hand position or vertically, with the girder on its side. Usually a double 'V' preparation consumes the least weld metal and, therefore, is most economical. Efficient back-gouging is necessary before the first run on the second side is commenced. Where large electrodes are used and down hand welding is limited to one side only, a single 'U' preparation is recommended. For site welds a single 'U' preparation is usually preferable, although a double 'V' assymetrical preparation can be comparatively economical [11].

9.4.3 Webs

The web in a plate girder can be of either constant or varying depth. A plate girder with a deeper web near the supports is termed a haunched girder. Increased shear near the supports may necessitate a deeper web at these locations. However, high fabrication costs have made haunched girders for simply supported spans rather unattractive. These are, in fact, economical for continuous girders and will be discussed in a subsequent section. The focus of discussion in this section, therefore, will be on girders with webs of constant depth.

As in the case of the flange, the thickness of the web can also be varied according to the shear along the length of the girder. However, the cost of steel, *vis-a-vis* cost of fabrication has to be considered first. Usually, for short spans, a uniform thickness in the web is preferred. In longer spans where joints in the webs cannot be avoided due to limits on length of plates or of transportation, varying the web thickness according to shear requirement may be considered.

The weld preparation for web plate may be double 'V' or single 'V', depending on the thickness of the plate.

9.4.4 Transverse Stiffeners

Basically, there are two types of transverse stiffeners:

1. Load bearing stiffeners
2. Intermediate stiffeners

Load bearing stiffeners

These are located at points of support to resist the end reactions and at points of concentrated loads. These are designed to act as columns with an effective area which includes a portion of the web, and an effective length equalling a portion of the depth between the flanges. Figure 9.9 shows a few typical load bearing stiffeners commonly used in bridges. In large bridges, when the plate girder is excessively deep, the web may need two pairs of bearing stiffeners. In such a case, the stiffeners should be placed sufficiently apart to permit proper welding and subsequent painting.

Connection of the stiffeners to the flanges can be done either by: (a) fitting the stiffener against the flanges sufficiently tight to transmit the load, or (b) butt welding the stiffeners to the flanges. The welds which connect the stiffeners to the web are to be designed to resist the total load.

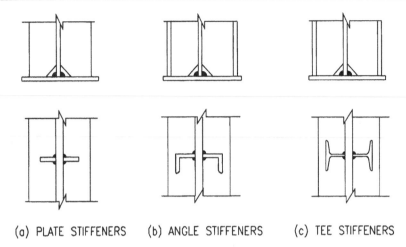

(a) PLATE STIFFENERS   (b) ANGLE STIFFENERS   (c) TEE STIFFENERS

Fig. 9.9   Typical load bearing stiffeners

Intermediate stiffeners

Loads on a plate girder cause bending moments along the length of the girder, setting up both horizontal and vertical shear stresses. These horizontal and vertical stresses combine and produce both diagonal tension and compression. If the web is deep and thin, the diagonal compression could cause the web to buckle (Fig. 9.10). Transverse intermediate stiffeners are provided to prevent such buckling of the web.

In a plate girder, however, the web is not an isolated plate. It is a part of a built up structure. Thus, when the critical buckling stress in the web is

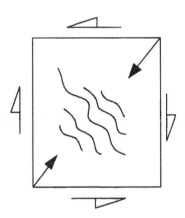

Fig. 9.10   Diagonal compression from shear stresses

reached, the girder does not collapse, as a new load carrying mechanism is developed. While the flanges carry the bending moment, the buckled web serves as an inclined tensile membrane stress field, and the transverse stiffeners become vertical compression struts. The load carrying mechanism in the post-buckling range of the plate girder becomes similar to that of a Pratt truss as shown in Fig. 9.11. The ultimate carrying capacity of the plate girder is greater under this analysis. Most of the modern codes are based on this concept. For the design and spacing of transverse intermediate stiffeners, the stipulations of the governing code are to be followed.

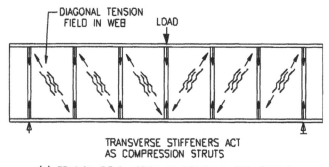

(a) TENSION FIELD ACTION IN INDIVIDUAL SUB-PANELS

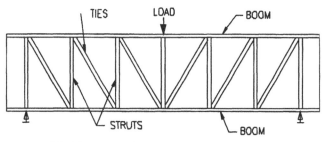

(b) TYPICAL PRATT TRUSS FOR COMPARISON

Fig. 9.11 Tension field action

Intermediate stiffeners may be connected on either one or both sides of the plate girder web. They may be plates, angles or tees. When stiffeners are used on only one side of the web, they should be welded to the compression flange to provide proper support to the flange. When a secondary member (e.g., cross frame or diaphragm) is connected to an intermediate stiffener, the possible effect of out-of-plane movement at that location should be examined.

The inside corners of the stiffeners should be suitably notched to allow the flange-to-web welds to be continuous. Fillet welding across the flow of stress in tension flange should be avoided, as this is likely to cause fatigue crack in the welds. Some commonly used ways of avoiding such welds have been shown in Fig. 9.12a, b, c and d. These are briefly discussed below:

Fig. 9.12a: The stiffener is to be stopped short of the tension flange plate, leaving a gap between the underside of the stiffener and the top face of the bottom flange.

Fig. 9.12b: The stiffener should fit into the flange sufficiently tight to exclude water after painting, eliminating the weld between the stiffener and the tension flange.

Fig. 9.12c: A small steel pad is welded on to the flanges using longitudinal fillet weld. The transverse fillet weld between the stiffener and the pad occurs on the pad.

Fig. 9.12d: A vertical plate is fillet welded on to the edge of the stiffener which is connected to the flange by weld in the longitudinal direction (in the line of the flow of stress).

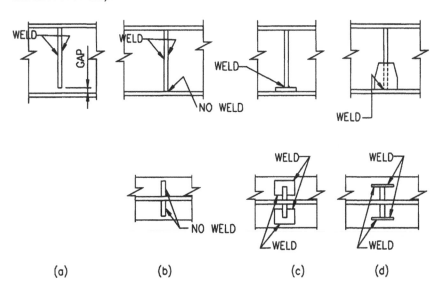

WELD
GAP
WELD
NO WELD
WELD
WELD
NO WELD
WELD
WELD
WELD
WELD
(a)
(b)
(c)
(d)
NOTE : NO TRANSVERSE FILLET WELD WITH BOTTOM FLANGE

Fig. 9.12 Typical details of stiffeners at tension flange

9.4.5 Longitudinal Stiffeners

A longitudinal stiffener is typically welded to one side of the web of a plate girder along the length of the girder. Incorporation of a longitudinal stiffener reduces the required thickness of the web plate considerably. A longitudinal stiffener also increases the shear and bending capacity of the girder, besides providing additional strength against lateral web deflection. Inspite of these advantages, however, longitudinal stiffeners have not found much popularity with the engineers because of their economic disadvantage by way of increased fabrication costs, particularly in spans below 60 m. In view of this, longitudinal stiffeners are generally not recommended unless the span length exceeds 90 m.

As mentioned above, longitudinal stiffeners are typically placed on only one side of the web plate. Some bridges have these on the inside faces of the girders, thereby leaving the outside faces of the girders smooth. If the transverse intermediate stiffeners are also placed on the inside faces of the girders, the longitudinal stiffeners are to be cut into short lengths and inserted between the transverse stiffeners. This detail results in increased fabrication cost. Placing the longitudinal stiffeners on the outside faces of the girders, and the transverse stiffeners on the inside faces, saves on fabrication time and costs. In such a case, automatic welding techniques may be used to weld the longitudinal stiffeners to the girder webs, thereby speeding the operation. Figure 9.13 shows the two alternative details.

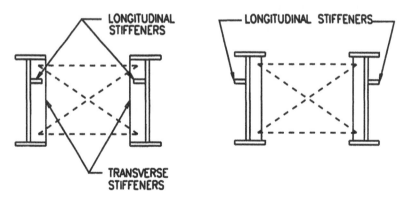

(a) LONGITUDINAL STIFFENERS ON INSIDE FACES OF THE GIRDERS

(b) LONGITUDINAL STIFFENERS ON OUTSIDE FACES OF THE GIRDERS

Fig. 9.13 Alternative arrangements for longitudinal stiffeners

9.4.6 Secondary Members

Secondary members are basically provided to prevent cross sectional deformation of the steel superstructure. They are also called upon to assist the structure to behave as a single unit, by distributing the loads between the girders evenly. For plate girders, secondary members may be classified under two heads: lateral bracings and cross frames (or diaphragms). Functions of these two types of bracings are discussed below:

Lateral bracings

Lateral bracings are placed in the horizontal planes at the top and bottom flange levels of a plate girder system to prevent lateral deformations of the primary members. In particular, lateral bracings provide lateral supports to the compression flange and also transfer the transverse wind and seismic loads to the ends of the bridge structure. Lateral bracings are not required at flanges which are attached to a rigid element like a concrete deck. In such a case, however, stability of the superstructure during construction should be investigated and if necessary, temporary lateral bracings may be provided. When precast concrete deck panels or timber planks (which are not securely attached to the girders) are used, lateral bracings should be provided.

Cross frames or diaphragms

Cross frames or diaphragms are secondary members placed in the vertical planes between the plate girders. These provide torsional stability to the system, help to evenly distribute vertical loads to the primary girders and transfer lateral loads such as wind and earthquke loads from the bottom level of the girder to the deck level (at intermediate locations) and from the deck level to bearings (at end locations). Normally, angle or tee sections in the form of 'X' or inverted 'V' are used for cross frames. The member size is designed to resist the lateral loads. For shallower girders rolled channels are commonly used. Cross frames are usually field bolted to transverse stiffeners welded to the girder webs. Spacings of cross frames are thus dependent on the locations of the transverse bracings. Figure 9.14 shows a few typical cross frame details.

9.4.7 Camber

Dead loads on a bridge girder can cause the primary members to deflect downwards, causing unsightly sagging. Plate girders are often cambered to minimise or eliminate this problem as well as to comply with the require-

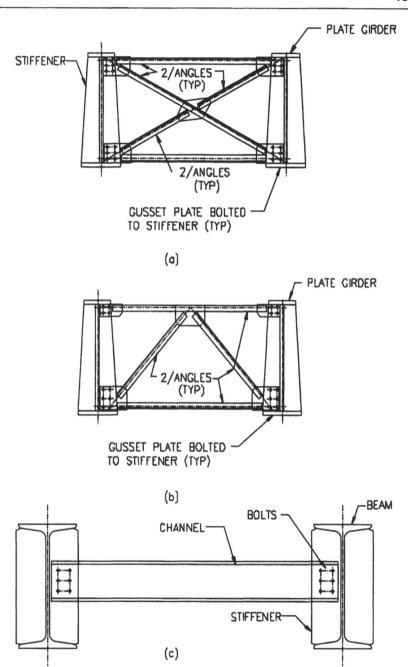

Fig. 9.14 Typical cross frame details

ments of vertical curve where applicable. Cambering of rolled beams can be done by local application of heat to the girder. However, the process is time consuming and, therefore, may not be cost effective. Alternatively, these can be cambered by cutting the web plates to the required profile and then welding the flanges to the web plates in that profile. For this purpose the designer has to indicate in the drawing the desired profile of the unloaded girder for the use of the fabricator.

9.5 CONTINUOUS PLATE GIRDERS

Since the introduction of moment distribution method for quick and easy analysis of continuous beams and frames by Hardy Cross in 1930, designers of steel bridges have increasingly been attracted towards continuous bridges. With the development of improved methods of analysis with grillage and finite element programs with computers, the desire for joint-free continuous bridges increased phenomenally. In some countries national codes even dictate 'multispan deck continuity' unless the possibility of significant differential pier and abutment settlement compels the use of a series of simply supported spans [12].

9.5.1 Advantages of Continuous Spans

Advantages of continuous spans are briefly discussed in the following paragraphs:

1. The most significant advantage of continuous bridges is economy, by way of lesser steel weight in the girders and elimination of deck joints at the piers. This latter item is very important from maintenance point of view as well, since leakage at deck joints is a constant source of problems related to maintenance of the steelwork underneath, including the bearings and the substructure. Experience reveals that decks of continuous spans are significantly more durable than those of multiple simply supported spans. Thus, by eliminating the need for joints at piers, a continuous span minimises the costs of maintenance and thereby reduces the life cycle cost of the bridge significantly.

2. Continuous spans result in reduction in the depth of the girder. In other words, the construction depth of the bridge is reduced. This can substantially reduce the costs associated with the approach ramps and earthwork. The overall bridge length can also get reduced. Figure 9.15 illustrates these points [12].

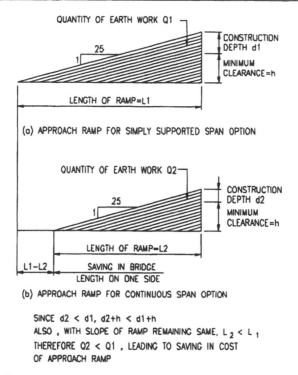

QUANTITY OF EARTH WORK Q1

CONSTRUCTION
DEPTH d1

MINIMUM
CLEARANCE=h

25
1

LENGTH OF RAMP=L1

(a) APPROACH RAMP FOR SIMPLY SUPPORTED SPAN OPTION

QUANTITY OF EARTH WORK Q2

CONSTRUCTION
DEPTH d2

MINIMUM
CLEARANCE=h

25
1

LENGTH OF RAMP=L2

L1–L2   SAVING IN BRIDGE
LENGTH ON ONE SIDE

(b) APPROACH RAMP FOR CONTINUOUS SPAN OPTION

SINCE d2 < d1, d2+h < d1+h
ALSO, WITH SLOPE OF RAMP REMAINING SAME, $L_2 < L_1$
THEREFORE Q2 < Q1, LEADING TO SAVING IN COST
OF APPROACH RAMP

Fig. 9.15 Saving in cost of approach ramp for continuous span vis-a-vis simply supported span

3. The economics achieved by adding continuity to constant-depth plate girders discussed in the preceding paragraph can be enhanced by further localised depth savings by using variable depth plate girders. By increasing the depth of the continuous girder at the supports, the sagging moment at mid-span will get reduced and the hogging moments over the supports will get increased. The degree of this 'shift' is dependent on the stiffness differential of the girder at these locations. Experience shows that inspite of the large reduction in the depth, the midspan bending stresses in a haunched girder are only a little more than the maximum stresses in a uniform depth girder. Thus, an overall economy can be achieved by using this arrangement [12].

Haunches are located over support piers and may be straight-tapered, circular or parabolic, the first option being the most cost effective from fabrication point of view. Modern trend, however, is to use girders with curved soffit, taking advantage of modern fabrication methods. This makes the structure elegant with pleasing lines.

4. For multiple simply supported spans, one set of bearings is required at the beginning of the first span at the abutment. For the first pier, one set is required for the first span and another set for the second span. Thus, every pier needs two sets of bearings—one for the preceding span—and one for the forward span. Continuous spans, however, require only one bearing at any pier, thereby reducing the number of bearings by half for every pier. Although these bearings would be of higher capacity, lesser number of overall requirement usually results in cost saving. Apart from this saving in the initial cost on bearings, maintenance of lesser number of bearings would again reduce the maintenance cost, and in turn, reduce the life cycle cost of the bridge.

5. Use of a single central row of bearings (instead of two rows) over each pier also means requirement of lesser width at the top. Additionally, due to the reduction in the dead load of the steel superstructure, and elimination of live load moments applied by off-centre sets of bearings (as in the case of multiple simply supported spans), the thickness of piers can be reduced substantially, leading to significant cost savings in pier foundations (Fig. 9.16).

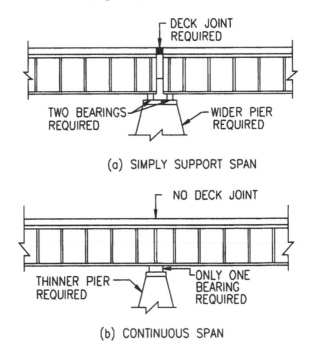

Fig. 9.16 Advantages of continuous span over simply supported span

6. A continuous structure has greater stiffness and also greater overload capacity. If overloads cause yielding at a particular location, loads are usually redistributed to adjoining locations which are not overstressed and the structure does not fail. Because of this feature, small settlements in supports have little effect on the ultimate strength of continuous spans. However, in sites where there is likelihood of large settlements, continuous spans are generally not recommended.

9.5.2 Structural Behaviour

In a simply supported span, maximum dead load moment occurs at midspan, and is positive. At this location the top flange is in compression and the bottom flange is in tension. In a continuous span, however, the maximum dead load moment occurs at the support, and is negative. At this location, the top flange is in tension and the bottom flange is in compression. The negative moment decreases rapidly with distance from the support and is reduced to zero at about quarter point of the span, generally called the point of contraflexure. The dead load moment between the two points of contraflexure in each span is positive and the maximum dead load moment at midspan is generally about half the negative dead load moments at the supports.

As regards live loads, in a simply supported span, the moments at each section are always positive. In case of a continuous span, however, maximum live load moments may cause reversal of stresses at some sections near the points of contraflexure, necessitating investigation of fatigue stresses at these locations.

In a continuous span, the total of dead load and live load moments is generally greater at supports than at midspan. As a result, usually the design criteria for a continuous span are controlled by the regions at supports. Interestingly, the sum of the dead load and the live load moments at supports of a continuous span is usually considerably less than the maximum midspan moment of a simply supported girder of the same span length. This contributes largely to the economy in using a continuous span in lieu of multiple simply supported spans.

In a composite construction, the deck slab at the support is considered to be cracked due to the tensile forces at the top of the section. This situation at the support offers two options to the designer, viz., either ignore the composite action, or take into account the effects of reinforing steel in the slab. The latter option means that the tensile properties of the reinforcing steel can assist the girder in resisting the tension in the top flange of the composite section. In the middle of the span, however, the behaviour of the composite section is the same as that for a simply supported span.

9.5.3 Span Lengths and Splices

Ideally, in a continuous bridge structure, the lengths of the end spans, which are simply supported at the abutments, should be about 0.75 to 0.80 of the penultimate span. In practice, however, due to site obstacles and restrictions, this stipulation is often not achieved and continuous bridges of irregular spans and skews are adopted. In such conditions, there can be even negative reactions in the bearings and supports due to reversal of stresses. These conditions should be adequately taken care of in the design.

The optimum span for plate girder continuous bridge structures in composite construction is about 45 m, with pier girders of about 18 m long to be site spliced after erection with span girders of about 27 m long. However, the maximum unit lengths for shop fabrication must take into account the limitations of transportation and handling capacities available at erection site. It is quite probable, that the unit lengths suggested above may have to be built up by splicing on the ground before erection to satisfy these limitations. Generally, 'pier girders' run over the piers and are site spliced near the points of contraflexure.

Site splices are formed generally with high strength friction grip (HSFG) bolts. If time permits, or where a smooth finish is required for appearance, site welded joints are recommended.

9.6 HYBRID GIRDERS

A hybrid girder is a plate girder in which high strength steel is used for top and bottom flanges while the web is built with comparatively lower strength steel. The primary advantage of this type of girder is economy in the size of the flange plates. Typically, flanges with steel having yield strength of 350 MPa to 700 MPa can be used, corresponding to webs with steel having yield strength of 250 MPa to 350 MPa. Studies, however, show that hybrid girders are marginally cost effective for spans upto about 60 m only [4].

9.7 JOINTLESS BRIDGES

Jointless steel bridges are continuous span bridges without expansion joints at either end. These are characterised by continuous spans where the steel superstructure is built integrally with shallow abutments. This type of construction is a modern development and has been adopted in a number of bridges in the USA during the last quarter of the 20th century. The design methods and abutment details vary considerably. Some authorities suggest details with the integral abutments supported on granular fills on which the abutments slide. The fill behind the abutment compresses to accommodate the movement. In an alternative scheme, the abutment rests on piles. In this case, the pile tops move to allow the horizontal movement of the abutment. Figure 9.17 illustrates these alternatives [9].

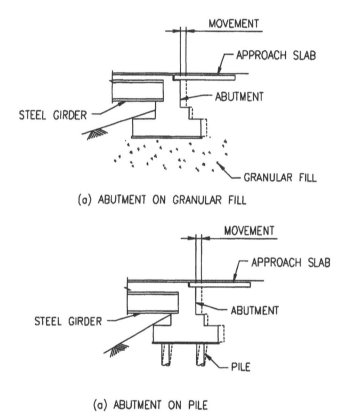

(a) ABUTMENT ON GRANULAR FILL

(a) ABUTMENT ON PILE

Fig. 9.17 Abutment details for jointless steel bridges [9]

Jointless bridges are generally considered as an answer to the chronic maintenance problems of the expansion joints of continuous bridges. It should, however, be noted that free movement of the concrete abutment may not be fully achievable, as the fill behind the abutment is bound to offer certain resistance to movement. Thus, secondary stresses are likely to be induced in the superstructure and substructure elements. This possibility needs to be addressed during the design stage. In any case, jointless bridges have yet to pass the test of time and their long-term performance and durability will determine their suitability for future construction.

9.8 COMPOSITE BOX GIRDER BRIDGES

Steel concrete composite box girders, i.e., steel box section with concrete deck can be used for longer spans because of their higher torsional rigidity and flexural capacity compared to I-shaped girders. They have been used extensively in the construction of urban viaducts and long span bridges. Their high torsional rigidity makes them attractive solutions for bridges with horizontal curved girders. Their closed shape makes them more efficient in corrosion resistance than plate girders, as the shape drastically reduces the exposed surface area, making them less susceptible to corrosion. In addition, box girders provide smooth lines and aesthetically attractive appearance.

Essentially, a composite box girder comprises of a concrete deck slab resting on an open top steel box section made up with two webs, two top flange plates, a bottom flange plate between the webs, and shear connectors welded to the top flanges and embedded in the concrete deck. The width of the top flanges should be sufficient to accommodate the required shear connectors and to provide adequate bearing for the conctete deck slab. The webs may be either vertical or inclined. The inclination is generally limited to 1 in 4. The webs are primarily designed to resist shear forces. The bottom flange, which resists bending, is often provided with longitudinal stiffeners, particularly in the regions with negative bending moment. The design of a box girder bridge should satisfy the requirements of the governing codes and specifications. While broad guidelines are briefly discussed in this text, the designer should study relevant literature for more detailed information.

A composite box girder bridge may have one or more boxes, depending on its width. Each girder may comprise of one or more cells. Boxes may be rectangular or trapezoidal. Figure 9.18 illustrates cross sections of some typical composite box girder bridges. Usually, the thickness of the concrete deck limits the spacing of the girder webs. However, thicker deck slabs may sometimes be justified to use wider girder cells for achieving overall economy.

In the case of a single box girder with multiple cells, the shear is shared by the multiple webs and the shear lag is reduced. Torsional rigidity of this system also assists the deck in better distribution of the load between the

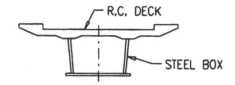

(a) SINGLE CELL SINGLE BOX

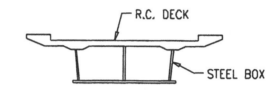

(b) MULTIPLE CELL SINGLE BOX

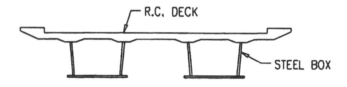

(c) SINGLE CELL MULTIPLE BOX

Fig. 9.18 Typical cross sections of composite box girders

adjacent girders. These characteristics make single box girders with multiple cells an economically attractive solution for long span bridges [2].

9.8.1 Top Lateral Bracings

In the hardened condition the concrete deck provides lateral support to the top flanges. However, before the concrete hardens, the top flanges may be subjected to lateral buckling. It is, therefore, necessary to provide a top lateral bracing system during erection so that the structure behaves as a closed box till the deck concrete is hardened and full torsional rigidity is achieved.

9.8.2 Diaphragms and Cross Frames

Diaphragms and cross frames are normally provided at the ends of a span and at intermediate locations. Diaphragms or cross frames at intermediate locations improve distribution of live loads to the girders, as also provide restraint to warping of the plates. They should withstand wind or seismic loads and also brace the compression flanges. They should also ensure that the distortion of the shape of the box girder is prevented during fabrication. They should be so detailed that access is available for final welding and post-construction maintenance.

REFERENCES

[1] Xanthakos, P.P. 1994, *Theory and Design of Bridges*, John Wily & Sons, Inc, New York, USA.

[2] Duan, L., Saleh, Y., and Altman, S. 1999, *Steel-concrete composite I-girder bridges*. In: Bridge Engineering Handbook, Wai-Fah, C. and Lian D. (eds.) CRC Press, Boca Raton, USA.

[3] Hedefine, A. and Swindlehurst, J. 1994, *Beam and girder bridges*. In: Structural Steel Designers' Handbook, McGraw-Hill Inc, New York, USA.

[4] Tonias, D.E. 1995, *Bridge Engineering*, McGraw-Hill Inc, New York, USA.

[5] Chatterjee, S. 1991, *The Design of Modern Steel Bridges*, BSP Professional Books, London, UK.

[6] Blodget, O.W. 2002, *Design of Welded Structures*, The James F. Lincoln Arc Welding Foundation, Cleveland, USA.

[7] Narayanan, R. 1994, *Plate girders*. In: Steel Designers' Manual, Owens, G.W., Knowles, P.R. and Dowling, P.J. (eds.), Blackwell Scientific Publications, Oxford, UK.

[8] Viest, I.M., Fountain, R.S. and Singleton, R.C. 1958, *Composite Construction in Steel and Concrete*, McGraw-Hill Inc, New York, USA.

[9] Parsons, J.D. 1994, *Continuous steel and composite bridges*. In: Continuous and Integral Bridges, B. Pritchard, E & FN Spon, (eds.) London, UK.

[10] Park, S.H. *Bridge Rehabilitation and Replacement*, New Jersey, USA.

[11] Brooksbank, F. 1952-53, *Welded Plate Girders*, Quasi Arc Ltd., Bilston, UK, Reprinted from Welding and Metal Fabrication.

[12] Pritchard, B. 1992, *Bridge Design for Economy and Durability*, Thomas Telford Services Ltd., London, UK.

Chapter 10

Truss Bridges

10.1 INTRODUCTION

Truss bridges or open web girders are used as simply supported girders, continuous girders, cantilever girders, arch bridges, as well as stiffening girders in suspension and cable stayed bridges. In this chapter, the various aspects to be considered for the design of truss bridges are discussed, considering a simply supported truss bridge as an illustrative example.

10.2 COMPONENTS

Principal components of a simply supported through type truss bridge are shown in Fig. 10.1 and those of a deck type bridge are shown in Fig. 10.2. These are briefly described below:

10.2.1 Deck System

Deck slab or similar structural system directly supports the wearing surface and carries the vehicular loads.

Stringers or longitudinal beams support the deck slab.

Cross girders or transverse beams support the stringers and transmit the deck loads to the main trusses at panel points.

10.2.2 Truss System

Chords are the top and bottom members of a truss system and resist tensile and compressive forces induced by bending.

Web members consist of diagonals and verticals. In parallel chord trusses, the diagonals resist the shear in the trusses. The verticals carry panel loads.

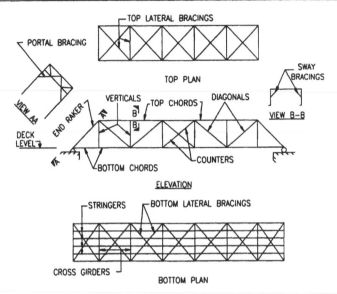

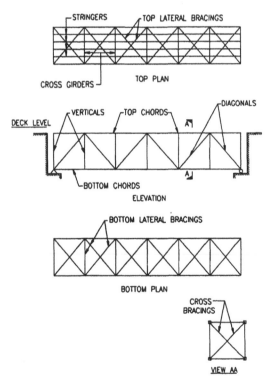

Fig. 10.1 Principal components of a through type truss bridge

Fig. 10.2 Principal components of a deck type truss bridge

Counters are a pair of diagonals placed in the form of an 'X' in a truss panel where the single diagonal would be subjected to stress reversals. These are generally found in older bridges.

End posts or end rakers are located at the ends of a truss to carry the lateral forces from the top chord level to the bridge bearings. For this purpose portal bracings are fixed onto them at the upper level.

Joints are points of intersection of different truss members. Joints at the top and bottom chords are often termed 'panel points'.

10.2.3 Bracing Systems

Lateral bracings are provided in the planes of the upper and the lower chords between the two trusses. They provide rigidity to the structure and also carry the transverse loads to the bridge ends.

Portal bracings are placed at each end of the span in the planes of the end posts (or end rakers) and transfer the horizontal transverse forces from the top lateral bracings to the bearings of the bridge.

Sway bracings are usually located at the top chord level at intermediate panel points for providing torsional rigidity to the truss frame, as also for distributing the transverse loads to the lateral system.

10.3   ACTIVITIES FOR THE DESIGN OF TRUSS BRIDGES

The following activities are arranged in a more or less logical sequence for the design of truss bridges.

- Layout and general proportions of the bridge.
- Design of deck system.
- Estimate of the dead load of the bridge.
- Analysis of truss (with assumed loads) to obtain forces in the truss members.
- Design of top chord members.
- Design of bottom chord members.
- Design of diagonals.
- Design of verticals (posts and hangers).
- Design of bracing system.
- Computation of the actual dead load of the truss and comparison with the dead load assumed in the analysis. In case the actual dead load

exceeds the assumed dead load, fresh analysis with the correct dead load is required to be done in order to obtain the final forces in the members. The design has to be checked accordingly.

- Design of joints, connections and details.
- Checking of deflections due to dead loads and live loads.
- Checking of secondary stresses in members, if necessary.
- Review of erection requirements.
- Review of aspects related to durability and post-construction inspection and maintenance.

10.4 LAYOUT AND ARRANGEMENT

Trusses may be of either single-plane or double-plane type. In a single-plane truss, all the gusset plates lie in one plane and each joint has only one gusset plate to which the converging members are connected. These members are normally single angle or channel, double angle or double channel, as shown in Fig. 10.3a, and are mostly used in lateral bracings. For the main structure, the members are normally of box or H shape. Gussets on two parallel planes are required for connection of these trusses, which are called double-plane trusses (Fig. 10.3b).

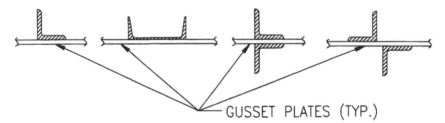

Fig. 10.3a Typical single plane truss members

Finalisation of the layout and arrangement of the bridge depends on a number of inter-related parameters.

10.4.1 Height-Span Ratio

Some codes specify generalised height-span ratios for truss bridges. However, the economic ratio depends upon many factors, such as the type of the truss, span length, loading requirements, user requirements, etc.

For simply supported highway bridges, some authorities suggest the ratio to be between about one-fifth and one-eighth. For continuous trusses, a

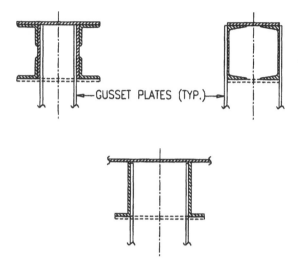

Fig. 10.3b Typical double plane truss members

ratio of around one-twelfth may be satisfactory [6]. Since railway loadings are heavier compared to highway loadings, normally railway bridges require deeper girders compared to highway bridges. However, these suggested ratios are not sacrosanct. User requirements such as minimum clearance diagram, may finally dictate the depth of the girder to be adopted.

10.4.2 Panel Dimensions

Panel lengths are influenced by the slope of the diagonals with the horizontal. The optimum slope may be taken at between 40 degrees and 60 degrees. When the span increases, the economical height also increases. In that case, if the same inclination of the diagonals is retained, the panel lengths will increase. If the panels become too long, the floor system becomes heavier and costlier. In order to reduce the panel lengths, often the panels are subdivided. However, subdivided trusses generally develop significant secondary stresses—a situation which should be guarded against.

10.4.3 Bridge Cross Section

The cross section to be adopted in a bridge should be decided quite early in the project—almost simultaneously with the finalisation of the depth of the truss, as these two items are more or less inter-related.

In a through girder system, the structure clearance diagram for roadway or railway bridges, as the case may be, normally determines the transverse spacing of the trusses. Also, if walkways are to be provided, suitable locations for these are to be considered and the spacing of the trusses are to be fixed accordingly. Provision for carrying utilities, such as water lines, power cables, etc., are to be considered, if necessary.

The truss configuration and panel length determine the spacing of the cross girders, because they are located at the panel points. The number and spacing of stringers depend on whether they are designed as simply supported beams between the cross girders or continuous over them, as also whether they are acting compositely with the reinforced concrete deck. The other alternative is to use an orthotropic plate deck system, in which case the stringers are integrated with the steel deck.

The structure clearance diagram along with the arrangement of sway bracings and portal frames determine the minimum height of the truss structure. Figure 10.4 illustrates a typical cross section of a through type truss road bridge.

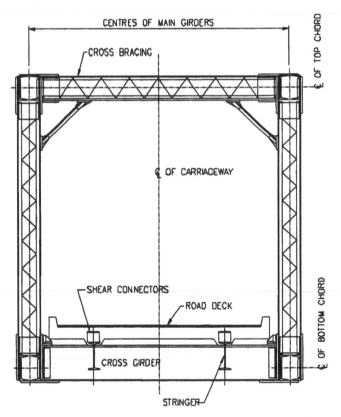

Fig. 10.4 Typical cross section of a through type truss road bridge

In the case of a deck type truss, the structure is located below the roadway deck or railway track. Therefore, the structure clearance diagram does not determine the transverse spacing of the trusses, which can be placed closer. This is likely to result in economy in the quantities of materials, because of shorter lateral and cross bracings as well as smaller piers. However, care should be taken to keep the transverse spacing of the trusses large enough to provide adequate lateral stability to the structure. Figure 10.5 illustrates a typical deck type truss road bridge.

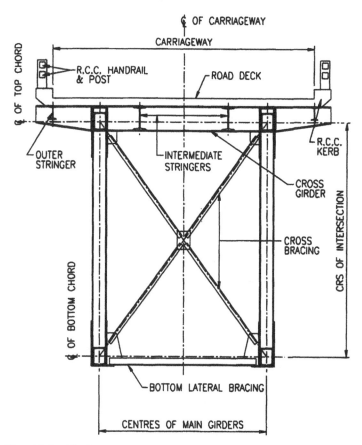

Fig. 10.5 Typical cross section of a deck type truss road bridge

10.5 DECK SYSTEM

In trusses, which are normally used in medium to long span bridges, the deck forms a sizable portion of the total dead load of the bridge. Thus, a light weight deck system can considerably reduce the total dead load, leading to substantial economy in the main structural elements and in the piers.

Railway bridges have either open deck structure or ballasted deck structure. In an open deck structure, the railway tracks are fixed on transverse timber or steel sleepers, which are supported by longitudinal stringers. These stringers span between cross girders, placed transversely at the panel points. In a ballasted deck bridge, the railway tracks are supported on ballast, which is carried by concrete deck supported on conventional steel stringer–cross girder system similar to the open deck structure. Dead load for ballasted deck structure is significantly more than that for open deck structure. However, because of advantages of comfortable ride and easy maintenance, ballast decks are preferred for most new bridges. Many railway authorities have their own standards for various items like walkways, trolley refuges, fixtures for rail tracks, etc., and the designer would do well to follow these standards while finalising his designs and details.

Bridge decks on curved railway tracks should allow for super elevation. For open decks, this may have to be built in the deck structure itself. In case of ballasted decks, this can be achieved by adjusting the concrete deck levels, or by varying the ballast depths at specific locations.

A conventional road bridge has a concrete deck supported by steel stringer–cross girder structural system, with the concrete slab acting compositely with the steel supporting girders. The deck slab may be transverse, longitudinal or a two-way slab, depending on the disposition of the longitudinal stringers and transverse cross girders. In case of a two-way slab, the most economical design solution would be when the spacings of the cross girders are almost equal to those of the stringers. However, this situation may not be achievable in actual practice and the designer may have to work on spacings as determined by other parameters.

10.6 ESTIMATE OF DEAD LOAD

Dead load on a truss bridge comprises the dead load on the deck system, the self-weight of the bracings and of the trusses. For the deck system, apart from the self-weight of the deck structure, the superimposed dead loads (such as wearing course, deck slab, walkways, railway tracks and their fixtures, etc.) are to be considered as well. It is preferable to design the deck and bracing systems first, so that the self-weights of these items can be estimated with reasonable accuracy. Prior to analysis and design of the main structure, its self-weight has to be assumed first. This is generally done using textbook empirical formulae, rough designs and accumulated office data. Experienced designers often have their own data, and the knowledge of where to look for somewhat similar structures and the capacity for shrewd guess work.

The weight of the structure, excluding the deck, may also be estimated by first determining the weight of certain components, and then approximating the weights of other members by referring to empirical charts giving average proportional weights of different members for a truss structure. One such chart [4] is given below :

Bottom chords	20%
Top chords	25%
Web members	25%
Bracings	10%
Connections	20%
Total	100%

The procedure is convenient, and is reasonably accurate for normal trusses, but can lead to misleading results in some cases. For example, a truss system with high 'depth-span ratio' will have proportionately more weight in the web members compared to one with low 'depth-span ratio', needing adjustments in the foregoing figures. The engineer should, therefore, use his judgment also while consulting such empirical charts. Nevertheless, even an approximate idea about these weights is bound to help the work of estimating the weight of the bridge.

10.7   ANALYSIS: PRIMARY AND SECONDARY FORCES

A truss is essentially a deep beam or girder, where the girder flanges are represented by the chords and the web plate is represented by the diagonal and vertical members of the truss. It is an assembly of straight members in a triangulated pattern, where the ends of each member are theoretically connected by means of pins which allow rotation of each member independent of other members.

The method of analysis of a pin-connected truss system was first developed in 1847 by Squire Whipple (see Chapter 1), and is still being used widely. In a pin-connected truss, as long as the loads are applied at the joints and no moments are applied or transmitted at the joints, the resulting forces in the members act axially along the members. These direct axial forces are termed *primary forces*, in contrast to bending forces, which are termed *secondary forces*.

The connecting straight lines between the intersections of a truss are generally termed 'working lines'. In order to avoid bending stress due to eccentricity, these lines should lie, as far as practicable, in the centroidal or gravity axes of the truss members.

Secondary forces in members of a truss may be produced by the following conditions:

- Rigidity of joints and consequent truss distortions which induce bending in the members.
- Transverse loads on members (e.g., self-weight of the members which produce bending moments).
- When the centroids of converging members do not meet at one point, producing moments due to eccentricity.
- When loads are applied on the truss from members not lying in the plane of the truss, producing torsional moments.

In the past, upto the early part of the 20th century, the system of pin-connected truss was used in many bridges and served well for the light loads then in use. However, as the loads kept on increasing with time, the system could not cope with the corresponding requirement of increased rigidity. Also the pins were never really friction-free and, therefore, some moments developed at the ends in any case, producing secondary forces. Thus, eventually, this system gave way to the modern practice of having the members rigidly connected at the joints by rivets, bolts or welds. Thankfully, the resulting restraint induces small moments, the magnitude of which vary, depending on the truss proportions and methods of fabrication and assembly. The secondary forces resulting from these small moments are generally neglected in the design. Also, the secondary forces produced by the bending moments due to self-weight of the member, being relatively small, are generally neglected. These self-weights are normally assumed to be applied at the panel points.

When a truss is analysed in a computer as a three-dimensional frame with moment resisting joints, it is generally found that virtually all members—primary as well as secondary members—share the effect due to application of live loads. Consequently, the stresses due to live loads in the primary members are reduced below those calculated by the conventional two-dimensional pin-connected truss analogy. However, in truss systems, which are normally used in relatively long span bridges, particularly in road bridges, the dead load stresses constitute a large proportion of the total load in many of the members. Thus, savings due to live loads from the use of three-dimensional analysis are likely to be rather small. It is, therefore, widely regarded that the conventional pin-connected analysis model is an adequate and reasonably accurate analysis tool. However, the following aspects need to be considered when this model is used:

- The structure should be fully triangulated both in the planes of the trusses and in the lateral planes of the chords.
- The converging working lines of the members should meet at a common point to avoid bending stresses due to eccentricity.
- Adequate cross bracings should be provided to prevent significant distortions in the transverse plane.
- Primary members should be properly cambered to eliminate secondary stresses in members in critical loading conditions. This topic will be discussed further later on in this chapter.

In cases where the working lines do not converge at a point, or where sway bracings and portals are eliminated due to aesthetic or any other reason, rigorous analysis of the truss should be carried out to ascertain bending moments of the members as well as the adequacy of the system from the point of view of rigidity.

In a conventional deck system, load from the deck is transmitted through the cross girders to the trusses at panel points only. This arrangement envisages that the deck system is designed to be structurally separate from the main supporting truss system, to which it is connected only at the panel points. When truss is analysed as a three-dimensional frame, the effect of eccentric loads from the deck system can be included in the analysis. However, in case of a two-dimensional truss analysis, the effect of these eccentric loads is not considered in the analysis, and therefore, needs to be examined separately. Generally, analysis of a cross frame consisting of cross girder, verticals of the trusses and bracing members in the frame should be sufficient to quantify the effects of the eccentric loads on the truss. If the deck slab is supported on the chords, the chords are likely to be subjected to bending in the longitudinal direction, as well as torsion in the transverse plane. The chords should be checked for this load condition also.

10.8 TRUSS MEMBERS

In modern bridges, the truss members consist of the following shapes:

Box sections

These are usually made with side channels, angles and plates. The side elements are connected at the top with solid plates and at the bottom with batten plates, lacings or perforated plates. Sometimes, the side elements are connected both at the top and bottom with batten plates, lacings or perforated plates. In modern bridges, four plate welded box members are used. For these, the flange plates are usually solid except at joint locations where access holes for bolting are provided.

I sections

These are either rolled or built up with angles and plates, or only plates welded to form the shape. The webs may be solid or perforated or with lacings and batten plates. I sections are easy to fabricate, easy to connect, and easy to maintain. However, they have one crucial disadvantage. They have a tendency for wind induced vibration. On the contrary, box sections are usually more structurally efficient and are widely used for top and bottom chords, as also for diagonals. The verticals, because of detailing reasons, are generally made of I sections.

In welded design of tension members, the individual elements that make up the section are shop welded and holes for connections are located *only at the ends*. In such cases, the tension member may be designed to their maximum capacity (considering their gross section) and the loss of area due to the holes may be compensated by welding additional plates locally to the connecting elements.

Salient features of different members of a simply supported truss bridge are discussed below:

10.8.1   Top Chords and End Posts

Figure 10.6 shows some of the cross sections used commonly for compression members forming the top chords and end posts. Generally speaking, these forms are structurally quite efficient and economical in terms of fabrication and are reasonably easy to connect to the gusset plates. These can be made of rolled or built up channels, turned either inwards or outwards. Connections may be welded, bolted or riveted. If the channels are turned inwards, the gussets are placed outside the chords. However, care should be taken to ensure adequate clearance between the tips of the channel for fixing the connections. Also, the bottom faces are connected by lacings, batten plates or perforated plates, which are not continuous members. This arrangement provides access for inspection and painting.

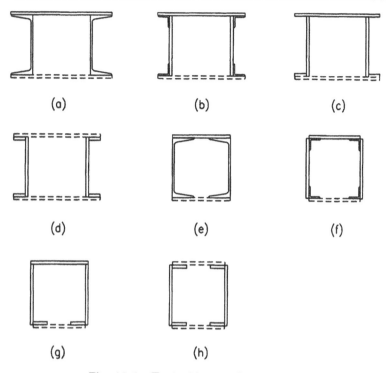

Fig. 10.6 Typical forms of top chords

Top chords are the most highly stressed compression members of the truss and need proper attention while proportioning and detailing, so that the desired strength, as well as satisfactory connections with other connecting members, can be achieved. The minimum thickness of the elements forming the members is subject to the requirements of the governing code. The maximum thickness of the webs is generally kept as thick as the economic fabrication allows. It is customary to use top flange plates in top chords; these plates connect the web plates and enhance the torsional rigidity of the section, as well as make the different elements to act together effectively as a single unit. Ideally, the depth and the width of the top chord compression members should be such that the radii of gyration about xx-axis and yy-axis are nearly equal. This condition is likely to provide an economical result.

In general, the top flange plate should be made as thin as possible, and as much material as possible should be concentrated in the webs. This would keep the centre of gravity of the section nearer to the middle of the member. It is desirable from detailing and fabrication points of view to have a uniform section of the chord throughout the length of the truss. For this purpose, the minimum section for the end panels should be designed first, and then the gross areas of the interior chords may be increased by providing additional plates in the sides of the webs, keeping uniform distance between the matching faces of the gussets.

10.8.2 Bottom Chords

Bottom chords are the most highly stressed tension members in the truss system and are normally made of rolled or built up channels. Some common shapes for bottom chords are shown in Fig. 10.7. Connections may be welded, bolted or riveted. It may be noted that the cross girder connections become simpler if the channels of the chords are turned inwards, as shown in Fig. 10.7d, e and f. In that case care should be taken to ensure adequate clearance between the tips of the channels for fixing the connections. When the flanges are turned inwards, the gusset plates are placed outside the chords. As in the case of top chords, in order to provide a uniform section throughout the length of the bottom chord, the minimum section of the end panels may be designed first, and then the area of the interior chords are increased by providing additional plates in the sides, keeping uniform distance between the matching faces of the gussets.

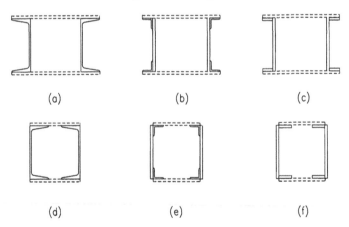

(a)   (b)   (c)

(d)   (e)   (f)

Fig. 10.7   Typical forms of bottom chords

10.8.3 Web Members

Depending on the type of the truss, some of the web members may be subjected to tension and some to compression. In a Pratt truss, the vertical members are in compression, while all the internal diagonals are in tension. In a Warren truss, alternate diagonals which are sloping away from the centre line of the truss are in compression, while those sloping towards the centre line are in tension. Vertical members may also be in compression or in tension; those in compression are termed 'posts' and those in tension are called 'hangers'.

Web members may be made from pairs of rolled channels (placed inward or outward), or built up from angles and plates. The two separate segments are connected at both faces by lacings or batten plates. Connections may be welded, bolted or riveted. The overall dimensions of the diagonal and vertical members are guided by those of the top and bottom chords, as these have to match with the connections at panel points. Some typical forms of diagonals and verticals are shown in Fig. 10.8.

In a conventional deck system, the cross girders are connected to the verticals at panel points. Thus, the sizes and forms of the verticals should be

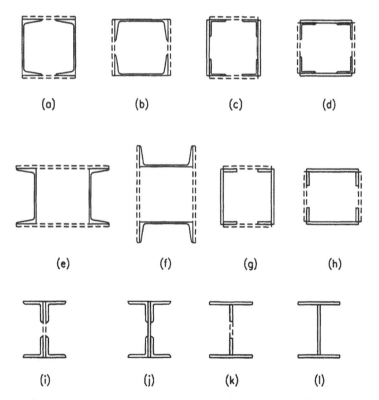

Fig. 10.8 Typical forms of diagonal and vertical members

so chosen that they are compatible with the cross girders to make proper connections at the joints.

10.9 BRACING SYSTEMS

Bracing systems in a bridge structure are considered as 'secondary' members, but they are, in fact, vital for the successful performance of the 'primary' members, both during erection and in service conditions.

Broadly speaking, bracings are designed to resist two types of forces, viz.,

- lateral forces, i.e., those acting transverse to the axis of the bridge
- longitudinal forces, i.e., those acting along the axis of the bridge

and to transmit these forces to the bearings of the bridge.

10.9.1 Lateral Forces

A bridge structure has to resist lateral forces due to :

- Wind loads
- Seismic loads
- Centrifugal forces
- Nosing action from locomotives caused by unbalanced behaviour of the locomotive mechanism (for railway bridges)
- Lurching movements of wheels against rails due to play between the wheels and the rails (for railway bridges)

Normally, three types of bracings are provided to transmit these lateral forces to the bearings :

- Lateral bracings between the two trusses in the horizontal plane
- Sway or cross bracings at intermediate locations along the length of the trusses
- Portal bracings at each end of the trusses

These bracings, acting in unison, provide lateral and torsional rigidity to the bridge structure.

Lateral bracings

Lateral bracings are placed between the top chords and bottom chords of a pair of trusses to transmit the lateral forces to the ends of the trusses. They are essentially horizontal trusses. The top and bottom chords of the main truss act as the chords of these bracings as well. Lateral bracings also serve to maintain the chords in line and to provide lateral support to the chords at intermediate points, thereby reducing the effective length in the plane of the lateral bracings. The overall effect stiffens the structure and prevents unwarranted lateral vibration. In a road bridge, although the deck slab can act as a stiffening member between the trusses, lateral bracings are required for providing rigidity during erection until the deck slab is constructed.

Although the top lateral bracings provide support to the compression chords, the supporting force is generally not considered in the design of the bracings. Furthermore, these bracings are often considered to act in tension only, because their long unbraced lengths tend to make these unsuitable to be treated as compression members.

A layout of the lateral bracing system which is commonly adopted in bridges is shown in Fig. 10.9a. In this case, the nodes of the lateral system coincide with the nodes of the main trusses and consequently interaction takes place, which needs to be taken into account in the design. This interaction can be significantly reduced if a 'Diamond' system of lateral bracing is used, as shown in Fig. 10.9b. In this detail, the nodes of the lateral system occur midway between the nodes of the main trusses. With this arrangement, when the chords are stressed, 'scissors-action' occurs and at the nodes of the lateral system, the chords deflect slightly laterally (outwards when the chords are in compression and inwards when in tension), thereby relieving the lateral system of additional stresses [8].

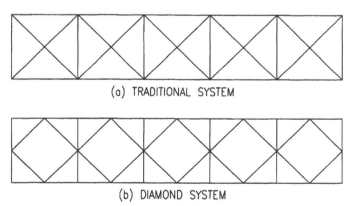

(a) TRADITIONAL SYSTEM

(b) DIAMOND SYSTEM

Fig. 10.9 Lateral bracing systems

Sway bracings

Overhead cross bracings or sway bracings are placed between trusses, usually in the vertical planes. In through trusses, these should be made as deep as possible and, in deck type trusses, should extend the full depth. Some codes require these bracings in every panel, athough in many existing bridges these are provided at alternate panels. Sway bracings are provided for distributing the transverse loads to the lateral system as also for providing torsional rigidity to the truss frame. It is, however, not easy to precisely analyse the proportion of the transverse load that can be carried by the sway bracings. Thus, the sway bracings are often provided with the real purpose of adding torsional rigidity to the truss system, rather than strength.

In general, if the sizes of the bracing members are selected to satisfy their respective slenderness ratio limits, it is most likely that their strength would also be adequate to resist the incoming forces.

Portal bracings

Portal bracings are located at the end posts or rakers and provide end supports to the top lateral bracing system. They are normally lattice frames with knee braces, and should be as deep as clearances will allow. The portal frames are generally statically indeterminate. In order that these may be sufficiently closely analysed by ordinary statics, it becomes necessary to make some simplifying assumptions. For example, the plane of the contraflexure may be assumed to be halfway between the bottom of the portal post and the bottom of the knee bracing (see Fig. 10.10). The end reaction on the plane may be assumed to be shared equally between the two end posts [7].

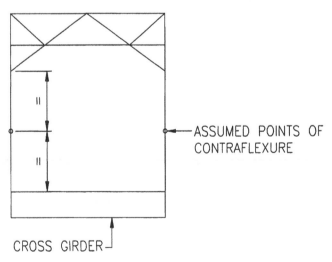

Fig. 10.10 Portal bracing

10.9.2 Longitudinal Forces

Starting or braking (acceleration or deceleration) of vehicular loads can impart longitudinal forces in bridges. In road bridges, these forces are negligible and are normally not provided for in the design. In railway bridges, however, these forces are significant and are to be considered in the design. These can occur in both the longitudinal (up and down traffic) directions. These forces are transmitted to the deck system through the rails.

Depending on certain factors such as rail continuity, rail anchorage and the deck connection to the span, a portion of the longitudinal forces is carried through the rails beyond the bridge structure. Some authorities recognise this fact and suitably adjust the forces to be considered in the design. The magnitude of the forces and their points of application are specified by the concerned authorities, which the designer has to follow.

Longitudinal forces are particularly significant in long span bridges and may induce severe bending stresses in the cross girders which are positioned transversely to the direction of the forces. It is, therefore, common for truss bridges to have traction or braking frames to divert the forces to the main trusses. In a single track bridge, transverse struts are normally provided in the level of the main lateral bracings between the points where these bracings cross the stringers in plan and struts are connected to both the stringers and the bracings, as shown in Fig. 10.11a. Since the plane of the lateral bracings may be below that of the underside of the stringers, it may be necessary to introduce suitable stools for connecting the concerned elements. The stools should be designed to transmit the longitudinal forces. For a double track system, a traction or braking truss of the type shown in Fig. 10.11b is normally fixed at each panel. The vertical distance due to the difference of levels between the bracings and the stringers may be made up by fixing suitable stools similar to those used in the case of single track system.

10.10 CAMBER

The term *camber* relates to the upward curve which a bridge truss is made to assume in its construction so that, under a specific condition of load, it will deflect and will regain its theoretical geometric shape and dimensions as determined by the final design drawings.

The need for camber arises from deflection, and deflection occurs when there are changes in the lengths of the truss members, i.e., lengthening of the tension members and shortening of the compression members. Thus, in

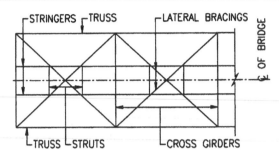

(a) SINGLE TRACK RAILWAY BRIDGE

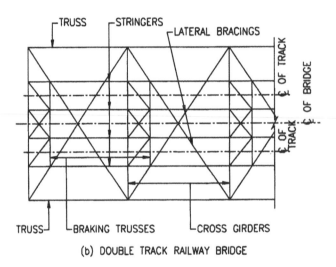

(b) DOUBLE TRACK RAILWAY BRIDGE

Fig. 10.11 Lateral bracings and braking truss arrangements

order to provide the required camber, it will be necessary to fabricate the tension members shorter by the amount to which they will subsequently lengthen under the specific load condition. Similarly, opposite action should be taken for the compression members. This method is normally adopted for longer trusses. For shorter trusses, an approximate method may be employed by lengthening the compression members by a specific amount (say, 1 mm for every metre in length), keeping the tension chords unaltered. The lengths of the web members are then computed to fit [7]. It is obvious that with the application of the specfic load condition the truss assumes its theoretical geometric shape; at any other loading state, the truss will be theoretically out of shape and the members will be subjected to secondary stresses. It is for this reason, that some national codes specify loading condition beyond dead load (for example, dead load + half live load) for computation of the cambered lengths, so that the secondary stresses in the members are kept minimal at higher load conditions.

For fabrication and erection purposes, normally a camber diagram is prepared. This camber diagram shows the amount by which the nominal lengths (i.e., the lengths which will *not* give camber) of members should be increased or decreased in order that the outline of the girder under the specific load condition assumes the nominal outline. Thus, the actual manufactured lengths of the members are to be the lengths 'with camber' given on camber diagram. The small variations in the lengths of members necessary to produce the required camber are made up, both in chords and web members, immediately outside the locations of the main gusset plates. Thus, the positions and angular setting out lines of all connection holes in the main gusset plates and also the positions of the connection holes in the chord joints and the machining of the ends should be exactly as shown in the design drawings. Furthermore, the groups of connection holes at the ends of all members are to be as per the design drawings, i.e., without any allowance for camber. Only the distance between these groups in each member should be altered by the amount of camber allowance in the member.

10.11 SKEWED BRIDGES

In a skewed bridge, abutments or piers are not perpendicular to the alignment of the bridge. Generally, the cost of steelwork for such a bridge is comparatively more than that for a normal bridge.

If the panel lengths can be made equal to the skew distance D tan φ, where D is the distance between the trusses and φ is the skew angle, the bridge can be arranged in the manner shown in Fig. 10.12. With this arrangement, the interior cross girders are at right angles with the trusses, and only the end cross girders are inclined in plan. In cases where the panel lengths cannot be made equal to the skew length, a similar arrangement may be made for the interior common panels and the difference in lengths may be made up in the end panels by adjusting the angle of inclination of the end rakers. Alternatively, the internal panel lengths may be varied for making the adjustment. In either case, the trusses may not be symmetrical about the centre lines and are not likely to be aesthetically pleasing.

Load distribution in a skewed bridge is a complex phenomenon. The pattern of loadings and their distribution through the deck system are affected by the unsymmetrical geometry of the bridge, resulting in unequal deflection of the trusses, and needing more in-depth analysis and

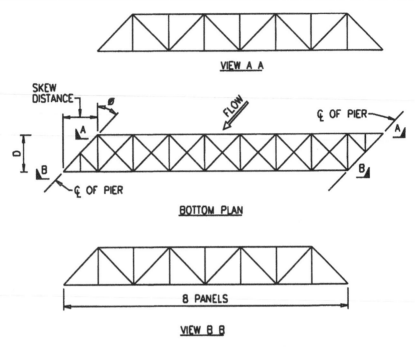

VIEW A A

SKEW
DISTANCE

FLOW

Ç OF PIER

D

Ç OF PIER

BOTTOM PLAN

8 PANELS

VIEW B B

Fig. 10.12   Typical skewed bridge

appropriate detailing of the connections of cross girders, lateral bracings and portals. Use of computer aided analysis and design should make the excercise comparatively easier.

Selection of the bearings also needs considerable attention. In a skewed bridge, the bearings are subjected to lateral rotation in addition to longitudinal rotation. The traditional bearings with rockers and rollers are designed to accommodate rotation only in one plane and are not suitable for such multi-directional rotation. Instead, it would be appropriate to use spherical, pot or elastomeric bearings which would cater for the abnormal rotation in a skewed bridge. The different types of bearings normally used in bridges are discussed in Chapter 14.

10.12   CONTINUOUS TRUSS BRIDGES

The principles discussed for a simply supported truss bridge in the preceding sections of this chapter are also applicable for continuous truss bridges. Analysis of forces in this case, however, is more complex as the structure is a statically indeterminate one. Use of computers in the analysis process makes the work much simpler. Once the forces in the different members are obtained, the design of the truss members are similar to those of a simply supported truss.

10.13 CONCLUDING REMARKS

Truss bridges have certain distinct characteristics, some of which are discussed below.

- The main members are subjected primarily to axial forces. By judicious design and detailing, secondary stresses can almost be eliminated.
- The increased depth inherent to an open web system provides greater rigidity in the plane of the truss, compared to an equivalent solid web girder. This rigidity results in reduced deflection.
- Because of its open web system and relatively large depth, a truss bridge can provide large carrying capacity for a comparatively small amount of steel. For eliciting economy, the area of steel furnished for individual components can be varied as often as required by the forces, since built up sections which form individual components can be specified with considerable exactitude. Using built up sections may add to the cost of fabrication, but overall economy can be achieved by saving on the total quantity of steel.
- Although the plethora of components associated with some truss bridges contributes to a somewhat unpleasing appearance, this construction system provides ample rigidity to the structure allowing it to behave like a space frame. In effect, a three-dimensional truss behaves much like a closed box structure, as its four planes are capable of resisting vertical, as well as lateral loads, and transmitting them to the end bearings through the end portals. Thus, in the unforeseen event of damage to a main truss member (from accident, war or terrorist attack), the bracing system can provide additional load paths to carry the loads around the damaged area. This situation was noticed in Hardinge railway bridge in Bangladesh which was damaged during a brief war in 1971. Although one bottom chord and some diagonals along with a portion of the deck in the central region of the 109-metre long ninth span were blown up, the bridge structure survived from collapse due to the presence of an extensive system of top and bottom lateral bracings along with sway bracings at every cross frame location (Fig. 10.13). The span was successfully rehabilitated later.
- Generally, truss bridges are comparatively easy to erect, needing only light equipment. Also, connections can be done with fasteners

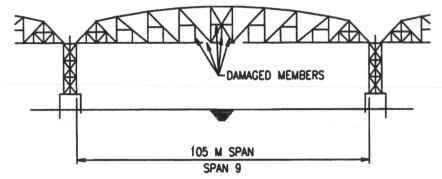

Fig. 10.13   Damaged Hardinge bridge, Bangladesh

(primarily bolts or rivets) without any sophisticated tools or equipment.

REFERENCES

[1] Xanthakos, P.P. 1994, *Theory and Design of Bridges*, John Wiley & Sons, Inc, New York, USA.

[2] Kulicki, J.M. 1999, *Highway Truss Bridges*. In: Bridge Engineering Handbook, Wai-Fah, C. and Lian, D. (eds.), CRC Press, Boca Raton, USA.

[3] Francis, A.J. 1989, *Introducing Structures*, Ellis Harwood Ltd., Chichester, UK.

[4] Bresler, B. and Lin, T.Y. 1960, *Design of Steel Structures*, John Wiley & Sons, Inc, New York, USA.

[5] Gaylord, E.H. (Jr.), Gaylord, C.N. and Stallmeyer, J.E. 1992, *Design of Steel Structures*, McGraw Hill Book Co. Inc, New York, USA.

[6] Kulicki, J.M., Pricket, J.E. and LeRoy, D.H. 1994. *Truss Bridges*, In: Structural Steel Designer's Handbook, McGraw Hill, Inc, New York, USA.

[7] Lothers, J.E. 1965, *Design in Structural Steel*, Prentice-Hall Inc, New Jersey, USA.

[8] *Commentary on BS 5400: Part 3 : 1982, Code of Practice for the Design of Steel Bridges*, The Steel Construction Institute, Ascot, UK, 1991.

[9] Sorgenfrei, D.F. and Marianos, W.N. (Jr.) 1999, *Railroad Bridges*. In: Bridge Engineering Handbook, Wai-Fah, C. and Lian, D. (eds.), CRC Press, Boca Raton, USA.

Chapter 11

Semi-through Bridges

11.1 INTRODUCTION

Semi-through bridges may be of plate girder or of truss construction. Figure
11.1 illustrates the configurations of both these types of semi-through
bridges.

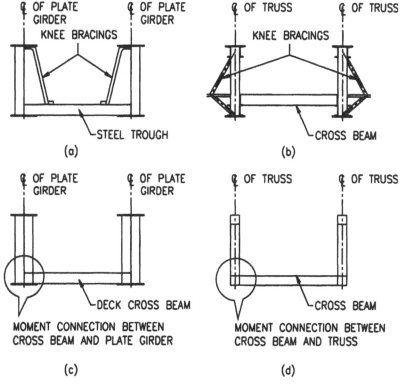

Fig. 11.1   Basic configurations of semi-through bridges

Essentially, semi-through bridges have the following characteristics:

- The deck lies at or just above the bottom flange (or chord) level of the longitudinal girders.
- Absence of top lateral bracings which are common in through and deck type bridges.
- The top flanges (or chords) are laterally restrained at the ends and at intermediate locations by either inclined knee bracings (as in older bridges, Figs. 11.1a and b) or by suitable moment connections between the deck system and members of the longitudinal girders (Figs. 11.1c and d). In modern bridges, the latter type of restraint is normally provided.

11.2   U-FRAME ACTION

When a strut is subjected to a progressively increasing compressive force, it will reach a critical stage when the bending action caused by any lateral displacement of the strut will lead to failure. This critical load is termed buckling load. Similarly, in a plate girder bridge, when the main girders are subjected to inplane bending, the bending compressive forces in the top flanges can reach a critical level, when these forces will cause the girders to buckle laterally. However, in a semi-through bridge system, the top (compression) flanges, though unbraced, cannot buckle freely due to the restraint provided by the bending tension forces in the bottom flanges and the connecting cross beams of the deck system which prevent relative rotation. Figure 11.2 shows a typical buckling mode for semi-through construction with rigid end frames. The device of connecting adjacent tension flanges of the girders and thereby increasing the lateral torsional rigidity of the girder system is commonly referred to as 'U-frame'. The webs of the plate girders are usually stiffened with vertical stiffeners placed at regular intervals along the span, and cross girders are commonly used to act compositely with the concrete deck to support the loads coming upon the deck. These cross girders are made rigidly connected with the vertical stiffeners, so that these elements can combine to form U-frames at discrete intervals. Where trusses are used instead of plate girders, the vertical and/or diagonal members of the truss can act as the 'vertical' of the U-frame.

Beam on elastic foundation (BEF) theory provides a suitable model for the analysis of the U-frame action, where the compression flange (or chord) may be considered as the beam deflecting transversely, and the restraints at the tops of the U-frames at discrete intervals provide the elastic foundations. The effective length of the compression flange (or chord) calculated on the basis

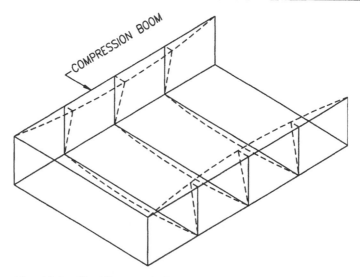

Fig. 11.2 Buckling mode for semi-through construction [5]

of this model can be used to determine the slenderness, and consequently the limiting stress in the flange (or chord). Theoretically, the stiffness property (rather than the strength) of the U-frame is the most significant factor in determining this effective length. Nevertheless, the requirement for adequate strength, particularly to resist transverse wind and other applied loads cannot be ignored. Moreover, the possibility of vehicle collisions with the vertical member of the U-frame must be considered in the design. As regards end frames, the traditional approach has been to provide almost rigid torsional end restraint to girders by means of very stiff end U-frames. For box girders (Fig. 11.3), which possess inherently high torsional rigidity, the requirement for restraint is significantly low.

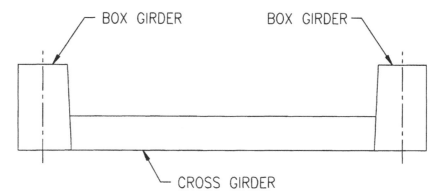

Fig. 11.3 Semi-through box girder bridge

The stiffness of the U-frame is determined by the aggregate stiffness of its following components:

(a) The deck and/or the transverse beams.
(b) The webs and vertical stiffeners of the plate girders (or the vertical and/or diagonal members of the trusses) which form the 'vertical' legs of the U-frame.
(c) The connections between (a) and (b).

11.3 ERECTION

In semi-through bridges, U-frames provide lateral supports to the compression flanges (or chords) of the completed structure, including the deck, which contributes significantly to the rigidty of the U-frames. The erection scheme should, therefore, recognise the need for adequate temporary supports to the incomplete structure during construction stage (i.e., before the U-frames are formed). The temporary bracing system should be adequate to ensure that the stability is maintained at every stage of construction, and that the deformations, if any, are kept within the permissible limits.

11.4 SKEWED SPANS

Choice of semi-through configuration for heavily skewed bridges is fraught with problems, particularly in interpreting codal provisions. For example, at the ends of the skewed span, 'L-frames' rather than 'U-frames' are formed to provide restraint to the compression flange (or chords), a situation usually not catered for by the code. This aspect should be considered at the conceptual stage of design, when the structural system is selected.

11.5 ADVANTAGES

As in the case of through bridges, semi-through bridges have an advantage where effective construction depth needs to be minimum. Thus, where the clearance under the bridge is required to be high, and the level of the roadway or railway track over the bridge needs to be as low as possible, this type of bridge is the most suitable solution.

The minimum construction depth also provides an economical advantage for semi-through bridges, by way of reduction of earthwork in the approach embankments.

Lastly, semi-through bridges are perhaps the only solution where the owner does not accept any clearance restriction over the bridge for transporting unusually tall cargo (e.g., approaches to power stations or similar utility units).

REFERENCES

[1] Evans, J.E. and Iles, D.C. *Guidance Notes on Best Practice in Steel Bridge Construction,* The Steel Construction Institute, Ascot, UK.

[2] Jeffers, E. 1990, *U-frame restraint against instability of steel beams in bridges.* In: The Structural Engineer, Vol. 68, No. 18, September 1990, Institution of Structural Engineers, UK.

[3] *BS 5400 : Part 3, Code of Practice for the Design of Steel Bridges,* British Standards Institution, 1982.

[4] *Commentary on BS 5400: Part 3: 1982 : Code of Practice for the Design of Steel Bridges,* The Steel Construction Institute, Ascot, UK.

[5] Owens, G.W., Knowles, P.R. and Dowling, P.J. (eds.) 1994, *Steel Designers' Manual* (fifth edition), Blackwell Scientific Publications Ltd., Oxford, UK.

Chapter 12

Arch Bridges

12.1 INTRODUCTION

Around 4000 BC, the Sumerians living in the Tigris-Euphrates Valley, discovered the advantages of the arch shape and started to construct arch portals and bridges using sun-baked bricks. By the time of the Romans, arch bridges were being constructed using stone instead of sun-baked bricks. Coming to more recent past, the first cast iron arch bridge was built in 1779, at Coalbrookdale, England, over the Severn River. By about the last quarter of the 19th century, steel established itself as a primary material for building bridges, and a number of steel arch bridges were built during the 20th century. Today, steel and concrete are the two primary materials for arch bridge building. However, considerations of size and weight of components would tend to favour steel. The present chapter presents different aspects of steel arch bridges.

12.2 TYPES OF ARCH BRIDGES

Structurally, arch bridges can be classified according to their degree of articulation. In a *fixed arch*, no rotation is possible at the supports and the structure is statically indeterminate to the third degree. In a *2-hinged arch*, rotations are allowed at the two supports, and the structure becomes indeterminate to the first degree. If, additionally, a hinge is provided at the crown (allowing rotation), the structure becomes statically determinate. This type of arch is termed *3-hinged arch*. 1-hinged arches are seldom used in actual practice. Figure 12.1 illustrates structural classification of arches.

In a *deck type arch* bridge, the deck carrying the traffic is located above the crown of the arch, and is supported by the arch. For this purpose, posts are placed on top of the arch along the length of the arch at different panel points. This type of arch is commonly known as *true arch*, in which both the horizontal and vertical components of the reaction or thrust are carried into

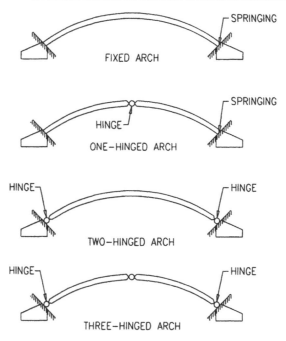

Fig. 12.1 Structural classification of arches

buttress situated at each end. In a *through type arch*, the deck is located at the spring line of the arch. Often the horizontal reactions of the arch are carried by a tension tie at the deck level of a through type arch. Such an arch is termed *tied arch*. This tension tie may be a single plate girder, a steel box girder, or a trussed girder, and may carry a portion of the live load as well. The proportion of the load carried is dependent on the proportion of the stiffness of the tie with that of the rib. Normally, a weak tie requires a deep arch rib, while a stiff tie requires a comparatively shallow arch rib. Thus, the sizes of the arch and the rib can be suitably proportioned to suit economic and aesthetic requirements. Deep trussed tie girders have often been used in combination with shallow solid web rib for double deck bridge construction. Both true and tied arches may be constructed with the deck located at an elevation between a deck arch and a through arch. These types are termed *part-tied*, *semi-through* or *half-through arches*. Figure 12.2 illustrates these different types of arches.

The framing for arches is normally one of the following three types:

- Solid rib
- Trussed rib
- Spandrel braced

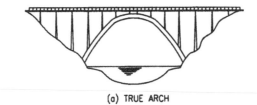

(a) TRUE ARCH

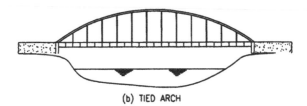

(b) TIED ARCH

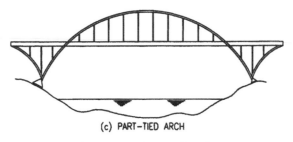

(c) PART-TIED ARCH

Fig. 12.2 Types of arches

Normally, *solid rib arches* are built with plate girders or box girders of uniform depth. However, variable depth arch bridges—with deep sections at the springings and tapering to shallower sections at the crown—have also been built for many bridges. The *trussed rib arches* consist of two parallel or nearly parallel chords at some distance apart connected by a truss system. An attractive trussed arch form is shown in Fig. 12.3. In this form, the bottom chord is a regular arch, while the top chord follows a somewhat reversed

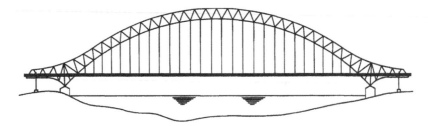

Fig. 12.3 Trussed rib arch

curve tapering from a deep truss at the end to a shallow depth at the centre. The trussed rib may also be of a crescent shape in the elevation (Fig. 12.4). Both solid ribbed arches and trussed rib arches can be used for deck, through or semi-through types of arch bridges. *Spandrel braced arches*, on the other hand, are applicable to deck type bridges only. This type of arch consists of the main arch rib (the curved bottom chord), a horizontal top chord at the deck level, and trussed web members, usually Pratt type (Fig. 12.5). A variation of a spandrel braced arch is the cantilever arch, as illustrated in Fig. 12.6. Cantilever arch can be well adapted in certain local site conditions. It considerably reduces the horizontal thrust on the arch support. It can also be erected with considerable ease by cantilever method.

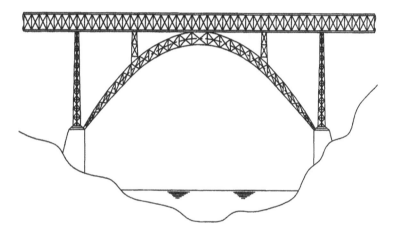

Fig. 12.4 Trussed rib of crescent shape

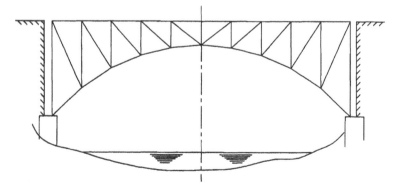

Fig. 12.5 Spandrel braced arch

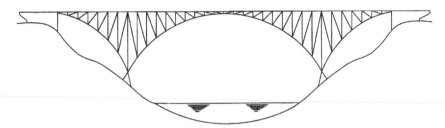

Fig. 12.6 Cantilever arch

Normally, arch bridges are designed with two planes of vertical arch ribs. However, single rib arch bridges with deck system cantilevered on either side of the rib have also been constructed. Arch bridges with two inclined ribs, widely spaced at the spring points, and closely spaced at the crowns have also been constructed. Recently opened Lupu bridge, the longest steel arch bridge in the world, over the Huangpu River in Shanghai, China, is of this type.

12.3   SELECTION OF ARCH TYPE

Arches are normally used on account of their superior aesthetic appearance as compared to simply supported truss bridges, as also for a possible economy for span ranges between 200 m and 500 m. Some of the important aspects which influence selection of the arch and its type are discussed in the following paragraphs.

12.3.1   Soil Conditions

Arches are eminently suitable for bridges across deep gorges with rocky sides, or for shallow streams with rock bottom and natural abutments. Foundation conditions should be suitable for construction of small and economical abutments or piers. In deep gorges with steep rocky walls, a deck-type arch bridge (true arch) is suitable. However, where the foundations are liable to settlement or lateral movement, it is not advisable to select true arch form as the superstructure. Often, need for placing the bearings above the high water level, may dictate selection of a semi-through structure to adopt a reasonable ratio of rise to span.

At locations where heavy reactions necessitate deep foundation, a single true arch span may not be economical. In such cases, a series of short spans, of equal or near-equal spans, may be feasible. The dead load thrusts at each intermediate pier would be nearly balanced, and the thrusts due to unbalanced live loads would not be large. Thus, even fairly deep foundations for the piers may be economical, or at least comparable to those for alternative simply supported or continuous truss spans. The other alternative would be to adopt tied arch construction, where the horizontal components of the reaction are carried by the central tie, while the vertical components are carried by the abutments. This arrangement would make the structure competitive with other types.

12.3.2 Rib Type

Selection of the rib type—whether trussed or solid—depends on various factors. For a bridge situated in a remote area with difficult access, trussed ribs may be economical, because trusses may be fabricated in small light weight components and easily transported to the site and erected. For bridges over 200 m span, particularly where heavy live loads are to be carried (e.g., railways), deflection requirements may dictate choice of trussed ribs in preference to solid ribs. Normally, for spans upto 230 m, solid rib arches are used for highway bridges. For bridges beyond 300 m spans, trussed rib arches are normally used. This stipulation, however, is not sacrosanct; there have been several exceptions.

12.3.3 Articulations

It is common practice to use 2-hinged construction for true arches, although there is hardly any particular advantage between fixed or hinged ends for such a construction. 2-hinged arches may be used for solid rib, trussed rib or sprandel braced types of arches. Some authorities suggest to let the arch act as 2-hinged under partial or full dead load and then fix the end bearings against rotation for additional loads. Similarly, the arch may be designed as 3-hinged under partial or full dead load, and also erected as 3-hinged arch and then converted to 2-hinged arch by making it continuous at the crown for subsequent loads. In this case, if the crown hinge is located on the bottom chord of the truss, and the axis of the bottom chord coincides with the load thrust line, conversion from 2-hinged to 3-hinged system would not change the load path pattern and the stresses in the top chord or web members would not change appreciably. Loads applied after the conversion process would only be carried by the top chord web members, which will consequently be relatively light. In the converted 2-hinged system, the bottom chord acts as the main load bearing arch. On the other hand, if the arch is designed as 2-hinged, the thrust under all loading conditions would be more or less equally shared between the top and bottom chords [1, 3].

Amongst the types of arches, the 3-hinged type is the least rigid, but is otherwise most satisfactory, in that the stresses caused by minor movements of the supports do not affect the stress condition of the structure appreciably.

12.3.4 Aesthetics

From an aesthetic point of view, particularly for road bridges, deck type arch is generally preferred by architects and engineers, as this type offers an uncluttered and clean view of the surrounding area to the bridge user. For modern through and semi-through bridges, solid ribbed arches are generally found desirable as the visible overhead structure presents clean and straightforward lines.

12.3.5 Form and Economy

In solid ribbed arches, the designer has to choose between two alternatives, viz., whether the rib should be curved throughout or constructed as straight between the panel points. No doubt, the former option would present a smoother appearance and be preferable from an aesthetic point of view. However, this would certainly be a more expensive option, because of increased material and fabrication costs. Material cost is increased as there will be some wastage in cutting the curved profile of the webs from rectangular plates. Also, more material will be required to cater for the secondary stresses due to bending induced by eccentricity between the curved axis of the rib and the thrust line which will be straight between the panel points (where most of the load is applied). A compromise solution would be to increase the number of panels and thereby reduce the panel lengths. As the panel length is reduced, the angular breaks at panel points would also be reduced and the segmental arch would look like a curved arch from a distance. A suggested upper limit of the panel length is about 1/15 of the span [1].

Often, instead of a constant depth rib, the rib is made to be deep at the springings, and gradually tapered at the crown. Fabrication cost of a tapered rib would be more compared to that of a constant depth rib. However, economy in material in tapered rib construction nearly balances the extra fabrication cost. Thus, overall costs of these two alternatives are almost always nearly equal.

12.4 DESIGN

In modern methods of analysis, quick results can be produced by computers using a three-dimensional finite element program. It is thus possible to carry out analysis and design of many alternative types and forms of arch bridges without much delay. While use of computers has speeded up the design process enormously, it must also be appreciated that, even before the advent of computers, analysis and design of arches were traditionally being carried out by experienced designers using slide rules and ordinary calculators to arrive at sound decisions after investigating a number of alternative layouts.

In this section, some traditional aspects relevent to design of arch bridges are briefly discussed.

12.4.1 Rise-to-span Ratio

The rise-to-span ratio of steel arch bridges vary widely. At one extreme, an arch can be semi-circular in elevation, and at the other extreme, it can be very shallow. Generally, this ratio varies between 1:4.5 and 1:6 in most arch bridges, both for solid rib as well as trussed rib arches. From an aesthetic point of view, shallow arches are desirable, particularly for through bridges because of pleasing appearance. Deeper arches are normally used for deck type bridges.

12.4.2 Panel Length

As already discussed in the previous section, for solid ribbed arches of segmental construction (i.e., with straight ribs between the panels), the panel length should not exceed 1/15 of the span. However, other considerations also need to be taken into account, such as economy of the deck construction, effect of additional moment due to eccentricity for comparatively long panel lengths.

12.4.3 Depth-to-span Ratio

For true arches without tie with constant-depth solid ribs, the depth-to-span ratio is generally in the range of 1:70 to 1:80. There are, however, instances of variations. For arches with variable-depth solid ribs, this ratio may be relatively small. Tied arches with solid ribs and deep ties require comparatively smaller depth of the rib. This is because a substantial portion of the bending moment is carried by the deep ties.

For trussed rib arches, the ratio of crown depth-to-span may be in the range of 1:25 to 1:50. In tied arches with trussed ribs, the tie primarily serves as a tension member to carry the horizontal component of the thrust. The arch truss carries substantial bending moment and is usually deeper with large moment of inertia [1].

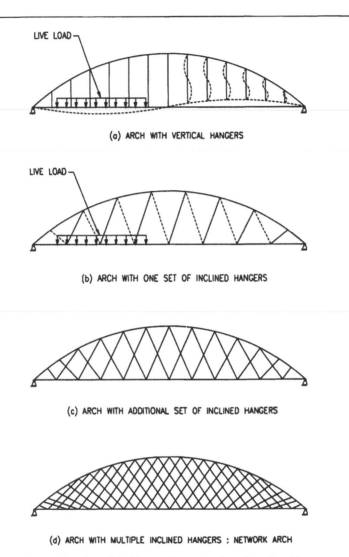

(a) ARCH WITH VERTICAL HANGERS

(b) ARCH WITH ONE SET OF INCLINED HANGERS

(c) ARCH WITH ADDITIONAL SET OF INCLINED HANGERS

(d) ARCH WITH MULTIPLE INCLINED HANGERS : NETWORK ARCH

Fig. 12.8 Steel arch bridge with vertical and inclined hangers

as predominantly compression members. With computerised analysis, the spacing of the inclined hangers can be so adjusted that there would be hardly any relaxation in the hangers due to any asymmetrical live load condition. Thus, all the hangers would be in tension or in 'near tension' condition. One other advantage of the system is that it results in smaller deflection. Even with increased number of hangers, the overall requirement of steel is likely to be reduced, making network arch system competitive with other structural systems for medium span bridges [4].

12.4.8 Bracings

For deck type arch bridges, there should be a top lateral bracing system along the deck, vertical sway bracing frames in the transverse plane between the posts and particularly effective lateral bracings between the ribs which are subject to high axial compression. Similarly, spandrel braced arches should be provided with adequate bracing system, such as top lateral bracings between the top chords, vertical sway bracing frames in the transverse plane between the posts, and effective bracing system between the bottom chords. In case of through or semi-through arch bridges, there should be adequate lateral bracings in the plane of the deck and also between the arch ribs. Also, adequate portal bracing system must be provided where the lateral bracings between the ribs are interrupted due to traffic clearance requirements.

Particularly special care should be taken to provide adequate bracing system to the arch ribs which carry high compressive forces and are liable to failure by buckling. Various arrangements can be used for these bracings. Normally, K-type or diamond pattern omitting cross struts at panel points are used. In some modern arch bridges, Vierendeel type bracings have been used. These provide an uncluttered and pleasing appearance. Figure 12.9 illustrates these bracing systems.

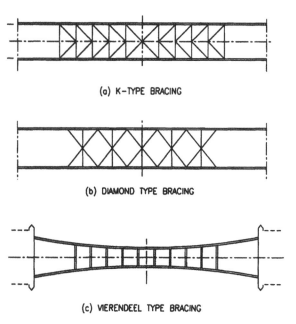

(a) K–TYPE BRACING

(b) DIAMOND TYPE BRACING

(c) VIERENDEEL TYPE BRACING

Fig. 12.9 Typical bracings for arch rib

REFERENCES

[1] Hedgren, A.W. (Jr.). 1994, *Arch bridges*. In: Structural Steel Designer's Handbook, Brockenbrough, R.L. and Merrit, F.S. (eds.) McGraw-Hill, Inc, New York, USA.

[2] Fox, G.F. 1999, *Arch bridges*. In: Bridge Engineering Handbook, Wai-Fah, C. and Lian, D. (eds.) CRC Press, Boca Raton, USA.

[3] Waddell, J.A.L. 1916, *Bridge Engineering*, John Wiley & Sons, Inc, New York, USA.

[4] Roy, B.C., et al. 2002, *Medium Span Bridges and Steel 'Network' Arch* : Conference Document of 7th International Seminar on Steel and Composite Bridges, held in Mumbai, India, Indian Institution of Bridge Engineers, 2002.

Chapter 13

Cable Supported Bridges

13.1 INTRODUCTION

Cable supported bridges are ideal for spanning large gaps, such as wide rivers, deep valleys or ravines where intermediate piers cannot be constructed. Modern cable supported bridges have been found to be competitive for span ranges from 250 m to 2,000 m, and beyond.

Majority of the cable supported bridges embody the following basic structural components, as indicated in Fig. 13.1:

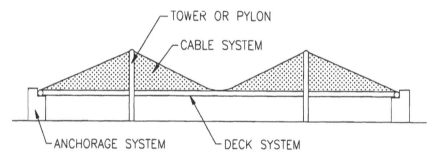

Fig. 13.1 Basic components of cable supported bridges

- The cable system supporting the deck system.
- The anchorage system to carry the forces from the cable system at the extreme ends (vertical and horizontal forces for suspension bridges, and vertical loads in the case of cable stayed bridges).
- The deck system consisting of the stiffening girders and deck structures that carry the traffic.
- The towers or pylons that support the main cables and transfer bridge loads to foundations.

13.2 CABLE SYSTEMS

13.2.1 Cable Configurations

Cable supported bridges can be classified as suspension bridges or cable stayed bridges, characterised by their distinctive cable configurations for supporting the bridge deck system. In suspension bridges, the deck is supported by vertical or slightly inclined cable suspenders hung from the main cables. The main cables, in turn, are hung across the span, supported on the top of the towers, with ends anchored firmly on either bank. The main cables, which are relatively flexible, take a profile shape that is a function of the magnitude of loading and position of hangers. Figure 13.2a illustrates the arrangement.

In cable stayed bridges (Fig. 13.2b) the main cables emanate from the towers and support the deck system directly. The inclined straight cables are in tension and provide relatively inflexible supports to the deck at several points. In effect, the inclined cables are tension members of a triangulated system, where the deck acts as the compression element. Thus, the deck performs the double duty of supporting the traffic, as well as serving as an important element in the main structural system.

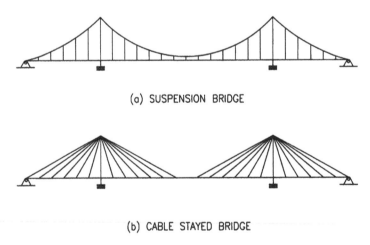

(a) SUSPENSION BRIDGE

(b) CABLE STAYED BRIDGE

Fig. 13.2 Cable supported bridge systems

Figure 13.3 illustrates an application of a combined system containing both the suspension system and the cable stayed system. This combination has been used in some old bridges, most notably in Brooklyn bridge in New York, where stay cables in the fan configuration (Section 13.6) have been used as a supplement to the main cable system.

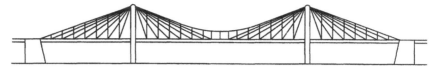

Fig. 13.3 Combination of suspension and cable stayed systems

13.2.2 Cables

Cables used for cable supported bridges can be composed of parallel wires, parallel strands or ropes, single strands, single ropes, locked coil strands, or solid hot rolled steel bars. Solid bars are, however, not suitable for suspension bridges because of the curvature requirements at the saddles (Section 13.2.4) on tower tops. These can be used in cable stayed bridges where saddles are not used and the cables terminate at the tower and are anchored to them.

Structural strands and ropes are normally available as prestretched by the manufacturer. Prestretching is generally done by subjecting the strand to a pre-determined load for a sufficient period of time at the manufacturer's workshop. Prestretching eliminates the constructional stretch and ensures that the product approaches a condition of true elasticity.

13.2.3 Suspenders

Generally, suspenders of suspension bridges are vertical. In some bridges, for example in Severn bridge in England and Bosphorus bridge in Turkey, inclined or diagonal suspenders have been used (Fig. 13.4). In the vertical suspender system, the suspenders do not contribute in resisting the shear resulting from external loading, which is resisted by the stiffening girder or by displacement of the main cables. In case of inclined suspenders, however, a truss action is developed and the suspenders participate in resisting the shear. Additionally, the inclined suspenders contribute to the damping properties of the system against aerodynamic oscillations.

13.2.4 Cable Connections

The connection between the cables and the towers is made either by means of saddles or by anchoring cables to the towers. In the former case, the cables

Fig. 13.4 Suspension bridge with inclined suspenders

are placed over the saddles fixed on the tower tops. This arrangement has been used in many suspension bridges. The saddles are normally made of cast steel or of fabricated steelwork. Typically, these are grooved to aid location of the cables. A cover plate is normally placed on top to protect the cables from corrosion. The whole unit is bolted down to resist movement. Figure 13.5 shows a typical saddle arrangement. Saddles may also be required at side piers to deflect the anchor span cables to the anchorages. These need to be designed according to site conditions. Splay saddles are often required at the anchorages.

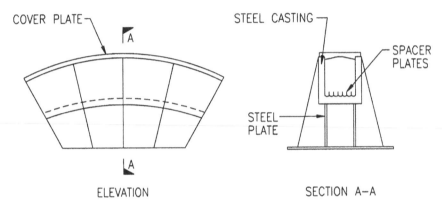

Fig. 13.5   Typical saddle arrangement

In a fan type arrangement of cables in cable stayed bridges (Section 13.6), where the cable stays converge on the top of the towers, and are continuous over them, saddles similar to those for suspension bridges are often used. In other types of cable arrangements, where stays are located separately along the tower length, similar but smaller saddles may be used. Where cables are not continuous over the towers, these are fixed on to the towers along their length. A typical arrangement is shown in Fig. 13.6.

In suspension bridges, cable bands are often used for attaching the suspender cables to the main cables. Typically, these are made of paired, semi-cylindrical steel castings

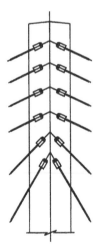

Fig. 13.6 Arrangement of cable anchors in a modified multi-cable fan system

with clamping bolts (Fig. 13.7). In smaller bridges, pre-fabricated steel twin hanger clamps (one above and one below the cables) may be used. In such cases, hanger ropes are suspended from the clamps by suitable eye bolt/socket arrangement. The suspenders may be attached to the cable bands by standard zinc-poured sockets. Attachments of the suspender cables to the girders depend on the girder detail. Often, standard zinc-poured sockets are used for connecting the suspenders to the girders. Where standard sockets are not suitable, the fittings may be specially developed and manufactured.

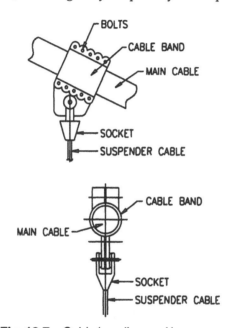

Fig. 13.7   Cable banding and hanger connection

13.3   ANCHORAGE SYSTEMS

In almost all suspension bridges, the forces on the cables are transferred to fixed anchorage systems at the two ends. These anchorage systems are broadly of two types:

- Rock anchorage
- Gravity anchorage

Rock anchorage involves drilling into the rock and grouting large bolt type anchors to which the cables are attached. The holes in the rock are often made to tapered shape in the slope of the connecting cable, so that the grouted bolt anchors are held by wedges of concrete in the tapered holes. For this type of anchorage, it is imperative that the adequacy of the geotechnical conditions are properly investigated. The second type, viz., gravity anchorage, is commonly used in most suspension bridges. This type is normally adopted where natural rock formation is not available at reasonable depths. In this type, the main cables are attached to steel anchor bolts embedded in massive concrete blocks. Both the vertical and horizontal components of the cable forces are resisted by a combination of overburden, dead weight and friction. An anchor house is often built over the mass concrete for inspection and maintenance of the anchorage. This construction adds to the dead load of the mass concrete. Figure 13.8 shows a typical gravity type anchorage.

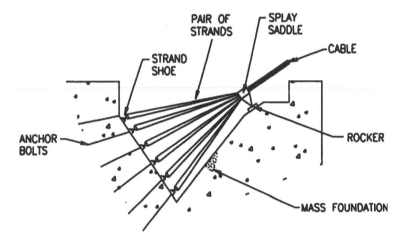

Fig. 13.8 Typical gravity type anchorage

In cable stayed bridges, the forces in the cable stays are typically transferred to the stiffening girders. In some cases, the cross sectional layouts of the girders allow direct connections of cables to such girders, as indicated in Fig. 13.9. In other cases, where the main girders are not located in the plane of the cables, transmission of cable forces are to be done by means of additional structural elements between the cable anchor points and the main girder. These additional structural elements are to be designed to cater for this purpose. The situation is shown in Fig. 13.10.

13.4 DECK SYSTEMS

For cable supported bridges constructed earlier, deck system was formed with a stringer-cross girder system, spanning between the stiffening girders

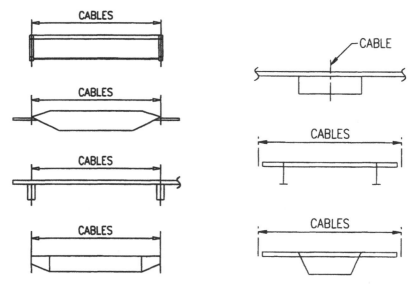

Fig. 13.9 Stiffening girders with direct connection to cables

Fig. 13.10 Stiffening girders with indirect connection to cables

on either side, which was held by the cables. The deck (steel/concrete) was laid on the stringer-cross girder grid. Subsequently, rectangular and aerofoil shaped steel box girders were introduced for minimising the effects of vortex due to wind and thereby ensuring aerodynamic stability of the system. Figure 13.11 shows a typical aerofoil shaped box girder.

Fig. 13.11 Typical aerofoil-shaped steel box girder

13.5 TOWERS

Towers are the tallest and most visible structural components of cable supported bridges. No wonder, designers invariably try to make them as attractive as possible. Towers carry all the loads transferred from the cables. Since the deck hangs from the cables and does not offer any lateral restraint to the towers, they are designed and proportioned to transfer such loads directly to the foundations. For bridges with bi-planer system of cables

(Section 13.6), tower shafts are provided with vertical or slightly inclined twin legs with the legs inter-connected with cross members in the form of diagonal bracings, or simply cross beams (portal system). Often, to improve the aesthetics, towers are made independent as single columns without any inter-connection between them. The transverse width of the tower at top should be wide enough to accommodate the cables. For economy, the transverse width of the tower shaft at the base should be minimum, considering, of course, the stability of the structure. Diamond shaped towers have also been used to support cable stayed bridges. Innovative ideas to improve the aesthetics of cable supported bridges have led to the design and construction of many attractive bridges during the last few decades. Figure 13.12 shows some different tower arrangements for cable stayed bridges.

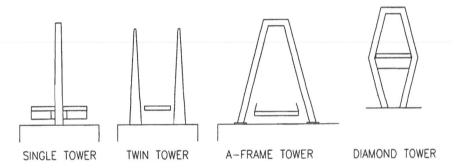

SINGLE TOWER   TWIN TOWER   A—FRAME TOWER   DIAMOND TOWER

Fig. 13.12   Typical tower configurations

In suspension bridges, the main cables are normally fixed at the top of the towers. With this arrangement, due to comparative slenderness of the towers, top deflections do not produce significant stresses. For relatively short spans, rocker towers are used with hinges both at the top and at the base. Towers fixed at the base and with roller saddles at the top have been used in some medium span bridges. Typically, such towers are tapered at the top as the required area decreases with height.

13.6   SPAN AND CABLE ARRANGEMENTS

Suspension bridge spans can be arranged in various ways. Figure 13.13 shows some such arrangements. The main cables are continuous from anchor to anchor for all the arrangements. Figure 13.13a shows a single span bridge with two towers and a hinged stiffening girder. A three-span bridge with two towers and hinged stiffening girders is shown in Fig. 13.13b. Figure 13.13c shows a three-span bridge with a continuous stiffening girder. A multispan bridge with three towers and hinged stiffening girders is shown in Fig. 13.13d. In a multispan bridge, the horizontal displacement of the tower tops may necessitate provision of horizontal ties at the top. Such ties were used in some French suspension bridges of the 19th century [2]. However, these are not pleasing from an aesthetic point of view. The multispan suspension bridge used in San Francisco-Oakland Bay bridge is essentially two three-span suspension bridges placed end to end with a central anchor pier.

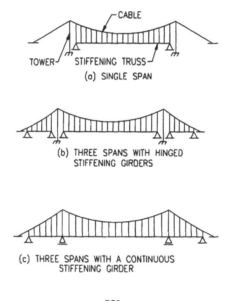

(a) SINGLE SPAN

(b) THREE SPANS WITH HINGED STIFFENING GIRDERS

(c) THREE SPANS WITH A CONTINUOUS STIFFENING GIRDER

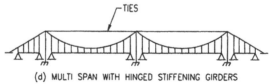

(d) MULTI SPAN WITH HINGED STIFFENING GIRDERS

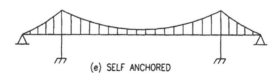

(e) SELF ANCHORED

Fig. 13.13 Suspension bridge arrangements

Most suspension bridges are externally anchored to earth embedded anchor system (Figs. 13.13a to 13.13d). In some bridges, where foundation conditions do not permit external anchorages, the ends of the main cables are fixed to the stiffening girders instead of earth anchorages, making the structure 'self-anchored' (Fig. 13.13e). Different anchorage systems have been discussed earlier in the chapter.

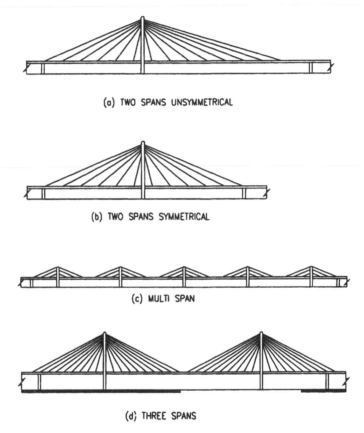

(a) TWO SPANS UNSYMMETRICAL

(b) TWO SPANS SYMMETRICAL

(c) MULTI SPAN

(d) THREE SPANS

Fig. 13.14   Cable stayed bridge arrangements

Cable stayed bridges may be two-span bridges, three-span bridges, or multispan bridges. Two-span bridges may be symmetrical or asymmetrical. In two-span asymmetrical bridges, the major spans are generally in the range of 60% to 70% of the total length of the stayed spans. In three-span bridges, the centre span is generally about 55% of the total length of the stayed spans, with the remainder 45% being equally divided between the two anchor spans. Multiplespan cable stayed bridges generally have equal internal spans, with symmetrical cable arrangements on each side of the towers. Often, the girder has 'drop in' sections at the centre of the span between the two forward stays, the length of which is about 8% to 20% of the length between the centres of the towers [2]. Figure 13.14 shows some two-span, three-span and multispan cable stayed bridge arrangements.

In cable stayed bridges, there are generally four different types of longitudinal arrangements of cables, known as fan, harp, modified fan and star. Figure 13.15 shows stay configurations of different cable arrangements.

In the fan arrangement (Fig. 13.15a), the stay cables radiate from the tower tops and consequently have maximum angles with the horizontal deck girders. This arrangement maximises the vertical components of cable forces resulting in smaller diameter of cables and smaller size anchorages. However, the high vertical forces make the towers heavy. Also, the cables, being concentrated at the top of the towers, present practical difficulties in anchoring these to the towers or saddles.

In the harp arrangement (Fig. 13.15b), parallel stay cables are used. This arrangement is often preferred in the case of double plane cable system, since the visual intersection of the cables, when viewed from an oblique angle, is the least. Also, spreading of the cable connections along the length of the tower, makes them easy to anchor. Furthermore, this arrangement results in a more efficient tower design.

A third arrangement—modified fan system—shown in Fig. 13.15c, is a compromise between the fan and the harp arrangements. In this arrangement, the anchor points at the tower top are spread out sufficiently to separate each anchorage, making the detail simple.

The fourth arrangement, known as star (Fig. 13.15d), has a unique appearance. However, this is not a very popular arrangement.

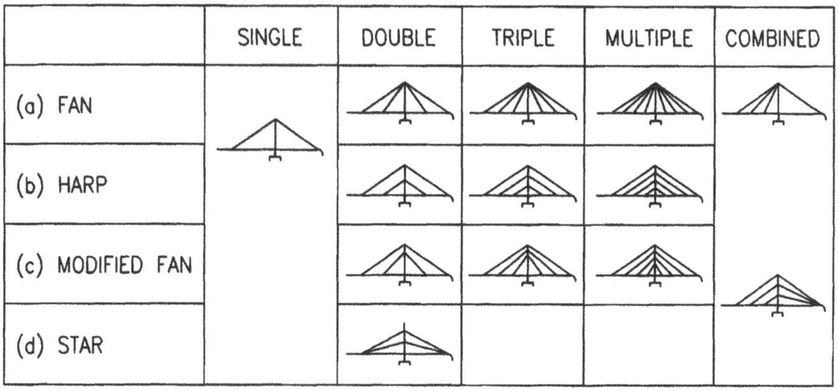

	SINGLE	DOUBLE	TRIPLE	MULTIPLE	COMBINED
(a) FAN					
(b) HARP					
(c) MODIFIED FAN					
(d) STAR					

Fig. 13.15 Longitudinal cable arrangements for cable stayed bridges

In cable stayed bridges, the number of stays to support the deck ranges from a single stay to multiple stays on each side of the tower. Increasing the number of stays leads to smaller spacings between the supports along the girder, smaller diameter stay cables, simpler connection details, and relative ease of erection. Therefore, overall economy is likely to be achieved if a large number of cables are used.

Transverse cable arrangements for cable stayed bridges may be of single plane, double plane—vertical, double plane—inclined, as also multiple plane. Figure 13.16 illustrates a few of the transverse cable arrangements. Figure 13.16a shows a single plane cable system at the longitudinal centre line of the structure. This arrangement requires a torsionally stiff box girder deck system to resist torsional forces due to unbalanced loading on the deck. The laterally displaced single plane system shown in Fig. 13.16b has often been used in pedestrian bridges. Systems shown in Figs. 13.16c, d and e have been used in a number of bridges in the past.

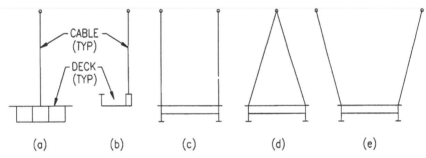

Fig. 13.16   Transverse cable arrangements for cable stayed bridges

13.7   DESIGN PRINCIPLES

Suspension bridges are characterised mainly by their lightness and suitability for longest of spans. However, their low dead load characteristic makes them relatively flexible, which can lead to large deflections under live load and make them susceptible to vibrations. Therefore, it becomes necessary to make an aerodynamic study for all major suspension bridges, and provide for adequate safety against wind-induced oscillations and avoid recurrence of the situation that caused the collapse of the first Tacoma Narrows bridge in 1940 (see Chapter 2).

Suspension bridges are normally stiffened by means of stiffening girders or trusses at the deck level. These stiffening girders equalise deflections due to concentrated live loads by distributing these loads to one or more main cables. Various factors may affect this distribution of live loads. These include:

- Rigidity of the stiffening girder. The stiffer the girder, better would be the distribution.
- Type of the stiffening girder, whether hinged at one, or more intermediate points, or continuous throughout.
- Rigidity of the cable system. This is determined by the dead loads, which establish the shape of the funicular polygon to which the cable conforms. An asymetrical live load would tend to distort this polygon, which would be resisted by the dead loads. Thus, larger these dead loads, better would be the rigidity of the cable system.
- The positions, lengths and the elastic properties of the hangers.
- Strain in the cables due to live loads or temperature variations.
- Deflection of the tower and variation of the tower height during service condition.

Suspension bridges which were constructed in the second half of the 19th century were designed with 2- or 3-hinged stiffening girders. The design rationally considered the interaction between the cables and the stiffening girders with the following assumptions:

- The entire suspension bridge is a continuous body, and the suspenders are closely spaced
- The cables are completely flexible
- The stiffening girder system is horizontal and straight, with uniform dead load and moment of inertia
- The dead load of the cables is uniform
- All dead loads are carried to the cable system
- The cables are of parabolic shape under all conditions of loading
- The stretch in the cables is relatively small

This theory, popularly known as elastic theory, did not take into account the displacement of the main cables under non-uniform traffic load. In order to cater for this condition, the deflection theory was developed by Melan in Vienna in 1888. Under this theory, equilibrium is more correctly established for the deflected system, instead of a system with the initial dead load geometry postulated in the elastic theory.

During the period when the deflection theory was introduced, calculation capacity of the designers was limited. Consequently, the complicated solution procedure for non-linear differential equations was rather tedious for the designers to use the theory in practice. Simplification had to be introduced in the form of charts, tables and correction curves, to obtain approximately correct results. Manhattan bridge in New York was the first major suspension bridge to have been designed with this approach.

The deflection theory was developed for vertical in-plane loading only. In 1932, a-three dimensional deflection theory was presented by L. S. Moissieff and F. Lienhard for calculation of the effects of lateral forces on suspension bridges. In this theory, the inclination of the cable planes caused by the lateral deflection of the stiffening girders were taken into account while calculating the moment and shear forces in the horizontal wind truss.

This new theory led to a substantial reduction in the lateral load in the stiffening girder. The reduction became more and more pronounced with increasing slenderness of the wind girder. These analytical methods were used by engineers, without intuitive understanding and with blind trust on the results of their calculations, to make the stiffening girders increasingly slender both in vertical and horizontal directions. This tendency was gradually leading to the aerodynamic instability of the structures. Ultimately, this aspect was brought to sharp focus by the unfortunate collapse of Tacoma Narrows bridge in 1940 (Chapter 2). Since then, the aerodynamic study and stability check became imperative for long span suspension bridges and new analytical methods were introduced for studying the dynamic behaviour of suspension bridges.

Of late, introduction of computers for engineering design has helped in making the suspension bridge design more rational. An entire bridge structure can now be analysed as a space frame by using finite deformation method. In this method, the actual behaviour of the bridge, such as elongation of suspender cables, which is disregarded in the deflection theory, can also be analysed [3].

In self-anchored cable stayed bridges, the taut inclined cables provide relatively stable point supports in the span. The cable forces are usually balanced between the central and the side spans. Because of the predominance of compression in the girders, the effects of asymmetric live loads requiring analysis by deflection theory are of relatively minor importance in such bridges. These bridges have, thus, greater stiffness properties than suspension bridge structures and with box girder deck system, they possess substantial torsional rigidity which make them eminently stable and resistant to wind and aerodynamic effects.

The basic structural system of cable stayed bridges is a series of overlapping triangles, comprising the vertical tower, the inclined cables, and the horizontal girder. All these members are predominantly subjected to axial forces with the cables under tension and the tower and the girder under compression. The girder is additionally subjected to vertical loads. However, it is essentially a continuous girder over the piers with additional intermediate elastic supports at the cable locations. All these components can be analysed satisfactorily with the help of computers.

REFERENCES

[1] Gimsing, N.J. 1997, *Cable Supported Bridges*, John Wiley & Sons, Chichester, UK.

[2] Podolny, W. (Jr.) 1994, *Cable suspended bridges*. In: Structural Steel Designer's Handbook, Brockenbrough, R.L. and Merritt, F.S. (eds.), McGraw Hill Inc, New York, USA.

[3] Okukawa, A., Suzuki, S. and Harazaki, I. 1999, *Suspension bridges*. In: Bridge Engineering Handbook, Wai-Fah, C. and Lian, D. (eds.), CRC Press, Boca Raton, USA.

[4] Tang, M.C. 1999, *Cable stayed bridges*. In: Bridge Engineering Handbook, Wai-Fah, C. and Lian, D., (eds.), CRC Press, Boca Raton, USA.

[5] Steinman, D.B. 1957, *A Practical Treatise on Suspension Bridges*, John Wiley & Sons, Inc, New York, USA.

[6] Podolny, W. (Jr.) and Scalzy, J.B. 1986, *Construction and Design of Cable Stayed Bridges*, John Wiley & Sons, Inc, New York, USA.

[7] Waddell, J.A.L. 1916, *Bridge Engineering*, John Wiley & Sons, Inc, New York, USA.

Chapter 14

Bridge Bearings

14.1 INTRODUCTION

Although small in size compared to other bridge elements, bearings are of significant importance for the proper functioning of a bridge. They are placed between the superstructure and the substructure and serve two principal functions, viz.,

- Transmitting the vertical and horizontal forces from the superstructure to the substructure
- Providing relative movements between the superstructure and the substructure.

14.2 FORCES ON BEARINGS

Bearings are subjected to a variety of forces, which include:

- Vertical forces
- Transverse forces
- Longitudinal forces
- Uplift forces

These are briefly discussed below.

14.2.1 Vertical Forces

These are mainly the dead load of the bridge superstructure and live loads from passing vehicles.

14.2.2 Transverse Forces

These consist of horizontal forces due to wind and earthquake forces acting across the centre line of the bridge. In bridges on horizontal curve, significant outward transverse force is also produced due to centrifugal effect from vehicles passing at high speed and/or applying brakes on the curve.

14.2.3 Longitudinal Forces

These forces include traction and braking forces, thermal forces, etc., and act parallel to the centre line of the bridge.

14.2.4 Uplift Forces

These forces are associated with reactions from transverse forces such as wind or earthquake, or from centrifugal effect of passing vehicles in bridges on horizontal curve.

14.3 MOVEMENTS IN BEARINGS

Movements in bearings include both translation and rotation. Translation in a steel bridge is mainly in the longitudinal direction. Change in temperature is the most common cause of longitudinal movement. Sharp skewed bridges or very wide bridges experience transverse movement as well. Rotations in bearings are associated with deflection of primary members, as also with uneven settlement of the foundation. Bearings are designed to cater for all these movements. If these movements are stopped or constrained, the bridge structure may be subjected to considerable additional forces, even in case of small spans.

14.4 TYPES OF BEARINGS

Bearings may be classified into two broad categories, viz., fixed bearings and expansion bearings. Fixed bearings resist longitudinal movement, but allow rotation. Expansion bearings accommodate both longitudinal as well as rotational movements. Both the fixed and expansion bearings are designed to transmit vertical as well as transverse forces to the substructure. Bearings may be of a variety of shapes and sizes, made of steel, as well as synthetic and elastomeric materials. The main components and functional characteristics of the principal types of bearings generally used are discussed in the following paragraphs.

14.4.1 Plate Bearing

This is the simplest, and perhaps, the oldest form of bridge bearing. It essentially consists of two steel plates, sliding at their interface to accommodate horizontal movements. The top (bearing) plate is fixed to the steel super-structure and the bottom (bed) plate is fixed to the bed block in the substructure. The sliding interface produces a frictional force. To reduce the coefficient of friction at the interface, bronze plates or lead sheets are often provided at the interface. Alternatively, grease or oil are also used as a lubricant. Plate bearings are generally used for smaller bridges (12 m to 20 m spans), where rotation at ends due to deflection of the girder is negligible, and therefore ignored. For longer spans, a bevelled bearing plate is welded to the upper bearing plate to allow for the rotation of the girder. Figure 14.1 shows a typical arrangement of this type of bearing.

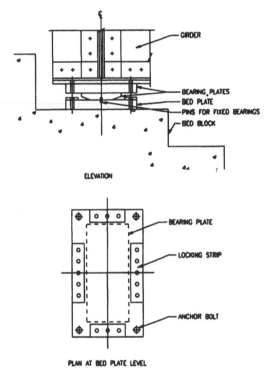

Fig. 14.1   Plate bearing

14.4.2   Rocker and Pin Bearings

Rocker and pin bearings are commonly used for spans beyond 20 m where the deflection of the girder due to live load becomes significant. These bearings can accommodate rotational movements at supports. These are typically made of steel and come in a variety of shapes.

Figure 14.2 shows a typical rocker type fixed bearing, where the rotational movement is provided by a saddle-knuckle arrangement. The bearing is connected to the substructure through a bed plate. Connection with the superstructure is made through a saddle by bolting or welding. The knuckle has projected ribs at both ends which resist transverse forces and prevent lateral movement of the saddle above.

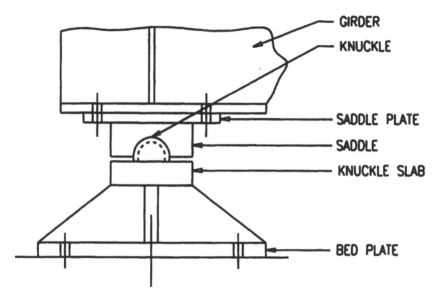

GIRDER

KNUCKLE

SADDLE PLATE

SADDLE

KNUCKLE SLAB

BED PLATE

Fig. 14.2   Rocker type fixed bearing

Figure 14.3 shows a typical pin bearing which consists of a solid shaft circular machined pin inserted between two pin plates with machined semi-circular recesses placed above and below the pin. Projected ribs at both ends of the pin prevent the pin from sliding off the seats.

Rocker and pin bearings of the types described above can be designed to support relatively large loads. However, they are suitable only where the direction of the rotation is in one direction only, viz., parallel to the axis of the bridge. It is necessary to protect the contact zones of these bearings from corrosion and deterioration. Regular inspection and maintenance are therefore essential for these bearings.

14.4.3   Roller Bearing

A roller bearing may be composed of one or more rollers between an upper and a lower steel plates. A single roller bearing can accommodate both

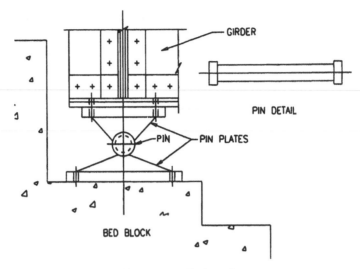

Fig. 14.3 Pin bearing

rotational and translational movements in the direction parallel to the axis of the bridge. However, a multiple roller arrangement can accommodate only translational movement in which case a rocker or pin arrangement has to be combined to cater for the rotational movement. Such an arrangement, shown in Fig. 14.4, allows longitudinal as well as rotational movements. Vertical tooth bars on the two sides of the rollers prevent lateral movement of the span. For multiple roller system, it is necessary that the rollers remain parallel and uniformly spaced during movement, otherswise these are likely

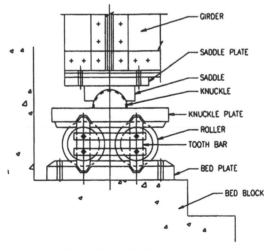

Fig. 14.4 Roller bearing

to get jammed. Generally, horizontal guide plates connecting all the rollers are provided for maintaining correct alignment. Also, since only the top and bottom surfaces of the rollers are used for movement, the unused side portions of the circular rollers are often eliminated to economise on space as well as material.

Like rocker and pin bearings, roller bearings are also susceptible to corrosion and deterioration. Regular inspection and maintenance are therefore required for these bearings.

14.4.4 Elastomeric Bearing

Elastomeric bearing is perhaps the simplest type of bearing, and has gained immense popularity in recent years. It is made of elastomeric (either natural rubber or synthetic material like neoprene) generally placed between two steel plates—one at the top and another at the bottom. It utilises the characteristics of the material and accommodates both translational and rotational movements (Fig.14.5a). Elastomer material is flexible in shear, but under vertical load, it expands laterally, leading to vertical displacement. In order to prevent this outward bulging of the material, it is common practice to introduce thin horizontal steel reinforcing sheets, vulcanised to the elastomer. These steel sheets are also encased completely within the

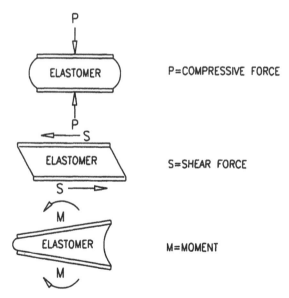

Fig. 14.5a Typical deformation for compression, shear and rotation of elastomer

elastomeric material for protecting these from corrosion. Figure 14.5b shows a typical elastomeric bearing with steel sheets. Reinforcing sheets may also be made of fibre glass or cotton. Elastomeric pad with fibre glass or cotton reinforcements can be produced in large sizes and cut to smaller required sizes for individual applications. This type of pad is comparatively cheaper than one with steel reinforcements, which has to be custom made for each application because of the end cover requirement for the protection of the steel reinforcing sheets from corrosion. Plain elastomeric pad (without any reinforcing sheet) can also be used for a short to medium span bridge where bearing stress is low. Because of the weather-resistant behaviour of the materials and absence of moving parts, an elastomeric bridge bearing provides virtually maintenance free service.

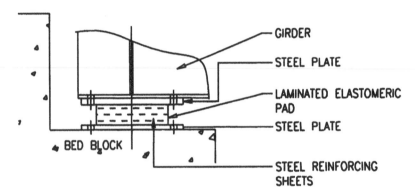

Fig. 14.5b   Typical elastomeric bearing with steel sheets

However, the following factors have negative influence on the durability of an elastomeric bearing and should be guarded against:

- Insufficient size
- Overloading
- Variation of the material properties over time
- Failure of bond between elastomeric material and the reinforcing plates
- Poor quality of manufacture
- Insufficient reinforcements
- Exposure to corrosive agents.

14.4.5   PTFE Bearing

PTFE bearing is the modern version of steel plate bearing. PTFE is the short term for *poly-tetra fluoro ethylene* and has an extremely low coefficient of friction with stainless steel surfaces. A thin stainless steel plate is seam welded to the underside of the top bearing plate. This stainless steel plate bears on the PTFE sheet placed on the top face of the bed plate and provides a very effective sliding surface. Since bonding property of the PTFE is very poor, it is customary to locate PTFE by confinement all around to prevent it from sliding away from the bearing. Figure 14.6 shows a schematic arrangement of a typical PTFE bearing.

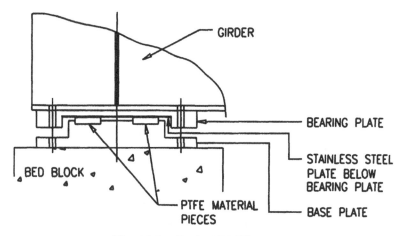

Fig. 14.6 Typical PTFE bearing

PTFE bearing is often used as a component of other type of bearing such as elastomeric bearing. Thus, while the rotation is accommodated by the elastomeric material, the horizontal movement is taken care of by PTFE.

14.4.6 Pot Bearing

Pot bearing is also of recent development and is used as an alternative system for a heavy type traditional steel bearing. It comprises a circular unreinforced elastomeric pad confined in a shallow steel cylinder or pot which prevents it from bulging and enables it to carry a much larger load than it would have normally done, if allowed to bulge, as in a conventional unreinforced elastomeric pad. The vertical load is transmitted to the elastomeric pad through a steel piston that fits closely to the steel cylinder (pot wall). When subjected to high compression forces, the unreinforced elastomeric pad behaves like a viscous fluid and allows the required rotation in any direction. A metallic (commonly brass) sealing ring is provided inside the cylinder to prevent leakage of the elastomer from the clearance between the piston and the cylinder.

Horizontal movements are restrained in a pure pot bearing, where the horizontal loads are transmitted through the steel piston bearing against the pot wall. To accommodate horizontal movement, a PTFE sheet is normally fixed on top of the circular steel piston, while a stainless steel plate is seam welded to the underside of the bearing plate.

A pot bearing can be conveniently used where rotation and horizontal movements in more than one direction are required to be catered for, such as in a skew bridge or a bridge on curve. Figure 14.7 shows a typical sliding pot bearing.

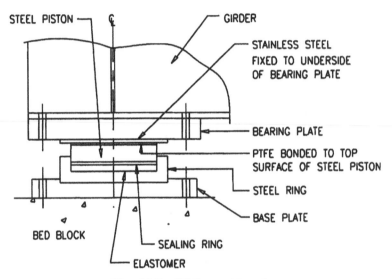

Fig. 14.7 Sliding pot bearing

A pot bearing has a very high vertical stiffness due to the nearly incompressible circular elastomeric pad confined within the steel cylinder. This characteristic makes pot bearing a very attractive solution for the bearings of high velocity railway bridges where the high vertical stiffness helps in preventing damage to the rails.

14.4.7 Spherical Bearing

A spherical bearing can also provide multidirectional rotations and horizontal movements at bridge supports. Generally, a spherical bearing has two sliding surfaces—one at the matching spherical surface which allows the bearing to rotate about any axis, and another flat surface which accommodates lateral movement in any direction. Both the sliding surfaces are required to be low friction surfaces, and PTFE sheets mated against stainless steel plates often provide such a condition. Figure 14.8 shows a typical spherical bearing.

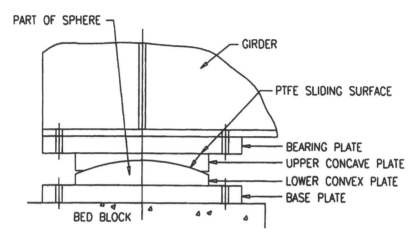

Fig. 14.8 Spherical bearing

14.4.8 Disk Bearing

Disk bearing is of relatively recent development and essentially consists of a hard elastomeric (polyether urethane) disk which supports the vertical load and a metal key at the centre, which transfers the horizontal load. A typical disk bearing is illustrated in Fig. 14.9. Rotational movements are provided by the deformation of the elastomer. The rotation, however, causes a shift of the axis of the load from the centre of the bearing. This shift has to be considered in the design. The basic disk bearing is a fixed type of bearing. To accommodate translational movements, a PTFE sliding arrangement is introduced.

14.5 SELECTION OF BEARING TYPE

Selection of the type of bearing to be used for a particular bridge depends primarily on its suitability to sustain the oncoming loads and movements and its overall cost. Thus, both technical as well as economical considerations govern the selection process.

Technical requirements consist of vertical and horizontal loads, rotational and transitional movements due to different loads. These should be collated and then compared with the characteristics of the various bearing types available in the market.

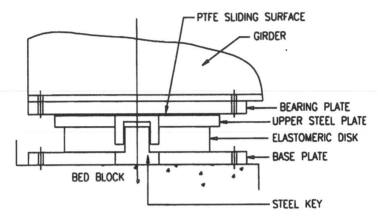

Fig. 14.9 Disk bearing

Considerations of available clearance between the underside of the bridge superstructure at the support locations is important for the maximum depth of the bearings to be chosen. Environment, ease of installation, maintenance costs and the client's preference are the other major considerations. Table 14.1 provides an approximate idea about the maximum load and movement capacities, as also relative costs of some common bearing types. However, these figures are purely indicative and are meant only to provide guidelines to the designer. The designer is expected to analyse and design a few alternative bearing types for obtaining quotations from the market for comparison

Table 14.1 Typical capacities and relative costs of bearing types [Source: Ref.1]

Type	Load Max (kN)	Translation Max (mm)	Rotation Max (rad)	Initial costs	Maintenance costs
Elastomeric pads					
Plain	450	15	0.01	Low	Low
(a) Cotton duck reinforced	1,400	5	0.003	Low	Low
(b) Fibreglass reinforced	600	25	0.015	Low	Low
(c) Steel reinforced	3,500	100	0.04	Low	Low
(d) Flat PTFE slider	>10,000	>100	0	Low	Moderate
Disc bearing	10,000	0	0.02	Moderate	Moderate
Pot bearing	10,000	0	0.02	Moderate	High
Pin bearing	4,500	0	>0.04	Moderate	High
Rocker bearing	1,800	100	>0.04	Moderate	High
Single roller	450	>100	>0.04	Moderate	High
Spherical PTFE	7,000	0	>0.04	High	Moderate
Multiple roller	10,000	>100	>0.04	High	High

purpose. Selection should be done keeping in mind the technical requirements and overall costs which should include future maintenance costs as well.

REFERENCES

[1] Feng, J. and Chen, H. 1999. Bearings. In: Bridge Engineering Handbook, Wai-Fah, C. and Lian, D. (eds.) CRC Press, Boca Raton, USA.

[2] Ramberger, G. 2002, *Structural Bearings and Expansion Joints for Bridges*, International Association for Bridge and Structural Engineering, Zurich, Switzerland.

[3] Ghosh, U.K. 2000, *Repair and Rehabilitation of Steel Bridges*, A.A. Balkema, Rotterdam, Netherlands.

[4] Tonias, D.E. 1995, *Bridge Engineering*, McGraw Hill, Inc, New York, USA.

Chapter 15

Welding Processes

15.1 HISTORICAL BACKGROUND

Welding is a process of joining together metal parts by application of heat resulting in fusion of the two sections along the line of joint.

The technology of forge welding is believed to have been used first by the Syrians as early as 1400 BC. Coming to the more recent past, until lately, the term 'welding' was generally associated with the joining of two metal pieces by a blacksmith by developing concentrated heat in his charcoal fire. The most common type of welding process now being used in structural steelwork is the electric arc process. The concept of using electric arc as a suitable source of heat was first put to practical application in the 19th century by producing an electric arc between a carbon electrode and the workpiece (UK Patent No.12984 of 1885 in the name of Benardos and Olszewski) [1]. Soon the carbon electrode gave way to a steel rod, which, apart from acting as a heat source, was also used to add deposition of molten metal on the workpiece. Since then, there have been phenomenal developments in welding technology and by the end of the 20th century a number of welding technologies were available for use in structural steelwork fabrication.

15.2 PRINCIPLES OF ELECTRIC ARC WELDING PROCESS

Essentially, in this process, a low voltage (15 - 35 V), high current (50 - 1000 Amps) electric arc operating between the end of a steel rod (electrode) and the components to be joined, melts metal from each steel component as well as the electrode to form a united pool of molten metal, which, on cooling, forms a solid bond between the components and provides a continuity of metal at the interface.

The composition of the electrode is usually so chosen that the resultant weld is stronger than the connected components. As the molten metal cools, heat flows into the adjoining parent metal, causing metallurgical changes in the structure of the steel upto a certain distance. This aspect is discussed in greater detail in Chapter 16.

15.3   COMMON ARC WELDING PROCESSES

The salient features of the welding processes which are commonly used in structural steelwork are discussed in the following paragraphs.

15.3.1   Manual Metal Arc (MMA) Welding

This process is also known as 'stick electrode welding', 'electric arc welding' or 'shielded metal arc welding'. In this system, the electrode is in the form of a hand-held steel stick coated with flux containing alloying elements such as manganese and silicon. The diameter of the steel core is generally 3.2 mm to 6.0 mm to match the level of the current used. The length of the electrode is generally 350 mm. As an arc is formed between the end of the electrode and the parent metals at the joint line, the heat melts both the parent metals and the electrode to form a weld pool. The molten flux layer and the gas generated by the flux protect the molten metal from oxidation and improves the mechanical properties of the weld. The welder has to move the electrode towards the weld pool to keep the arc gap at a constant length in order to produce a uniformly wide weld deposit. The welder has also to move the electrode forward with a uniform travel speed to achieve continuity of the weld deposit. Thus, the quality of weld in MMA welding depends, to a considerable extent, on the skill of the welder in adjusting the feed rate of the electrode and at the same time its travel speed. When the electrode is reduced to a length of about 50 mm due to melting or 'burning off', the arc is extinguished and the solidified flux (or 'slag') is removed. A fresh electrode is then used to continue the welding. A schematic diagram of MMA welding is shown in Fig. 15.1.

Apart from the skill of the welder, the quality of weld in MMA welding also depends on certain other factors such as the choice of electrode, voltage, current, etc. Furthermore, because of the limited length of the electrode which can be handled in the MMA process, the welder cannot deposit 'continuous' welds beyond a certain length. This situation presents some problems while executing long lengths where uninterrupted weld deposit is preferred. However, in spite of these apparent limitations, MMA technology has

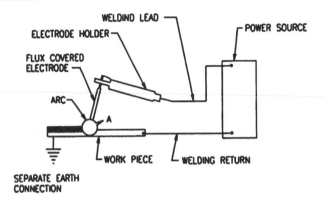

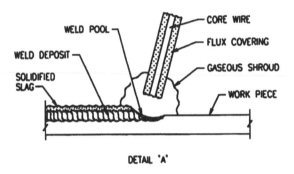

DETAIL 'A'

Fig. 15.1 Schematic diagram of manual metal arch (MMA) welding

gained immense popularity amongst structural fabricators because of a number of advantages it offers. The capital cost of a unit is comparatively low and it can be moved to almost any location, both within the workshop as well as in the construction sites. The system also offers welding to be performed in any position (down-hand, horizontal or overhead) without any trouble and has a lead of upto about 20 m from the power supply point, making it extremely useful, particularly at work-sites.

15.3.2 Metal Active Gas (MAG) Welding

This process is also known as 'Metal inert-gas' (MIG) welding, 'metal arc gas-shielded' (MAGS) welding, 'gas metal arc welding' (GMAW), or CO_2 welding. This is a manually operated system, but can also be used with a mechanical traversing system. In this process, the electrode consists of bare wire of 0.9 mm to 1.6 mm diameter containing the alloying elements and is fed at a constant speed by a motorised unit. The arc and the molten weld metal are shielded by a gas which does not react with the molten steel. The common practice is to use pure carbon dioxide or a mixture of argon and carbon dioxide in the proportion of 80 : 20. Flux is not necessary in this process; however, in some special cases, a flux-cored wire is used to maintain the required weld profile, as in case of fillet welds in the horizontal-vertical positions. The current is determined by the pre-set speed of the wire feed and the arc length is determined by the pre-set power supply unit. The welder has to maintain the required height of the nozzle above the weld pool as also the travel speed. A schematic diagram of MAG welding process is shown in Fig. 15.2.

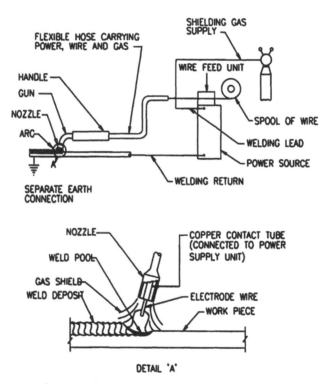

Fig. 15.2 Schematic diagram of metal-active gas (MAG) welding

Unlike MMA process, MAG welding does not require changing of electrodes and, therefore, is a continuous process. Consequently it is a comparatively faster process. However, it needs a certain degree of sophistication in the execution and requires considerable skill of the welder to produce good quality welds. The process is generally useful in the controlled environment of a workshop and is not quite 'handy' in site operations where it may not be possible to maintain adequacy of the gas shield due to possible high wind conditions.

15.3.3 Submerged Arc Welding (SAW)

This is a mechanised process in which the arc is completely covered by a mound of granular flux (hence the name). A bare electrode wire of 2.4 mm to 6 mm diameter, coiled on a spool, is fed continuously into the arc by mechanically operated drive rolls. As an arc is formed between the end of the electrode wire and the parent metal, the flux is deposited continuously from a hopper around the wire on the surface of the joint. The voltage and the current are controlled automatically at the preselected values.

The electrode and the drive assembly are traversed along the joint by a mechanical drive system. Alternatively, the work can also be moved in relation to the welding head. The electrode feed rate is so maintained that the pre-set arc length remains constant throughout the operation. The arc melts a portion of the flux which covers the weld-pool as it cools, solidifying into a slag. On completion of the operation, the slag is removed from the weld and the unfused flux is collected for recycling. A schematic diagram of submerged arc welding system is shown in Fig. 15.3.

SAW may be used as a fully automatic system or as a semi-automatic system, depending on the nature and extent of the work. The process is particularly useful for welding long joints. Its continuous operation and high welding speed offer very high productivity rate. Also, mechanical operation of the process and the effective shielding provided by the flux produce uniformly smooth high-quality welds. Consequently the process is used extensively in the structural workshops the world over.

15.3.4 Stud Welding

For composite girders, steel stud shear connectors are very popular. Studs are end-welded on the top surface of the top flange of the girder by means of a mechanised device or 'gun'. The stud, which is normally with a head on top, acts as an electrode and is held in the chuck of the gun which is connected to the power supply. The operator holds the gun in the welding position with the stud pressed firmly on the surface of the steel with a ceramic ferrule around the stud. As soon as he presses the trigger, the current is switched on and the stud is moved away automatically from the steel surface to establish an arc. The heat melts the end of the stud and the welding surface to form the molten pool, and the stud is automatically

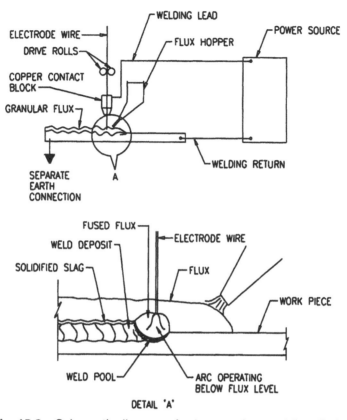

Fig. 15.3 Schematic diagram of submerged arc welding (SAW)

plunged into the pool when the current is switched off. The molten metal is formed into a fillet around the stud.

Direct current is used for stud welding. While some skill is required initially to establish the weld parameters such as current, voltage, arc time and force, the operation of the equipment is relatively simple and offers an accurate and fast method of welding shear connectors on to the girders. Usually, the welding gun welds only one stud at a time. For major bridge work requiring a large number of studs, multi-stud set-ups are used.

15.4 CONTROL OF WELDING PARAMETERS

For achieving the required strength in any joint, it is necessary that the bonding between the weld metal and the components are of acceptable quality throughout the length of the joint without any interruption. This implies that the heat input in the arc has to be maintained at a minimum level throughout the welding process. Heat input, in turn, depends primarily on the following parameters:

(a) Arc voltage
(b) Arc current
(c) Travel speed
(d) Arc length
(e) Electrode feed rate

In other words, these parameters are to be controlled in order to achieve a good quality continuous weld.

15.5 SELECTION CRITERIA OF WELDING PROCESS

In the preceding sections, the salient features of different welding processes have been discussed. No doubt, while choosing a process, the primary objective is to achieve the required quality of weld at the lowest possible cost. In practice, however, one has to take a holistic view of the situation and consider many other factors before finally selecting the appropriate process for a particular application. Some of these factors are discussed in the following paragraphs.

15.5.1 Location of the Works

This factor plays an important role in the selection of the process. Controlled and protected environment of a fabrication shop provides an ideal location for SAW and MAG welding processes. Level of accuracies in alignment and fit-up required for these processes are generally available in fabrication shops equipped with various jigs, fixtures, positioners (manipulators), etc., which may not be readily available at sites. MMA welding, on the other hand, can be effectively used in the open and unprotected environment of sites and is very popular for site operations.

15.5.2 Welding Position

Mechanised processes like SAW or MAG welding are eminently suitable for continuous and downhand welding. For joints in the vertical or overhead positions, mechanised systems are not suitable. In such cases, manipulators are generally used to rotate the job suitably to enable the welding head of the mechanised process to perform downhand welding. At work sites where such facilities are normally not available, MMA welding is, perhaps, the most suitable option for performing the obligatory overhead welding. Adequacy of access for the electrode, welding gun or welding head also need to be looked into while selecting a particular system.

15.5.3 Composition of Steel

It has to be examined whether a particular metal needs special procedures to eliminate the risk of crack formation in the weld. In this context, hydrogen content in the weld metal can be better controlled in MAG and SAW processes than in MMA process [2]. Aspects related to risk of crack formation have been discussed in greater detail in Chapter 16.

15.5.4 Availability of Welding Consumables

In this respect MMA welding has an advantage. Electrodes with a variety of flux combinations to suit a wide range of applications are generally available in the market.

15.5.5 Costs

Costs in an arc welding operation comprise the following elements :

- Direct welding labour
- Associated labour
- Cost of consumables
- Equipment capital cost
- Maintenance cost of the equipment
- Overhead expenses

In addition to the welder-time spent on actual welding (i.e., arcing time), direct labour costs also include the idle time due to electrode change (for MMA welding) or aligning the joint (in SAW or MAG welding), etc., and also waiting time between successive operations. The arcing time expressed as a percentage of the total time is termed 'duty cycle':

$$\text{duty cycle} = (\text{arcing time} / \text{total time}) \times 100$$

The process which can offer maximum duty cycle is obviously the most attractive one. Thus, SAW and MAG welding offer best advantage on long joints where the idle time and waiting time can be kept very low. Also, the faster operation in these processes leads to increased productivity. This, however, is not the case where short weld runs at different locations are required. In these cases, MMA welding offers the best advantage in spite of its high idle times.

15.5.6 Availabilty of Skilled Welders

In general, welding work requires reasonably skilled workmen. Therefore, availability of workmen with sufficient skill must be ensured for any process to be adopted for the welding work. However, for MMA welding, this problem may not be so acute as in the case of more sophisticated SAW and MAG welding processes. This aspect assumes considerable significance in remote areas where welders with sufficient skill may not be available. Importing of skilled welders from other locations may have to be considered in such cases.

REFERENCES

[1] Houldcroft, P. and John, R. 1988. *Welding and Cutting*, Woodhead-Faulkner Ltd., Cambridge, UK.

[2] Gourd, L.M. 1995, *Principles of Welding Technology*, Edward Arnold, London, UK.

[3] Easterling, K. 1992, *Introduction to Physical Metallurgy of Welding*, Butterworth-Heinemann Ltd., Oxford, U.K.

[4] *Procedure Handbook of Arc Welding Design and Practice*, 1959, The Lincoln Electric Co. Cleveland, Ohio, USA.

Chapter 16

Welded Joints

16.1 TYPES OF WELDS

In structural fabrication work, two types of welds are commonly used to form welded joints. These are:

16.1.1 Fillet Weld

In fillet weld, the weld metal is deposited outside the profile of the joining elements. Figure 16.1 shows a few typical fillet welds.

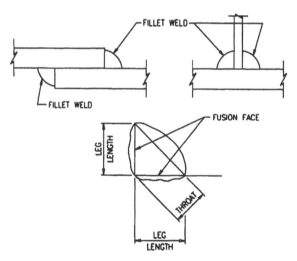

Fig. 16.1 Fillet welds

Although the size of the fillet weld is specified by the leg length, the throat thickness is considered for computing its strength. In a fillet welded joint, the layout of the weld is important, since stress distribution depends not only on the load, but also on the layout of the weld or group of welds.

16.1.2 Butt Weld

In a butt welded joint, the edges of the members are butted against each other and joined by fusing the metal to produce a continuous joint. Depending upon the current used, the arc can melt the metal upto a certain depth only. If the thickness of the members to be joined is more than this depth, the edges of the members are required to be 'prepared' to form a groove along the joint line, so that the joint continuity through the thickness can be achieved. The prepared groove is then filled by weld metal from the electrode. Figure 16.2 shows a few common types of butt welds.

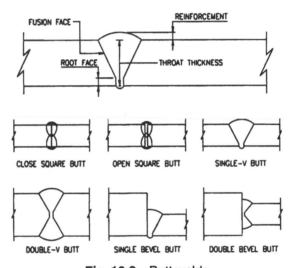

Fig. 16.2 Butt welds

Properties of the parent metal, edge preparation, selection of the electrode and welding parameters (current, speed, voltage, etc.) play a vital role in developing the strength of a butt welded joint.

16.2 TYPES OF WELDED JOINTS

Using the two types of welds, viz., fillet weld and butt weld, a variety of joints can be formed. These joints can be made up from the four basic configurations described in the following paragraphs.

16.2.1 Butt Joints

Butt joints are commonly used to join lengths of plates, in solid web girders, as shown in Fig. 16.2.

16.2.2 Tee Joints

Depending on the service requirement, these joints can be formed either by fillet weld or by butt weld. Typical examples of such welds are at flange-to-web connections, stiffeners welded to the web of solid web girders. Figure 16.3 shows some examples of the use of Tee joints in bridge work.

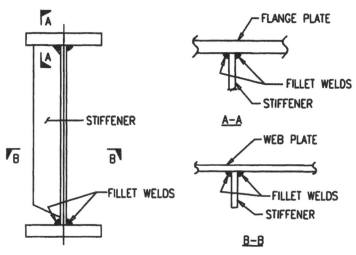

Fig. 16.3 Tee joints

16.2.3 Corner Joints

These joints are commonly used in welded box girders and can be either fillet-welded or butt-welded as shown in Fig. 16.4.

Fig. 16.4 Corner joints

16.2.4 Lap Joints

Lap joints are commonly formed by fillet welds. Bondings in lap joints are at the interfaces of the fillet welds only. These are widely used in the structural fabrication work, such as in connecting gussets to main members, as shown in Fig. 16.5.

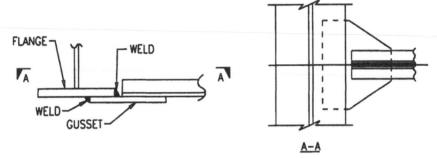

Fig. 16.5 Lap joint

16.3 HEAT AFFECTED ZONE (HAZ)

The solidified weld metal which joins the components is essentially a mixture of the molten parent metal, and steel from the electrode. It is generally made stronger than the components by judicious adoption of the composition of the electrode. As the molten metal starts cooling, only a small amount of heat escapes through the surface of the weld pool, leaving the majority of the heat to flow through the parent metal on either side of the joint. As a result, steel upto a certain distance is sujected to a severe thermal cycle—heating and then rapidly cooling. This causes change in the micro-structure and properties of the steel in the region. This region of the parent metal is called 'heat affected zone' (HAZ). The parent metal near the fusion boundary is subjected to a maximum temperature close to the melting point. This maxumum temperature gradually falls as the distance from the fusion boundary increases until the outer boundary of the HAZ is reached where the temperature is below the range for metallurgical change. The phenomenon is shown in Fig. 16.6.

As the weld pool cools, the metallurgical structure of the steel in HAZ is changed from a ductile to a hard form. When the hardness is above a critical level, the metal becomes prone to cracks. The level of hardness in the HAZ depends primarily on two factors, viz.,

- Chemical composition of the parent steel
- Rate of cooling of the weld.

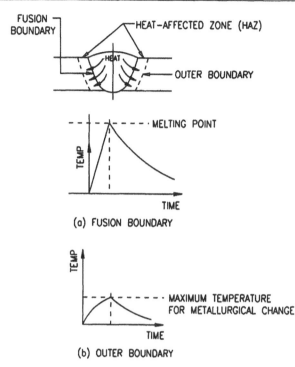

Fig. 16.6 Heat affected zone: typical temperature gradients at fusion boundary and outer boundary [1]

16.3.1 Chemical Composition of Steel

Carbon is primarily the strengthening element in steel. However, increased carbon content impairs ductility and weldability of the steel. Therefore, in order to attain better physical properties in steel (while keeping the carbon level low), other admixtures or alloys are generally added during the process of steel making. Some of these, (such as manganese and chromium), however, increase the hardness of the metal and consequently the risk of crack in the HAZ.

The relative influence of chemical contents on the weldability of a particular steel is guided by the value of 'Carbon Equivalent' (CE) which is derived from the following empirical formula :

$$CE = C + (Mn/6) + (Cr + Mo + V)/5 + (Ni + Cu)/15$$

where the chemical symbols represent the percentage of the respective elements in the steel. It may be noted that the amount of each alloying element is factored according to its contribution towards hardening of the steel. Several other variants of this formula are available and are recommended by different authorities. The formula given in the present text has been adopted by the International Institute of Welding and is widely followed in different countries.

In structural steel, the values of CE range from 0.35 to 0.53. As a general rule, however, a steel is considered weldable when the CE is less than 0.4. With increased value of CE, use of low hydrogen electrodes and pre-heating of the components to be joined become important.

The relationship between the cooling rate and CE can be depicted by a curve AB shown in Fig. 16.7. Faster cooling rate for material with low CE may be tolerable as the risk of cracking is less. Thus, higher the CE, the lower will be the tolerable cooling rate and consequently, the harder and more brittle will be the HAZ and more susceptible will it be to cracking.

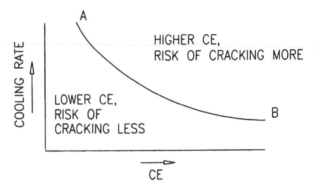

Fig. 16.7 Relationship between cooling rate and CE [1]

The level of hardness in steel is measured by Vicker's Pyramid Number (VPN). In this system, an indenter is forced into the surface of the steel and the size of the impression is compared with a pre-set standard. VPN for steel to be used for fabrication should ideally be in the range of 190 to 200. This should be taken into account in the design stage while specifying the material to be used.

16.3.2 Rate of Cooling

The moving heat source used in arc welding process induces steep temperature gradients around the melt zone. A typical form of isotherm and temperature gradient along and across the line of movement of arc is illustrated in Fig. 16.8. It may be particularly noted that the movement of the arc results in the piling up of the isotherms at the leading edge.

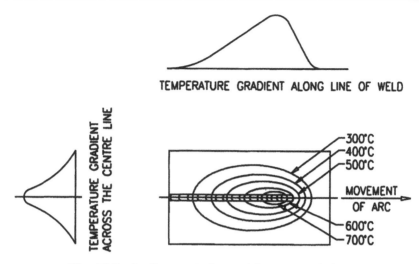

Fig. 16.8 Isotherms during welding on steel plates

The pattern of flow of heat from the source area depends on the thickness of the plates. In the case of thick plates, the flow of heat moves not only horizontally but also in the vertical direction. In the case of thin plates, however, the flow of heat is only in the horizontal direction. This phenomenon is illustrated in Fig. 16.9.

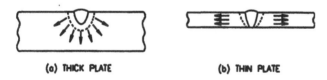

(a) THICK PLATE   **(b) THIN PLATE**

Fig. 16.9 Heat flow in welding (Reproduced from [6] with permission from Elsevier)

Rapid cooling makes steel comparatively hard and brittle. The relationship between cooling rate and cracking is illustrated in Fig. 16.10.

Rate of cooling in the HAZ depends on a number of factors:

- High heat-input entails a slower cooling rate and reduces the risk of cracking in the HAZ.
- A thick section cools more rapidly than a thin section and is thus more susceptible to cracks in the HAZ.

It is a common practice to pre-heat thick components in order to reduce the temperature gradient between the weld and the adjoining metal. This would reduce the level of hardening of the metal and also the chance of consequent

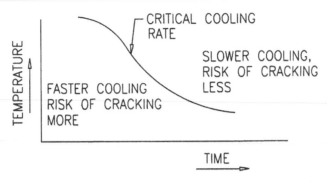

Fig. 16.10 Relationship between cooling rate and cracking [1]

cracks in the HAZ. Pre-heating is also useful in dispersing hydrogen from the weld metal and thereby reducing the risk of its embrittlement. Other uses of pre-heating are removal of surface moisture in humid conditions and also maintaining ambient temperature in a cold environment.

16.4 LAMELLAR TEARING

Lamellar tearing is the separation of parent metal caused by through thickness strains induced by weld metal shrinkage. It is associated with highly restrained welded joints with heavy members under high stress. As the weld metal cools, it contracts. When the resultant stress is carried in the through thickness direction, any lack of cohesive strength of steel material in this direction will allow the plate to be separated (see Chapter 4). The separation has a step-like appearance comprising of a series of terraces parallel to the surface. Figure 16.11 shows a typical lamellar tearing in a cruciform welded joint.

Studies reveal that proneness of steel towards lamellar tearing is governed by a number of factors. Shape of non-metallic matter in the steel plays an important role in the lamellar tearing in the member. Long sharp inclusions make the plate more susceptible to tear than small round inclusions. The rolling temperature in the mills also affects the behaviour of the inclusions in the steel. Furthermore, the welding process itself often contributes to the tendency of lamellar tearing. The factors that influence the phenomenon include welding sequence, pre-heat, size of weld, the weld restraint in the through thickness direction, etc.

To avoid lameller tearing, it is advisable to design and detail joints in thick plates avoiding weld contraction in the direction normal to the surface

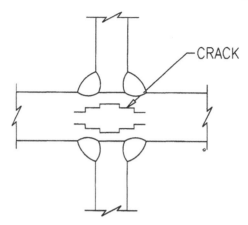

Fig. 16.11   Example of lameller tearing

of the plates. Also, heavy fillet welds should be avoided in the design. For fabrication of major members, suspect plates with risk of lameller tearing should be replaced by plates with adequate through thickness properties.

16.5   CRACKING AND FRACTURE IN WELDED JOINTS

The problems of cracking and fracture in welded joints in bridges have been causing concern to the designers and scientists during the past few decades. This has led to extensive research programmes on the subject around the world. Of late, there has been a greater understanding of these problems and designers are now able to design welded bridges with greater confidence. Some of these features will be discussed in the following paragraphs.

The cracks in welds can be classified into two broad groups, viz., those which are related directly to weld defects, and those which are related to external phenomena occurring during the service life of the bridge, such as fatigue cracks.

16.6   WELD DEFECTS

In the design stage, while assessing the strength of a particular joint, it is assumed that the weld metal is free from any defect. In actual practice, however, this ideal result is rarely achieved, and in most welds a few defects are present in spite of best efforts. Some typical defects which are noticed in welds are discussed in the following paragraphs.

16.6.1 Undercut

Undercut (Fig. 16.12a) is a potential danger for initiating fatigue cracking. This should not be allowed in a bridge structure which is generally subjected to fatigue stresses. Possible causes of this defect are:

- Current too high
- Welding speed too rapid
- Arc length too long
- Incorrect manipulation

To eliminate this defect, the welding parameters need to be reviewed.

16.6.2 Porosity

Porosity (Fig. 16.12b) in the weld metal may not be a particularly damaging defect. However, its presence in the surface of the weld may lead to tiny

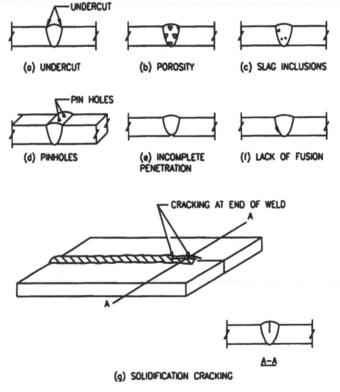

Fig. 16.12 Typical weld defects

notches on the weld face and initiate premature fatigue cracks in the weld. Possible causes of porosity are:

- Faulty electrode
- Welding speed too rapid
- Current too low
- Presence of sulphur or other impurities in the metal

Use of low-hydrogen electrode may reduce the possibility of this defect.

16.6.3 Slag Inclusion

Particles of flux left behind in the weld pool form slag inclusions in the weld metal (Fig. 16.12c). This defect may be particularly damaging when such large inclusions or a group of small inclusions lie across the direction of stress which might initiate fracture in the weld metal. Possible causes of this defect are:

- Weld temperature too low
- Cooling too rapid
- Included angle of the joint too narrow
- Viscosity of the molten metal too high
- Inadequate cleaning of the slags between multi-run welds

Pre-heat of the components is recommended to reduce the possibility of this defect. Also, in multi-run welds, care should be taken to ensure that the slags are removed and the weld surface is cleaned before the succeeding runs are deposited.

16.6.4 Pin Holes

Probable causes of pin holes (Fig. 16.12d) are:

- Damp electrodes
- Presence of rust, scale, paint, etc., on the work

The defect may be eliminated by ensuring dry electrodes. Use of ovens for drying the electrodes is recommended. Also, all contaminants on the work must be carefully removed prior to taking up welding operation.

16.6.5 Incomplete Penetration and Lack of Fusion

The possible causes of incomplete penetration and lack of fusion (Figs. 16.12e and f) are:

- Incorrect penetration
- Welding speed too rapid
- Current too low
- Electrode too large

To eliminate these defects, the profile of the preparation needs to be checked. Also, welding parameters may need review.

16.6.6 Solidification Cracking

Solidification cracking (Fig. 16.12g) occurs during the process of cooling, predominantly at the center-line of the weld deposit. This type of crack is noticed particularly at the ends of weld deposits. This is a very serious defect and is caused by:

- Cooling too fast
- Wrong choice of electrode
- Incorrect edge preparation
- Welds stressed during welding operation

The possibility of cracks in a weld can be reduced by a number of methods. Slower cooling can be ensured by pre-heating as well as post-heating. Use of correct electrodes (e.g., low-hydrogen electrodes) and correct profiles of the preparation are essential features of good and crack-free welds. To reduce stresses in the weld during the welding operation, it may be necessary to review the design and arrangement of the jigs and fixtures and also pre-heat the components, if necessary.

It may be noted that the above faults are not due to the characteristics of the parent metal. Rather, the faults are related to the choice of various welding parameters as also, to a large extent, to the level of skill of the welder engaged in the work. This aspect should be kept in mind while trying to improve upon the quality of the weld.

16.6.7 Effects of Weld Defects

Primary effects of weld defects is that they set up discontinuity in the path of the tensile stress. It may appear that the main concern should be the impairment of strength of the joint owing to the reduction of the effective load bearing area. However, in practice, defects due to porosity, slag inclusion, lack of fusion, under-cut or cracks, occur locally and are seldom concurrent at a particular cross section. Therefore, chances of significant loss of area along the entire cross section are rather remote. Nevertheless, acceptability of such defects depends on the size and the nature of the defects in a particular location and has to be examined for each individual case. In cases of incomplete penetration, however, the defect is likely to occur along a considerable portion of the weld and therefore, needs to be examined closely. Figure 16.13 illustrates stress paths and locations of typical failures in plates joined by butt welds.

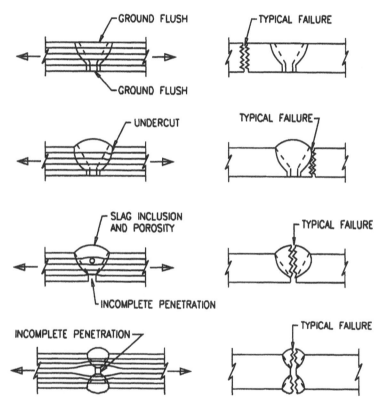

Fig. 16.13 Stress paths and locations of typical failures in plates joined by butt welds [8]

A designer's primary concern regarding weld defects is the phenomenon of concentration of stress at the locations of discontinuities of the stress path. In general, the chances of stress concentration for smooth or rounded discontinuities such as porosity, and slag inclusion, are much less than those for sharp notches like under-cuts or cracks. These have significant impact on fatigue strength of the joint and have been discussed in greater detail in a separate section.

16.7 SHRINKAGE AND DISTORTIONS

All fusion welding methods involve movement of a molten weld pool which creates a 'heating-cooling' cycle along the axis of the joint. Heat flows from the pool into the adjoining parent metal, thereby affecting the properties of the metal and causing dimensional changes to the elements being joined. The former (i.e., the change in the properties of the metal) has been discussed in Section 16.3. In this section, the characteristics of the dimensional change due to thermal expansion and contraction during welding will be briefly discussed.

The 'heating-cooling' cycle in the molten weld pool involves three distinct phenomena :

- Contraction of the liquid metal (thermal)
- Change in volume of the weld pool on solidification (density), and
- Contraction of the solidified metal (thermal)

As the heat source moves forward along the joint centre-line, progressively melting the parent metal at the leading edge, the weld pool metal at the trailing end solidifies, starting from fusion boundary and progressing towards the centre-line. As the density of the solidified metal is higher than that of the liquid metal, the solidified metal at the lower portion (Fig. 16.14) occupies a smaller space than its predecessor (liquid metal). Simultaneously, the liquid metal, during cooling and solidification process also shrinks due to thermal contraction. The reduction of the metal on solidification (due to the combined effect of density and thermal contraction) is made up by fresh molten metal being fed from the leading edge of the weld pool and the additional metal contributed by the electrode.

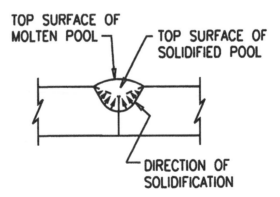

Fig. 16.14   Shrinkage during solidification

It has been observed that during the process of intense heating and rapid cooling cycle, the mechanical properties of the materials in and around the weld pool can change drastically. Thus, the yield stress of steel at high temperature may suffer a drastic change at temperatures as low as 500 degrees to 600 degrees Centigrade.

As the electrode moves forward, the heat flows from the weld into the surrounding material which results in progressive cooling of the joint. During this stage, the weld material contracts, causing the joint to shrink. This contraction is prevented by the surrounding cooler material inducing stresses in the weld and producing plastic deformation. As long as these stresses are above the yield point of the metal at the prevailing temperature, the deformations are of permanent nature. The various instances of deformations and the methods of controlling these are discussed in the following paragraphs.

Essentially the metal experiences shrinkage both in the direction of the weld (longitudinal) and at right angles to it (transverse). The longitudinal shrinkage of the weld causes the adjoining plates to be reduced in length along the line of joint. Similarly, due to transverse shrinkage of the weld metal, the overall width of the welded plate is reduced. A typical example of distortion in a butt welded joint is illustrated in Fig. 16.15. Due to the combined effect of longitudinal and transverse shrinkages of the weld, the joint takes a somewhat distorted shape. Since contraction is proportional to the length of the metal being cooled, in asymmetrical welds such as single-V butt welds, shrinkage in the transverse direction will be greater at the top

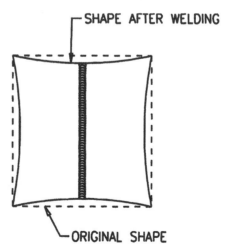

Fig. 16.15   Changes in shape due to shrinkage

surface than the bottom (root). This will produce a rotational (angular) distortion as shown in Fig. 16.16.

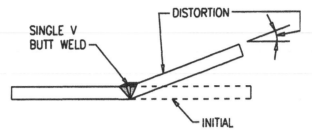

Fig. 16.16 Rotational (angular) distortion

It is necessary to keep the amount of the angular distortion within an acceptable limit. There are a number of methods for achieving this. One method is to clamp the pieces to restrain the movement. This method is likely to produce 'locked in' stresses at the joint. An alternative method is to adopt a weld preparation in which the shrinkage of the plates across the thickness of the weld would be more or less same. Generally a double-V joint may provide such a solution (Fig. 16.17). However, since, in a multi-run weld, the first run is likely to produce more angular rotation than the subsequent runs, symmetrical double-V preparation may not produce a completely distortion-free result. A slightly asymmetrical preparation (Fig. 16.18) may produce satisfactory results.

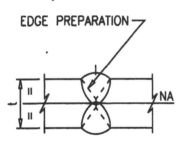

Fig. 16.17 Symmetrical double-V joint

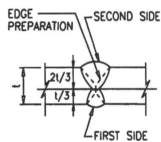

Fig. 16.18 Asymmetrical double-V joint

In the longitudinal direction, the forward moving 'heat-cool' cycle on the joint induces asymmetrical shrinkage, which produces a longitudinal bow in the joint (Fig. 16.19). This can be best tackled by resorting to a pre-determined sequential and directional welding in short lengths along the entire joint. A typical example of such a joint has been shown in Fig. 16.20.

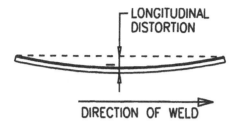

Fig. 16.19   Longitudinal distortion in a butt joint [1]

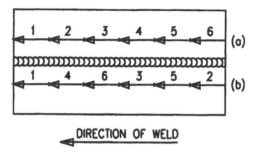

Fig. 16.20   Typical sequences (a) and (b) and direction of short length welds (¬) in a butt joint to reduce longitudinal distortion [1]

Angular rotations have often been noticed in flanges of welded plate girders (Fig. 16.21a). The distortion can be controlled by pre-cambering the flange plate prior to welding. (Fig. 16.21b). An alternative solution is to deposit welds in a pre-determined sequence. Since the angular distortion is maximum in the first weld (on one side of the joint), the second weld on the

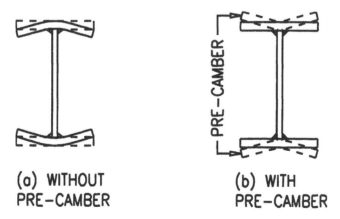

(a) WITHOUT PRE-CAMBER

(b) WITH PRE-CAMBER

Fig. 16.21   Method of pre-cambering to correct distortion

other side cannot pull the plate back fully to the normal position. As a result there will be a residual distortion in the joint. To minimise this defect, the joint can be initially set at a pre-determined angle so that at the end of the second run, the required angle of 90 degrees is achieved. Figure 16.22 illustrates the procedure. The method of pre-setting the joint requires considerable experience to achieve the desired result.

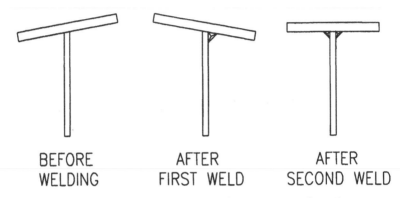

BEFORE WELDING   AFTER FIRST WELD   AFTER SECOND WELD

Fig. 16.22   Method of pre-setting to correct distortion

Another alternative method of straightening flange plates distorted during welding, is by utilising the simple concept of thermal expansion and contraction. Figure 16.23 illustrates this principle. When a narrow band on the top face of top flange is heated by applying concentrated heat from a

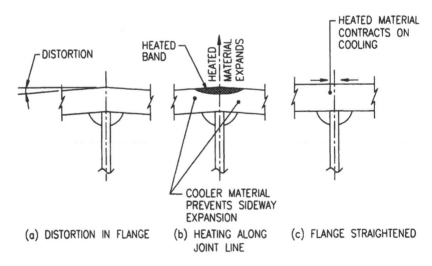

(a) DISTORTION IN FLANGE   (b) HEATING ALONG JOINT LINE   (c) FLANGE STRAIGHTENED

Fig. 16.23   Flame straightening [1]

blow pipe, the heated portion will tend to expand sideways. This expansion will be prevented by the surrounding cooler material and the relatively weak heated part is forced to expand at right angles to the surface. When the area cools, the metal contracts. This contraction causes the band to straighten in a manner similar to which caused the distortion due to weld. Although the method appears to be quite simple, it needs sufficient field experience and practice to be successful. As a result, though very effective, the method has not gained popularity amongst technicians.

For the prevention and control of distortion, one or a combination of the following simple rules may be followed:

1. Reduce the effective shrinkage force

This can be achieved by keeping the welds to the minimum size to meet the service requirement. For example, in a fillet weld the strength is determined by the effective throat thickness. Making the weld profile too convex does not increase the strength, but certainly increases the effective shrinkage force. A flat or a concave profile with adequate throat thickness will reduce the shrinkage force without impairing the strength.

Similarly, selection of proper edge preparation in a butt welded joint, will also reduce the effective shrinkage force. The aim should be to obtain the proper fusion at the root of the weld with a minimum of weld metal. A bevel upto 30 degrees should satisfy the requirement. The gap between the two pieces should also be kept to the specified minimum, so that the least amount of weld metal is deposited, thereby keeping the shrinkage force to the minimum.

Distortion in the lateral direction can be controlled by using less passes with the largest suitable size of the electrode, bearing in mind plate thickness, welding position, length of run, etc. A suitable run sequence may also reduce the effective shrinkage force (Fig. 16.20).

2. Make shrinkage force work to reduce distortion

Distortion can be countered by pre-setting or pre-bowing the parts out of their final position. The amount of pre-setting or pre-bowing has to be determined by previous experience (Figs. 16.21 and 16.22).

3. Balance shrinkage forces with other forces

This can be achieved by employing proper welding sequence for depositing weld metal at different points about the structure, so that, shrinkage of weld metal at one location will counteract the shrinkage forces of welds already made at another location. A simple example of this situation is welding alternately on both sides of the neutral axis of a double-V preparation in a butt welded joint.

Peening of individual weld run in heavy butt joints immediately after deposition stretches the weld bead, thereby counteracting the tendency to contract and shrink as it cools. Peening should, however, be used with great care as excessive peening may damage the weld metal. Final layers should never be peened.

Use of jig and fixtures to hold the work piece in a rigid position during welding is another example of balancing the shrinkage forces of the weld with sufficient counterforces to minimise distortion. In effect, the balancing forces of the jig and fixtures cause the weld metal itself to stretch, thereby reducing the distortion.

16.8   RESIDUAL STRESSES

As discussed in the preceding section, the weld metal undergoes plastic deformations during cooling. As the weld cools and the plastic deformation is completed, there will be some stresses locked up in the joint. These are termed residual stresses. This stress pattern after the weld cools and the plastic deformation has ceased, is shown in Fig. 16.24. It will be noted that, moving out from the centre-line of the joint, the residual tension is reduced to zero and then there is a zone of compression. This residual stress is present in most of the welded joints.

Presence of residual stresses in a great many applications such as low pressure pipe work, storage tanks, etc., does not present appreciable problems in the behaviour of the joints during service conditions. Consequently, joints in such cases are not required to undergo any post weld treatment for relieving the residual stresses. However, in certain situations it may be necessary to consider the effects of residual stresses. As for example, presence of residual stresses in a very cold environment may embrittle the material causing brittle fracture. Also, fluctuating loading pattern during service condition (as in a bridge), may lead to fatigue cracks due to the presence of residual stresses. Furthermore, in ceratin types of environments, some steel materials are susceptible to corrosion in the tension zone and presence of residual tension stress in the welded joints located in the tension zone enhances the risk of cracking due to stress corrosion. In such cases it is advisable to subject the welded joints to stress relieving treatment.

The most common method of stress relieving a welded joint is by heat treatment. As already stated, yield stress of steel decreases with the

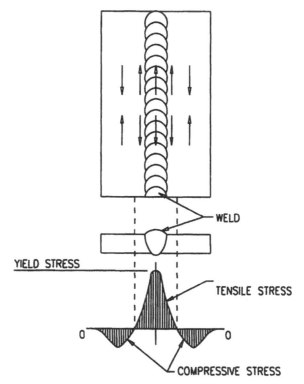

Fig. 16.24 Residual stresses in welds [2]

application of heat. Thus, if a welded joint is so heated that its temperature reduces the yield stress to a value which is below the residual tensile stress, a localised plastic deformation will occur and the tensile stresses will be reduced. Simultaneously, the compressive stresses will also be reduced to balance the equilibrium of the heated zone. The approximate temperature for effective stress relieving depends on the relationship between the temperature and the type of steel (alloy content) on which depends the corresponding yield stress. Typical stress relieving temperatures corresponding to various steel types is shown in Table 16.1.

A note of caution needs to be highlighted here. As thermal treatment involves localised heating, there is always a danger of differential expansion and contraction causing new residual stresses. Therefore, heating and cooling need to be carefully controlled. This can be achieved by maintaining the temperature distribution between the weld centre-line and the outer limits of the specified width of the heated zone to a uniform pattern throughout the length of the joint.

Table 16.1   Stress relieving temperatures for different types of steel [1]

Type of steel	Stress relieving temperature (degrees Centrigrade)
Low Carbon	580
Carbon-Manganese	580
Carbon - 0.5% Molybdenum	650
1% Chromium - 0.5% Molybdenum	650
2.25% Chromuim - 1% Molybdenum	690
5% Chromium - 0.5% Molybdenum	725

16.9   INTERACTING VARIABLES

Fusion welding is a process which involves a number of interacting variables, any one or more of which may influence the final properties of the joint. These variables are listed in the following paragraphs.

16.9.1   Composition of Base Metal, Electrode and Flux

The weld metal consists of a mixture of materials obtained from the base metal, the electrode as well as material from the flux. All these materials have important effects on the final properties of the solidified weld.

16.9.2   Welding Process

Welding process determines the size of the weld pool and its geometry.

16.9.3   Environment

During welding process, moisture and gases such as oxygen, nitrogen and hydrogen are likely to penetrate into the weld pool. Of these, hydrogen makes the weld prone to cracks.

16.9.4   Speed of Welding

Speed of welding has a direct influence on the solidification speed of the weld, which affects the properties of the final weld.

16.9.5   Thermal Cycle of Weld

The pattern of thermal cycle also affects the properties of the weld.

16.9.6   Size and Type of Joint

In case of thick plates, multi-run welds are preferable as these are likely to reduce residual stress. This entails consumption of more time leading to costlier (but better) joints.

16.9.7   Manipulation of Electrodes

In order to produce good weld, it is necessary to manipulate the electrode properly so as to avoid craters at the edge of the plates, which may lead to cracks in the future.

REFERENCES

[1]   Gourd, L.M. 1995, *Principles of Welding Technology*, Edward Arnold, London, U.K.

[2]   Dowling, P.J., Knowles, P.R. and Owens, G.W. (eds.) 1988, *Structural Steel Design*, The Steel Construction Institute, London, UK.

[3]   Ghosh, U.K. 2000, *Repair and Rehabilitation of Steel Bridges*, A.A. Balkema, Rotterdam, Netherlands.

[4]   Ghosh, U.K. 2000, *Quality assurance and training for design and detailing of durable steel structures*. In: The Bridge and Structural Engineer, Journal of the Indian National Group of the International Association for Bridge and Structural Engineering, New Delhi, India, Feb-Mar 2000 issue.

[5]   Ghoshal, A. 2000, *Quality assurance and training for fabrication and erection of durable steel structures*. In: The Bridge and Structural Engineer, Journal of the Indian National Group of the International Association for Bridge and Structural Engineering, New Delhi, India, Feb-Mar 2000 issue.

[6]   Easterling, K. 1992, *Introduction to Physical Metallurgy of Welding*, Butterworth-Heinemann Ltd., Oxford, UK.

[7]   Houldcroft, P. and John, R. 1988, *Welding and Cutting*, Woodhead-Faulkner Ltd., Cambridge, UK.

[8]   Xanthakos, P.P. 1994, *Theory and Design of Bridges*, John Wiley & Sons, Inc., New York, USA.

Chapter 17

Bolted and Riveted Joints

17.1   INTRODUCTION

Phenomenal progress in the use of bolting and welding during the recent decades has significantly reduced the importance of riveting in structural steelwork. However, riveting has been the traditional method of connecting metal components for centuries, and is still being used in some countries in fabrication and erection of steel bridges, particularly railway bridges, where fatigue effects are predominant. In the present text, therefore, riveted connections are also briefly covered.

The modes of transmission of loads in the bolted and riveted joints are somewhat similar, and connections using these two types of fasteners are discussed in the current chapter.

17.2   BOLTED JOINTS

17.2.1   General

Bolted joints are a very simple and sufficiently reliable method of connecting steel elements. They do not require particularly sophisticated equipment and are generally quicker to install compared to other types of fasteners. Because of these advantages, bolted connections have become very widespread in field operations.

17.2.2   Physical Properties of Bolts and Nuts

Bolts used in structural connections are generally supplied to International Standards Organisations (ISO) specifications. The designation system of the strength grade number of the bolts consists of two figures. The first figure in the grade number represents one tenth of the minimum tensile strength of the material in kilograms force per square millimetre and the second (decimal) figure is the factor by which the first must be multiplied to give the yield stress. The two most commonly used bolts in structural connections are grade 4.6 mild steel bolts and grade 8.8 high strength bolts. The computation of the respective ultimate tensile stress and the yield stress of these grades are shown in Table 17.1.

Table 17.1 Ultimate tensile and yield stresses of bolts

Grade	Ultimate tensile stress Kgf / sq. mm	Yield stress Kgf / sq. mm
4.6	$4 \times 10 = 40$	$40 \times 0.6 = 24$
8.8	$8 \times 10 = 80$	$80 \times 0.8 = 64$

There is a similar designation system for nuts also. This is a single figure number and represents one tenth of the ultimate tensile stress of the steel. Thus, grade 8 nut has an ultimate tensile stress of 80 Kgf/sq.mm. and can be used with a grade 8.8 bolt. Often, higher grade of nuts are used to reduce the risk of thread stripping.

17.2.3 Types of Bolts

Broadly, there are two types of bolts which are used in bolted connections, viz., bearing type bolts and friction type bolts. Bearing type bolts are those which transmit load from one member to another either by shear in the shank or by bearing against the members. Friction type bolts, on the other hand, transmit the load from one member to another by means of friction developed due to the clamping action produced by tightening the bolts to a high degree of tension. The first type includes unfinished bolts and, turned and fitted (precision) bolts. The latter category mainly comprises High Strength Friction Grip Bolts. The characteristics of these bolts are discussed in the following paragraphs.

Unfinished bolts

These bolts, sometimes termed 'ordinary' or 'black' bolts, are forged from rolled steel round bars which have not been finished to accurate shank dimensions. Consequently, they have large dimensional tolerances on shank and thread and are used in 'clearance holes' with diameters normally 1.5 mm to 2.0 mm greater than the diameters of the shanks. Steel washers are normally used under the nuts to distribute the clamping pressure on the bolted members, and also to prevent the threaded portions of the shanks from bearing on the connecting members.

These bolts are used where slippage or vibrations do not matter. Therefore, in steel bridges, where stress reversals may occur and slippage is undesirable, these bolts are usually not recommended for permanent connections of main members of bridges.

These bolts are generally manufactured in mild steel, but can also be of high strength steel.

(a) Load transmission In a joint with unfinished bolts, the load is considered to be transmitted by shear or bearing. The friction between the interface of the members due to tightening of the bolts is ignored. As the load is applied, the members slip and the bolts bear against the edges of the clearance holes. It is at this position that the load can be transferred from one member into the bolts and, in turn, be transmitted from the bolts to the connecting member by shear in the shank or bearing on the bolt. Figure 17.1 illustrates the phenomenon. The shank may have one, two or more shear planes as shown in Fig. 17.2a, b and c respectively. Figure 17.3 shows a case where the load is transmitted through bearing between the surfaces of the components and the bolt head and nut, producing tension in the shank. In this case, the root area of the threaded portion of the shank (not the gross shank area) will have to be considered for the transmission of the load. For bolts subjected to combined shear and tension, the effect is allowed in national codes by a linearised interaction formula.

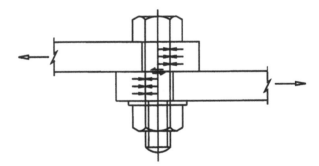

Fig. 17.1 Joint with unfinished bolt

(b) Shear stress Under elastic condition, the shear stress across the gross section of the shank is not uniformly distributed. However, for design purpose, this stress is assumed to be uniformly distributed. This inaccuracy is taken care of by the safety factor in the design process.

In a properly designed and installed connection, the threaded portion of the bolts does not bear on the connecting members. In such a case, the shear-

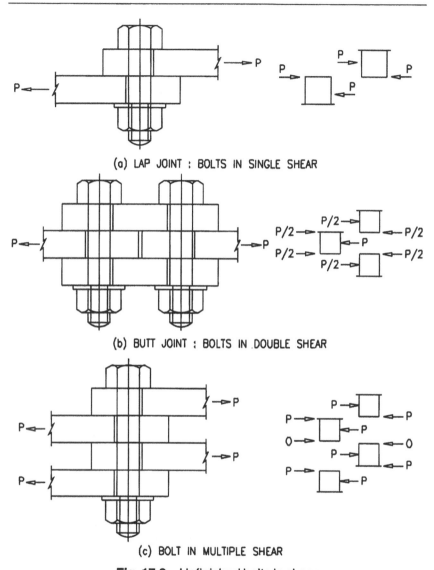

(a) LAP JOINT : BOLTS IN SINGLE SHEAR

(b) BUTT JOINT : BOLTS IN DOUBLE SHEAR

(c) BOLT IN MULTIPLE SHEAR

Fig. 17.2 Unfinished bolts in shear

ing stress of the bolt is computed by using the gross sectional area of the bolt. However, if the threaded portion of the shank bears on the connecting members, the net section at the root of the thread is to be considered for computing the shearing stress. While designing the connections, it may be expedient to consider only the net section for computing the number of bolts as the designer will have little control over the actual position of the threaded portion of the bolt shank in the steelwork at work site.

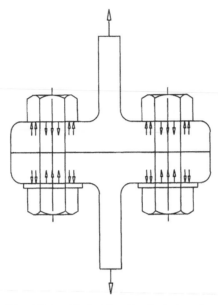

Fig. 17.3 Unfinished bolts in tension

(c) Bearing stress The bearing between the bolt and the member is concentrated at the edge of the member and is likely to produce high local stresses, until plastic deformation takes place.

(d) Spacing requirements National codes normally specify both minimum and maximum limits for the spacing between the centre-lines of adjoining bolts and also between the centre-line of a bolt and the edge and end distance of the member.

Limits for the minimum value of space between the centre-lines ensure adequate space for tightening of these bolts. Maximum values serve two distinct purposes: (a) they control the unsupported distance between two bolts and guard against possible local buckling of the plies of compression members, and (b) they prevent corrosion due to ingress of water between the plies.

The minimum value of the end distance is determined to guard against shear out failure of the plate as shown in Fig. 17.6d.

(e) Sharing of loads in long joints It is traditionally assumed that bolts in a group of long joints, when subjected to direct load, will share the load equally. This traditional concept is based on the assumption that the connecting plates are perfectly rigid and the bolts perfectly elastic. Consequently, pure translation of one plate relative to the other produces

equal deformation and shearing strain in all the bolts. Thus, if the cross section of the bolts are same, the load in each bolt would also be same. This is the basis of the 'rigid plate theory' commonly adopted in the design.

However, in actual practice, the plates are not absolutely rigid and the elongations between the bolts are not same. Therefore, the bolts do not share the load equally. In fact, experiments have revealed that the bolts at the outer rows (ends) tend to carry more loads than those in the interior rows. This concept is usually termed 'elastic plate theory' and underlines the desirability of arranging bolts in a large joint, as far as possible in a compact manner, so as to avoid uneven distribution of load amongst the bolts. In the plastic range, as the outer bolts deform without taking additional load, the loading will redistribute causing a more uniform sharing of the load until all the bolts are stressed to the yield point. This behaviour provides the justification for using the rigid plate theory as the basis of design. Some national standards (e.g., BS 5400), however, recognise the unequal distribution of the load between the bolts in long joints and recommend reduction in the strength (capacity) of such bolts in long joints by a suitable multiplying factor.

(f) Tensile stress concentration in the member   Distribution of tensile stresses across a member with holes is not uniform within the elastic range. Stresses are dependent on the size and location of the holes. Typical shape of stress diagrams are shown in Figs. 17.4a and b. The high stress concentration would cause the fibres adjoining the hole to reach yield point first. As the load increases, these fibres will deform and the stresses in the next fibres will increase. The process will continue and the distribution of stresses across the member will become more and more uniform until finally the load reaches the ultimate strength.

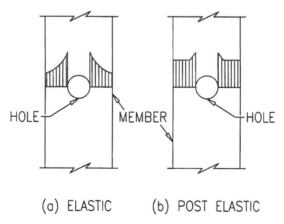

(a) ELASTIC   (b) POST ELASTIC

Fig. 17.4   Stress distribution in member across the hole

(g) Prying forces When bolts are loaded in tension such that the line of the applied force is eccentric to the axes of the bolts, an additional tension is induced due to prying action. The situation is best illustrated in a moment connection at the end plate of a cross girder and the flange of a vertical member in a truss bridge. A relatively flexible end plate would bend under the applied load P and compressive reactions Q between the outer edges of the end plate and the flange of the vertical are set up. These reactions are termed prying forces which add to the tension in the bolt (Fig. 17.5). Thus:

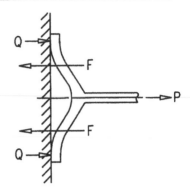

Fig. 17.5 Prying action in bolts

$$\text{Tension in a bolt } F = P/2 + Q$$

The magnitude of the prying force can be reduced by increasing the thickness of the connecting end plate (by reducing its flexural deformity) and/or by decreasing the eccentricity (by placing the connecting bolts nearer to the web). A number of semi-empirical formulae are available to determine the values of prying forces. Some national codes (e.g., BS 5400) recommend design procedure with formation of plastic hinges at the root and, if required, at the bolt line. However, provided the connecting plate is relatively thick, the spacings of bolts are reasonably small, and the edge distances are sufficiently large, the prying forces are very small and often neglected.

(h) Failure modes of a bolted joint A bolted joint normally deforms appreciably before failure. Failure of a joint may take place in any one of the following modes, depending upon which one is the weakest. Different failure modes have been illustrated in Fig. 17.6.

Shear failure of a bolt Shear failure of a bolt occurs across one or more planes between the members it connects, depending on whether the bolt is in single or in multiple shear.

Bearing or crushing failure of a bolt The mode of failure occurs at half circumference contact surface of the bolt and the hole in the member. This may occur due to bearing failure of the member or of the bolt or of both.

Tension failure or tearing of the member This mode of failure may occur when the cross sectional area of the member is inadequate to transmit the load.

Shear out failure of the member This mode of failure occurs when there is insufficient edge distance in the member along the line of the load. This type of failure can be avoided by providing adequate edge distance.

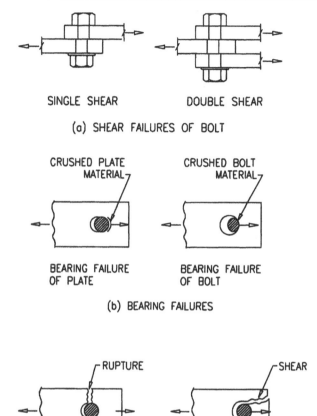

Fig. 17.6 Modes of failure of a bolted joint

Turned and fitted (precision) bolts

These bolts, with nuts and washers, are usually used in structures—such as in bridges—where slip-free connections are desired. These bolts are made from rolled steel bars and turned down to closer tolerances and are used in close tolerance holes generally produced by reaming. The tolerances vary according to different national standards. Commonly used tolerances are: + 0.000 mm, – 0.125 mm for the shanks and + 0.125 mm, – 0.000 mm for the holes. These bolts are generally manufactured in high strength steel; but can also be of mild steel. These bolts are inserted into the holes by means of light blows of hammers. The faces under the head of the bolt and the nut are usually machined. The washers are also machined on both faces. In order to avoid threads encroaching into the steelwork connected, the length of the thread should be clearly specified for each bolt type.

In some turned and fitted bolts, the threaded portion is made of lesser diameter than the shank or barrel diameter. With this variation, there is lesser chance of the thread getting damaged during insertion of the bolt in the hole and thus makes site operation that much easier. These bolts are often called turned barrel bolts.

When members are joined by turned bolts fitted in a reamed close tolerance hole, the slip in the joint is minimal, and the load is transmitted directly by shear and bearing (Fig. 17.7), as in the case of a riveted joint. The high local bearing stresses associated with unfinished bolts in clearance holes are also minimal. Behaviour of a joint with turned and fitted bolts or turned barrel bolts is thus considered to be much better than that of a joint with unfinished bolts.

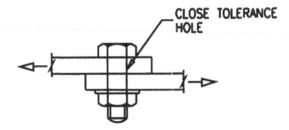

Fig. 17.7　Joint with turned and fitted or turned barrel bolt

In the design, bending in short bolts are normally ignored. However, in bolts with grips exceeding five times the diameter, the effect of bending should be checked in the conventional manner.

High Strength Friction Grip (HSFG) bolts

The term 'High Strength Friction Grip' bolt relates to a high strength bolt with a high strength nut and hardened steel washer(s) installed in a clearance hole, and tightened in a controlled way to a specified shank tension in order that the members (plies) joined are clamped together between the bolt head and the nut.The clampling force thus developed, transfers the load in the connected members by friction between the parts, and not by shear or bearing as in the case of other types of bolts.

Figure 17.8 illustrates the mechanism of a High Strength Friction Grip bolt. As the nut is tightened to produce the tensile force T in the shank of the bolt approaching yield point, the connecting members are clamped tightly together, developing a frictional resistance F. The maximum value of F is given by:

$$F = \mu T$$

where μ is the coefficient of friction between the members at the interface. The value of T should be approximately 80% to 90% of the yield strength of the bolt. As long as the external load P does not exceed F, no slip will occur and the joint will transmit the load.

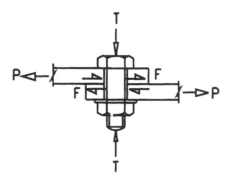

Fig. 17.8   Joint with friction grip bolt

The coefficient of friction is also termed 'slip factor'. In a multi-bolted connection, slip factor is defined as :

$$\text{Slip Factor} = F/(n \times T \times b)$$

where,   F is the slip load, i.e., the force causing displacement between the two interfaces,

n is the number of interfaces through which each bolt passes (excluding the packing pieces),

T is the specified minimum shank tension in one bolt, and

b is the number of bolts in the joint.

The values of slip factor are affected by the joint geometry, surface treatment, etc., and are generally specified in the code for conventional situations. Slip factor is normally taken as 0.45 for surfaces free from any paint or other applied finish, oil, dirt, loose rust or mill scale. For unusual situations, slip factor should be found out from standard tests with simulated assembly conditions of the actual job. Safety factor, which takes into account certain uncertainties in the friction grip bolts, is to be considered in the above equation in accordance with the standards being followed in the design.

Since the strength of a joint with HSFG bolts is derived from the clamping force caused by the tension in the shanks of these bolts, any additional external tension in the bolts will reduce the clamping force. If the external tension is further increased, the clamping force will also decrease further. In order to avoid complete neutralisation of the clamping force (thereby causing separation of the interfaces), the national codes specify the limit to which the external tension can be applied.

When a joint with HSFG bolts is subjected to a combination of tension and shear (as in the case of connections of stringer to cross girder or cross girder to vertical member in a truss bridge), the behaviour of these bolts is not simple. However, a simple linearised interaction formula is specified in national codes to balance the interaction of the shear and the external tension.

(a) Control of bolt tension As the behaviour of the friction bolted joint depends largely on the tension produced in the shanks of the bolts, it is imperative that the bolts are tightened adequately to develop the desired tension. There are a number of methods to ensure the correct shank tension in the field condition, such as part turn method, use of proprietary devices like calibrated wrenches, load indicating bolts or washers, twist-off bolts, etc.

Part turn method: In this method, the nut is initially tightened to a 'snug-tight' bolt condition to bring the joining components together. This initial condition is recorded by making a mark on the nut and the protruding thread portion of the bolt. From this point the nut is finally turned with respect to the bolt shank an additional half to three quarter turn depending on the diameter and length of the bolts as specified by the codes. This method does not require any sophisticated tools (ordinary spanner may suffice) to achieve the required tension and is also easy to inspect. However, the efficiency of this method depends on the correctness of the initial tightening operation prior to giving the final turn.

Torque control method: In this method, a calibrated torque wrench is used, which may be operated either manually or by power. The calibration in the wrench needs to be checked at regular intervals. This type of torque wrench needs calibration for different sizes and grades of bolts and may become rather unwieldy where a number of grades and sizes of bolts are to be tightened.

Load indicating devices: Several types of load indicating bolts have been in use. Typically such a bolt has its head so shaped that before tightening, it (the bolt head) makes contacts with the steel at its four corners only. This leaves a gap between the steel and the underside of the bolt head. With the tightening of the nut, the parts of the bolt head in contact with the steel yields, causing gradual reduction of the gap. A pre-determined minimum gap indicates that the desired bolt tension has been achieved. This gap can be checked by using feeler gauges. Another simple device is the load indicating washer. The washer has a number of protrusions on one of its surfaces and is normally fitted under the bolt head, with the protrusions in contact with the underside of the head, leaving a gap between the general surface of the washer and the underside of the bolt head. The nut is tightened until the protrusions are crushed and the gap is reduced to a pre-determined value (checked by feeler gauge), which indicates that the desired bolt tension has been achieved (Fig. 17.9). There is also another system of 'twist-off' bolts, in which an extended portion of the threaded shank is automatically sheared off when the desired tension in the shank is achieved.

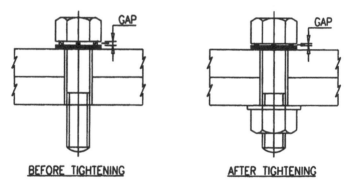

Fig. 17.9   HSFG bolt with a load indicating washer

(b) Factors affecting preload   In spite of using these devices, however, it may not be possible to achieve the intended tension in all the bolts in a particular group, as the final pre-load is dependent on some other factors, some of which are discussed in the following paragraphs.

Surface preparation: The surfaces at the interface of the connecting members should be free of defects which are likely to prevent proper seating of the part, especially dirt, burr and similar other foreign material. Also, the interface should be free from oil or grease, paint, lacquer, galvanising or similar finishes, which might prevent the development of the friction.

Sequence for installation: In a large joint, there is always a chance of bedding down of the connected plies during installation of these bolts; as a result, tension in some of the bolts tightened first may have been reduced by the time the latter bolts are tightened. To avoid this effect, the bolts should be tightened incrementally, in a staggered pattern. Also, the bolts should be tightened in two stages—first, when the members are brought together to a just tight (snug tight) position, and the final stage when the nut is turned to achieve the required tension in the bolt shank.

Pressure distribution: The pressure between the plies caused by bolt tightening is distributed around the bolt hole over a limited circular area. This pressure is maximum at the edge of the hole and gradually reduces to zero at the periphery of the pressure area (Fig. 17.10). The portions of the plies outside the pressure area have a tendency to warp away from the interface and may lead to separation of the plies at the outside edge of the joint. This effect is deleterious from corrosion point of view and needs to be considered while finalising the corrosion prevention regimen for the bridge. This effect is also important for the understanding of the fatigue behaviour of the joint with friction grip bolts.

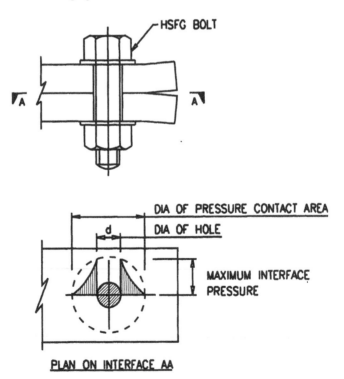

Fig. 17.10 Interface pressure distribution with a tightened HSFG bolt

Creep: Friction grip bolts suffer creep and consequent loss in bolt tension due to high local stresses in the threads and contact surfaces, mostly within the first few days after tightening. The effect is least with a smoother surface (e.g., grit-blasted). Loss in bolt tension may also be caused by the tension in the joining members due to Poisson's ratio effect, which tends to reduce the thickness of the plies and consequently the pretension strain in the bolt.

Furthermore, presence of prying effect of friction grip bolts subjected to combined shear and tension may also aggravate the situation. These aspects need due consideration while prescribing the desired pretension in the bolts.

Other aspects :   Other aspects, such as spacing (e.g., pitch, edge distance, end distance), long joints in shear and effects of prying, which were discussed earlier in the section of ordinary bolts, are equally relevant for friction grip bolts also and should be considered as such.

(c) Advantages of HSFG bolts   Compared to other types of bolts, HSFG bolts have many advantages. Some of these are:

- HSFG bolts produce very rigid joints, since the load is transmitted by frictional resistance and without any slip.
- As the phenomenon of frictional resistance is effective outside and around the hole area, only a portion of the load is transmitted at the net section of the connecting components. This reduces the risk of failure occurring at the net section.
- Since alternating loads have little effect on the stresses in the bolts, these joints command a very high fatigue strength. Consequently these bolts are very useful for connections in bridges.
- Furthermore, the tension in the bolts prevents the nuts from loosening —a problem generally associated with other types of bolts.
- The bolts are tightened in accordance with set specifications, tools and equipment. It is, therefore, rather easy to train a person to follow these specifications and achieve fairly uniform joints.
- The noise level is very low during installation of these bolts, and the effect on the environment is very friendly.

17.3   RIVETED JOINTS

17.3.1   General

Riveting is a forging process, in which a rivet with one head and shank is inserted into pre-aligned oversized holes on the components to be joined, with the head tightly pressed by means of a bucking bar with a head die, against one of the components. The shank which protrudes through the hole to the other side is shaped into a driven head by pressure from a rivet driving machine. The process is illustrated in Fig. 17.11. Rivets used in structural steelwork are almost always heated to a minimum temperature of about 1,000 degrees Centigrade in a furnace before they are inserted in the hole.

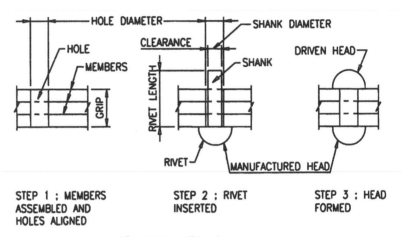

STEP 1 ; MEMBERS ASSEMBLED AND HOLES ALIGNED

STEP 2 ; RIVET INSERTED

STEP 3 ; HEAD FORMED

Fig. 17.11   Riveting process

17.3.2   Behaviour of Riveted Joints

The modes of transmission of load in a riveted joint are somewhat similar to those in a bolted joint, viz., shear, bearing, tension or friction. However, the actual manner of load transmission in a riveted joint varies significantly from that in a bolted joint. These distinctive features in a riveted joint are dicussed below:

Frictional resistance

As the rivet is being driven, the pressure from the riveting machine forces the hot shank to fill the clearance in the oversized hole. Subsequently as the shank cools, it tends to shrink longitudinally as well as diametrically. The longitudinal shrinkage in the shank is resisted by the connected components. This shrinkage induces a high tension (close to elastic limit) to be developed in the shank, while the connecting components are subjected to considerable compression. This compression produces a frictional resistance to sliding in the interface of these components. The phenomenon of initial shrinkage of the rivet and the resultant friction between the connected components make a riveted joint to behave in a way somewhat intermediate between a friction type connection and a bearing type connection. For design, however, the frictional resistance is ignored and only a bearing type joint is considered.

Shear stress distribution

As in the case of bolts, the shear stress across the gross section of the rivet is assumed to be uniform for the purpose of design.

Bearing stress distribution

Figure 17.12a shows the radial nature of bearing stresses in a rivet. For the purpose of design, however, an equivalent nominal bearing stress on a diameter plane through the rivet is assumed as shown in Fig. 17.12b.

(o) ACTUAL BEARING STRESS
IN RADIAL DIRECTION

(b) NOMINAL BEARING STRESS

Fig. 17.12 Bearing stress distribution in a rivet

Bending

Theoretically, a rivet is subjected to bending stress in addition to shear and bending stress along the length of the rivet. However, in practice, the bending is likely to be prevented by the friction between the connecting components, and as the effect of bending will be only marginal, it is normally neglected.

Sharing of loads in long joints

The behaviour of a long riveted joint is much the same as that of a long bolted joint which has been discussed already. However, for the purpose of design of long riveted joints, the concept of rigid plate theory is generally followed and a uniform sharing of the load is traditionally considered.

Filling of the rivet hole

For the purpose of design, it is assumed that the high pressure from the driving machine would force the heated rivet to completely fill the oversized hole. This assumption may not be correct as some of the rivets may fail to fill the holes completely. Therefore, these rivets will receive the load only after the other rivets in the group become fully loaded and a redistribution of the load takes place.

REFERENCES

[1] Dowling, P.J., Knowles, P.R. and Owens, G.W. (eds.) 1988, *Structural Steel Design*, The Steel Construction Institute, London, UK.

[2] Bresler, B., and Lin, T.Y. 1960, *Design of Steel Structures*, John Wiley & Sons, Inc, New York, USA.

[3] Grinter, L.E. 1965, *Design of Modern Steel Structures*, The Macmillan Co., New York, USA.

[4] Brockenbrough, R.L. and Merrit, F.S. (eds.) 1994, *Structural Steel Designer's Handbook*, (second edition), McGraw Hill Book Co Inc, New York, USA.

[5] Owens, G.W. and Knowles, P.R. (eds.) 1994, *Steel Designers' Manual*, (fifth edition) Blackwell Scientific Publications Ltd., Oxford, UK.

Chapter 18

Connections: Design Principles

18.1 INTRODUCTION

Connections play a very important role in steel bridges as they are not only required to transmit the forces from one member to other members and *vice versa*, but they must also ensure that the completed structure behaves in the manner that was envisaged by the designer in his analysis. It is necessary, therefore, that design of connections should be given adequate attention during design stage. In fact, the basic conception of connections to be adopted in a particular bridge should be decided at an early stage of the project, as the layout and arrangement of the bridge structure may, indeed, be influenced by the choice of connections.

No doubt, from the foregoing, it would seem to be advantageous if the design and detailing of the connections are done by the same agency entrusted with the structural layout and design of the member sizes. Unfortunately, however, this is not always the case. Therefore, it is desirable that the designer of the structure should have a working knowledge of steelwork detailing process, so that he is in a position to convey in his design drawings the salient features of the structure and its intended behaviour for the benefit of the detailer.

18.2 GENERAL CONSIDERATIONS

Much of the overall costs of the structure depends on the fabrication and erection costs. Here, again, the type of connection plays an important role. A connection which is easy to fabricate at shop and simple to install at site would reduce the overall cost of the structure considerably. In other words, the cheapest connection is likely to be that which involves the minimum shop and site labour.

In order to achieve economy, a few points related to connections should be sorted out early in the design stage:

First to decide would be the locations of the connections. These largely depend on the sizes in which materials are available from the mills. Next to decide would be which of these connections are to be done at the shop and which ones at the site. Obviously, as much as possible of the joining should be done at the shop, leaving the number of site joints to the minimum. Ease of handling, location and condition of the site and type of the route to be negotiated for transporting fabricated units from the workshop to the site may govern the number of site connections.

Selection of joining system—welded, bolted or riveted—would be the next important decision to be made. In general, the preferred practice nowadays is to use welding for most of the shop connections and, either bolted or welded joints for site connections. In case of bolted site connections, the type and grade of bolts to be used will also have to be decided.

As stated earlier, in bridgework, use of bearing bolts in clearance holes is recommended by many authorities to be restricted to other than the main structural connections. Thus, for main connections, HSFG bolts are the preferred choice. In some countries, rivets are used for main as well as secondary connections, mostly for railway bridges. However, the popularity of riveted connections is fast declining.

18.3 BOLTED VS WELDED SITE CONNECTIONS

For bridge steelwork, bolted connections are not only easier to install compared to welded connections, but they are also less expensive. In case of bolted connections, correctly fabricated components are almost self aligning and do not need very sophisticated equipment or highly skilled workmen like those required for welded connections. On the contrary, in welded connections, components are required to be firmly held in position while the welding is performed. Also, welded connections are susceptible to distortions and, therefore, need considerable care during welding process. Because of their possible internal defects, welded connections need a sophisticated inspection system, whereas bolted joints are comparatively easier to inspect.

While bolted site connections are generally considered to be advantageous, in some situations they do have their limitations. Welding can provide continuity and smooth stress flow in a joint. It provides rigidity to connections where the design considerations demand such rigidity. Also, unlike bolted joints, welded joints do not require any holes. Thus, deduction in the cross-sectional area is avoided, making the member more cost effective. One other point in favour of welded joints is that they are aesthetically more pleasing than bolted joints.

18.4 DESIGN METHODOLOGY

Methodology for design of connections of a bridge structure may be described as follows:

1. Calculate the forces in a particular joint and identify the correct load paths.
2. Based on the above, make a preliminary sketch showing the proposed layout.
3. Select the type of fastener or weld. The criteria for selection of these have been discussed earlier.
4. Compute the number and diameter of fasteners or the size and length of weld required from the forces already calculated and arrange them suitably in the layout sketch. This should be done considering maximum efficiency of the connection, making it as compact as possible to economise on use of material.
5. Check adequacy of the cross-sectional areas of the connecting components.
6. Finalise the layout sketch, considering practicability for fabrication, and ease for execution (e.g., access for welding, or insertion and tightening of bolts).

18.5 DESIGN CRITERIA

In the development of connection design, the criteria that need careful consideration are:

1. The connection should be as direct and as simple as possible. It should be complete and structurally in equilibrium, and should satisfy the requirements of the governing codes (such as spacing of bolts/rivets, edge distance, etc.).
2. The design of the connection should be generally in line with the assumptions made in the analysis of the structure. Thus, if continuity over a pier is assumed in the analysis of a solid web girder bridge, the connection details over the pier should be designed for the moments and shear forces at that location.

3. Behaviour of local connection elements (other than rivets, bolts or welds) which participate in the functioning of the connection needs to be considered and approriate details introduced to make the connection strong and durable.

4. In the splice location, the centroidal axis of the splice materials should coincide with that of the components joined, so as to avoid any secondary stresses being developed. If this is not practicable, the effect of eccentricity should be taken into account in the strengths of both the spliced components and the splice materials.

5. In open web bridge girders, care should be taken to ensure that the centroidal axes of the converging members meet at a single point, so as to avoid secondary stresses in the joint. In cases where the centroidal axes do not meet at a point, the eccentricity should be considered in the design of the connection and also of the concerned members. This aspect should be taken into account while finalising the form of the girder configuration.

6. In any connection, load should not be considered to be shared between fillet welds and bearing type bolts in clearance holes, the reason being that the deformation characteristics of fillet welds and bearing type bolts are not compatible. In fact, fillet welds, being stiffer than bearing type bolts, would initially carry most of the load. However, because of their limited ductility, they are likely to fail as the load increases, when the entire load will be transferred to the bolts which may not be able to carry it and finally fail.

7. Details at connections should avoid sudden change in sections as they are likely to become the focal point of stress concentration. This aspect is important in bridge structures which are normally subjected to fluctuation of stresses leading to fatigue fracture.

8. The overall size and degree of squareness of a shop fabricated component are always subject to manufacturing tolerances, which include mill tolerances for the sections and plates, as also fabrication tolerances allowed in the codes. These aspects should be carefully considered while detailing a connection.

9. Design of connection should consider easy access for inspection during fabrication and erection, as also for inspection and maintenance of the structures during the service life of the bridge. This aspect is very important from the point of view of durability.

10. Possibility of accidental damage during handling, transportation and erection at site should attract the attention of the detailer during

finalisation of any connection. As an example, abrupt projection of any shop connected gusset should be avoided.

18.6 ANALYSIS OF TYPICAL CONNECTIONS

The objective of this section is to present the basic methods of analysis for a few typical connections which are commonly used in the design of steel bridges.

18.6.1 Bolted Connections

Concentric shear connection

A concentrically loaded shear connection is shown in Fig. 18.1. Essentially, it consists of two tie plates spliced by two cover plates placed on either side of these connecting plates and secured by bolts of same diameter.

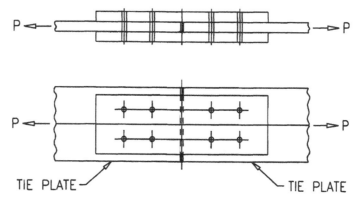

Fig. 18.1   Concentric shear connections

The layout of the connecting bolts is such that the centre of area of the group of bolts on either side of the joint lies in the line of the forces. This is defined as a direct or concentrically loaded connection. As the plates are assumed to be rigid, the force is considered to be shared equally by the bolts in the group. Thus, the force F in each bolt would be:

$$F = P/n$$

where P is the force to be transmitted and n is the number of bolts in the group.

In case bolts of different diameters are used in the same group, it is normally assumed that the bolts would share the load proportional to their cross-sectional areas. In such a case, the centre of area of the bolt group must coincide with the line of the force in order that the connection can be considered as a concentrically loaded one. For the computation of the centre of area of the bolt group, the differing areas of the bolts are to be duly considered.

Figure 18.2 shows the details of splice in the flanges of a girder where the flange forces (due to moment) are transmitted by cover plates secured to the flanges by bolts. If M is the moment at the section and d is the distance between the the centres of the flanges, the force in the flange will be M/d. The geometry of the connection is such that the splice can be analysed as a concentric shear connection. Analysis for web splice to transmit the shear has been discussed in the following paragraphs. Contributions of the bolts in the web splice to bending resistance is generally ignored, except in the case of deep girders.

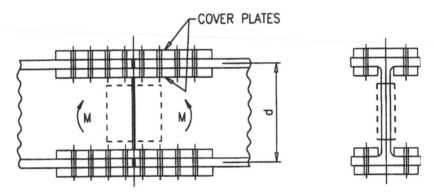

Fig. 18.2 Flange splice

Eccentric shear connection

Figure 18.3a shows a bracket connection in which a load P is applied in the plane of the connection at an eccentricity e from the centroid 'G' of the bolt group. This eccentric load is equivalent to a concentric force P passing through 'G' and a moment $P \cdot e$ which tends to rotate the side plate about 'G' (Fig. 18.3b).

For the purpose of analysis, these two load conditions may be treated separately and then the results can be superimposed to get the combined effect.

(a) Concentric force As the plate is assumed to be rigid, and the line of the force P passes through the centroid of the bolt group, the force will be shared

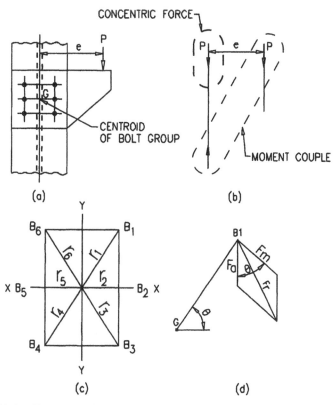

Fig. 18.3 Eccentric shear connection (bolt group in shear and torsion)

equally by each bolt. Thus, if there are n number of bolts in the group, the shear on each bolt is Fa = P/n.

(b) Moment   To calculate the force on a bolt due to the moment, it is assumed that the shearing force Fm (due to the moment) on any bolt is proportional to its distance from the centroid of the bolt group 'G' and that the force Fm acts in a direction normal to the line joining the concerned bolt and 'G'. Thus, the bolt farthest from the centroid of the bolt group will carry maximum load.

Considering the group of bolts in Fig. 18.3c, the bolts B_1, B_2, B_3, etc., are respectively at distances r_1, r_2, r_3, etc., from the centroid 'G' of the group. The force on bolt B_1, which is farthest from 'G' is say, Fm_1. In consideration of the assumption that the force in a bolt is proportional to its distance from 'G', it follows that the force on B_2 (located at a distance r_2 from 'G') will be $Fm_1 \times r_2/r_1$. Similar will be the forces for other bolts also.

Thus the moment of reistance (MR) of the bolt group will be:

$$MR = Fm_1 \cdot r_1 + Fm_1 \cdot \frac{r_2}{r_1} \cdot r_2 + Fm_1 \cdot \frac{r_3}{r_1} \cdot r_3 + \cdots$$

$$= \frac{Fm_1}{r_1} (r_1^2 + r_2^2 + r_3^2 + \cdots)$$

$$= \frac{Fm_1}{r_1} \Sigma r^2 = \frac{Fm_1 \Sigma r^2}{r_1}$$

Equating this to the applied moment,

$$P \cdot e = \frac{Fm_1 \Sigma r^2}{r_1}$$

or
$$Fm_1 = \frac{P \cdot e \cdot r_1}{\Sigma r^2}$$

Fm_1 is the force in the most heavily loaded bolt B_1

(c) Combined effect   The two forces *Fa* due to direct load and *Fm* due to moment have different directions. The resultant force Fr can be found graphically as shown in Fig. 18.3d. The algebraic formula can be derived as follows:

$$Fr = \sqrt{(Fa)^2 + (Fm)^2 + 2 \cdot Fa \cdot Fm \cdot \cos\theta}$$

where θ is the angle between the forces Fa and Fm.

Figure 18.4 illustrates a typical web splice in a girder. This is a common example of eccentric shear connection, which, in effect, consists of two eccentric shear connections with a common shear force V acting at an eccentricity e from the centroid of the bolt group. Analysis of the bolt groups can be done in the same manner as has been shown for a bracket in the previous paragraphs.

Another common example of eccentric shear connection, is the cleated end connection of longitudinal stringer girder to transverse cross girder in a bridge deck system. There are mainly two possible behaviours of the connection which would affect the analysis, viz., the cross girder is either free to rotate or is prevented from rotation. As an example of the first possibility, Fig. 18.5a shows a one-sided connection of the stringer to the cross girder with low torsional rigidity. The connection is assumed to rotate with the application of shear load from the stringer. The shear V is considered to be transmitted to the web of the cross girder by means of

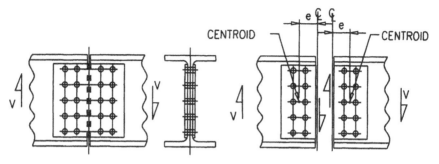

Fig. 18.4 Web splice

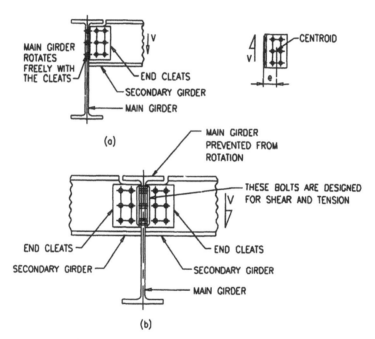

(a)

(b)

Fig. 18.5 Connections with end cleats

double angle cleats. The bolt group connecting these cleats to the stringer is analysed for eccentric moment due to the shear V acting at an eccentricity e from the centroid of the bolt group, as shown in the figure. The bolts connecting the cleats to the web of the cross girder are required to resist the vertical shear V only. A variation to the previous example is shown in Fig. 18.5b, where two stringers from two sides are connected to the cross girder. With this double sided connection detail, the cross girder is prevented from rotation. If the cleats are sufficiently stiff and do not rotate under the applied shear load from the stringer, the bolt group connecting the cleat to the web of the cross girder will be required to resist not only the vertical shear, but also the tension due to the moment $V \cdot e$. Analysis of moment connection with bolts in shear and tension is discussed in the following paragraphs.

Connection with bolts in shear and tension

Apart from the example described in the previous paragraph, a few more cases of connections subjected to combined shear and tension in bolts are shown in Fig. 18.6. In all these cases, the applied moment is not in the plane of the bolts and tends to rotate the joint across the plane of the bolts, thereby inducing tension in the bolts in addition to shear. These connections can be treated somewhat similar to the typical bracket connection shown in Fig. 18.7a, in which a load P is applied at an eccentricity e from the edge of the bracket. The conservative approach for analysis is to consider the shear to be shared equally by all the bolts and check the top-most bolts for combined shear and tension.

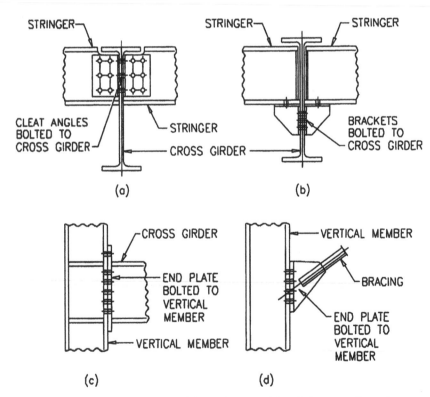

Fig. 18.6 Connections subjected to combined shear and tension

In a simplified method of analysis, the centre of rotation is assumed to be at the bottom bolt of the group, the loads varying linearly as shown in the figure. If T_1, T_2, T_3, etc., be the tensions in bolts 1, 2, 3, etc., located respectively at distances r_1, r_2, r_3 etc., from the bottom-most bolt, then, by similar triangles:

$$\frac{T_1}{r_1} = \frac{T_2}{r_2} = \frac{T_3}{r_3}, \text{etc.}$$

If M_1, M_2, M_3, etc., are moments shared by the bolts marked 1, 2, 3, etc., respectively, then, considering two rows of bolts, the total moment resisted by the bolt group is:

$$M = 2\,(M_1 + M_2 + M_3 + \cdots)$$

$$= \frac{2\,T_1\,\Sigma r^2}{r_1} = P \cdot e$$

From the above equation, the maximum bolt tension in the top-most bolt is:

$$T_1 = \frac{P \cdot e \cdot r_1}{2\,\Sigma r^2}$$

The load Fs due to direct shear is:

$$Fs = P/\text{No of bolts.}$$

The top-most bolts are to be checked for combined shear and tension as per the provisions of the code.

In an alternative method, only the bolts located at the top flange are considered to take the tension along with their share of the shear. This method is illustrated in Fig. 18.7b. The tension in the top group of bolts is:

$$n \cdot T = M/d$$

where n is the number of bolts considered as tension bolts, T the tension in each bolt, M is the applied moment, and d the distance between the centres of the flanges. These bolts are to be checked for combined shear and tension.

18.6.2   Fillet Welded Eccentric Connections

Two types of eccentric connections are considered here:

(a) Load lying in the plane of welds, and

(b) Load not lying in the plane of welds.

In both these types the fillet welds are subjected to shear due to direct load and moment.

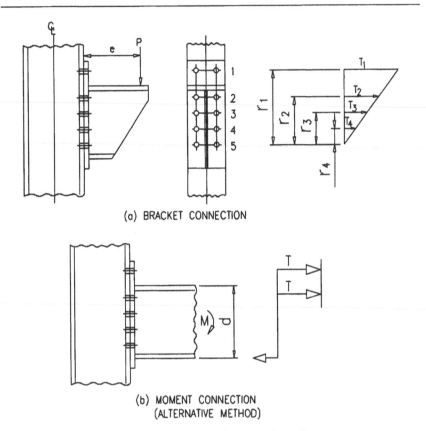

(a) BRACKET CONNECTION

(b) MOMENT CONNECTION
(ALTERNATIVE METHOD)

Fig. 18.7 Bolts in shear and tension

Load lying in the plane of welds

The theory for analysis of eccentric shear connection for bolt groups can be broadly applied in the welded joint also (Fig. 18.8). While in a bolt group, bolts are discrete individual connections, in a welded joint the weld is a continuous connection and the analysis has to be done accordingly. The connected members are assumed to be rigid. The eccentric load P is applied at an eccentricity e from the centroid 'G' of the weld. This eccentric load is equivalent to a concentric force P passing through ' G ' and a moment P · e, which tends to rotate the side plate about 'G'. These two load conditions are treated separately and then the results are superimposed to get the combined effect. The shear due to the concentric force P is assumed to be uniform throughout the weld. The force in the weld due to the moment is considered to be directly proportional to its distance from the centroid of the weld. Thus the weld farthest from 'G' will carry maximum load. The combined effect of the direct load and the moment can be obtained by adding the two results vectorially.

Assuming the weld to be of unit leg length and uniform throughout, the shear force Fa in the weld due to concentric force P is given by:

$$Fa = P/L$$

where L is the total effective length of the weld.

Torsional moment $P \cdot e$ produces bending forces Fm in the weld about an axis passing through 'G' and perpendicular to the plane of the weld, and is given by:

$$Fm = P \cdot e \cdot r / Ip$$

where r is the distance of a point under consideration from 'G' and Ip is the polar moment of inertia of the weld and is given by:

$$Ip = I_{xx} + I_{yy}$$

$$I_{xx} = 2 \cdot \frac{y^3}{12} + 2 \cdot x (\tfrac{y}{2})^2$$

$$I_{yy} = 2 \cdot \frac{x^3}{12} + 2 \cdot y (\tfrac{x}{2})^2$$

where x and y are the lengths of weld along XX- and YY-axes respectively.

In Fig. 18.8a, 'A' is a point farthest from the centroid of the weld and is the maximum loaded weld. The distance r of 'A' from 'G' is given by:

$$r = \sqrt{\left(\frac{x}{2}\right)^2 + \left(\frac{y}{2}\right)^2} = \frac{1}{2}\sqrt{x^2 + y^2}$$

The resultant force Fr at 'A', the position of maximum shear, is given by:

$$Fr = \sqrt{(Fa)^2 + (Fm)^2 + 2 \cdot Fa \cdot Fm \cdot \cos\theta}$$

where θ is the angle between AG and XX-axis.

If the weld layout is not symmetrical (Fig. 18.8b), the centroid of the weld has to be located first to determine the values of eccentricities and the polar moment of inertia. The stresses can be obtained as above.

Load not lying in the plane of welds

A simple bracket connection is shown in Fig. 18.9. In this case also, the eccentric load P, applied at a distance e from the plane of the weld, can be considered equivalent to a direct force P, passing through the weld and a

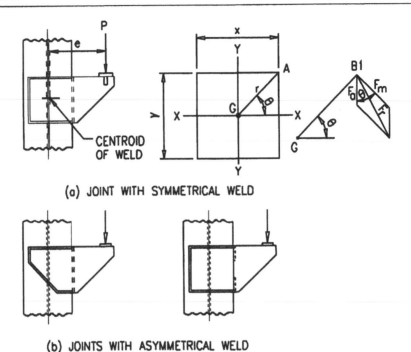

(a) JOINT WITH SYMMETRICAL WELD

(b) JOINTS WITH ASYMMETRICAL WELD

Fig. 18.8 Eccentric welded connections (load lying in the plane of weld)

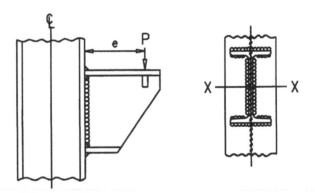

Fig. 18.9 Eccentric welded connections (load not lying in the plane of weld)

moment P·e which tends to rotate the joint across the plane of the weld. The load on the weld can be obtained by applying beam bending formula.

Assuming the weld to be of unit leg length and uniform throughout, the shear force in the weld due to the direct force P is given by:

$$Fa = P/\text{length of weld}$$

The load due to moment is given by:

$$Fm = P \cdot e \cdot y/I$$

where I is the moment of inertia of the weld and y is the distance of the farthest weld from the neutral axis of the weld layout.

The resultant force Fr is given by:

$$Fr = \sqrt{(Fa)^2 + (Fm)^2}$$

18.7 CONNECTIONS IN TRUSSES

The converging members of trusses are generally joined by means of gusset plates at joints where these members meet. Although, for analysis, pinned connections are commonly assumed, this idealised condition is hardly attained in actual practice, where the members are connected to the gusset plates by bolting, riveting or welding. A typical connection which is normally used in trusses is shown in Fig. 18.10. As discussed earlier in this chapter, care should be taken to avoid secondary stresses in joints by having the centre-lines of the converging members meet at a single point.

At any joint, for maximum force in any member, or for maximum force across a joint, there is a set of co-existing forces in all other members converging at a joint. Theoretically, these co-existing forces should be considered in the design. However, as per usual practice, only the maximum forces in members are calculated and tabulated in the stress sheet and, with moving loads, these do not occur simultaneously in all members. Although the co-existing forces for all the joints against various load combinations are available from the computer, working with so much data becomes laborious. It is, therefore, customary to design a connection from usual stress sheet data which shows only the maximum forces in the members for different load combinations.

18.7.1 Types of Connections

In an open web girder system, connections may be broadly classified into four categories:

1. *Internal joints:* These joints usually connect the web members (e.g., vertical and diagonal members) to continuous chords. Figure 18.11 shows a typical example of this type of joint.
2. *Site splices:* Fabricated units sent from the shops are assembled and spliced at site to make up the completed structure. Very often these

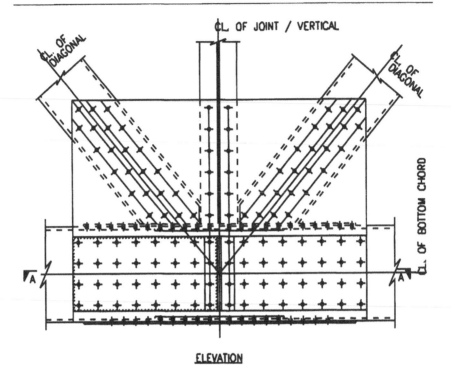

CL OF JOINT / VERTICAL

CL. OF DIAGONAL

CL. OF DIAGONAL

CL. OF BOTTOM CHORD

ELEVATION

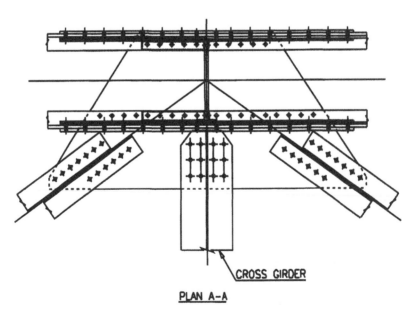

CROSS GIRDER

PLAN A-A

Fig. 18.10   Joint detail for bottom chord

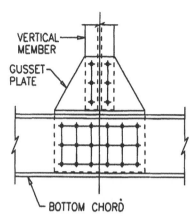

Fig. 18.11 Internal joint

splices are located at the node points. Figure 18.10 illustrates an example of this type of connection detail for bottom chord.

3. *Bracing connections:* These are generally required to connect the lateral bracing systems with the vertical girders. These include the portal bracings, sway bracings, etc. Figure 18.12 shows some typical details of these connections.

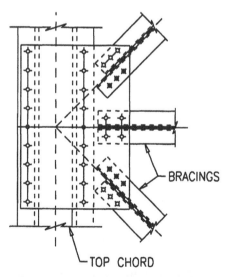

Fig. 18.12 Bracing connection

4. *Bearing connections:* Open web girders transmit the load to the substructure through bearings placed at the ends of the girders.

18.7.2 Transmission of Force in Chords

In a truss joint, there are two distinct types of force transmission in chords:

1. Where the chord is continuous through the gussets, the force in the chord is directly transmitted within the chord itself and only the difference of chord forces is transmitted through the gussets. If chord splices are required, they are made outside the joint. This arrangement is often used in heavier trusses in order to relieve the gusset plates of large forces.
2. Where the splices of chords are located right at the joints, the gussets are subjected to heavy forces, since they, along with the splice cover members, transmit the entire chord forces. In case of compression chords, where the bearing surfaces are milled, a substantial portion of the force is transmitted by direct bearing from one chord to the other, leaving the balance to be transmitted by the gussets and splice cover plates. In case of tension splices, however, the splice cover plates and the gussets carry the entire force from one chord to another.

Normally, the centroid of the splice cover materials approximately coincides with that of the chords. In such a case, it may be assumed that the cover materials act as one and in line with the chord. The centroid of the gussets, on the other hand, is usually some distance away from that of the chords, and consequently the gusset plates are subjected to bending across a section at right angles to the chord. Gusset plates should be checked for such bending. When the free edges of the gusset plates are subjected to compression, they are usually stiffened by plates or angles.

18.7.3 Gusset Plates

Truss members are joined by the gusset plates where the members meet. Each web member (diagonal and vertical in Fig. 18.13) is connected to the gusset for the maximum force in the member and the gusset plate is made large enough to accommodate the connections. A minimum plate thickness is necessary to develop the full strength of the bolts or rivets. If the rivets or bolts are in single shear, a smaller bearing thickness may be adequate, whereas if the rivets or bolts are in double shear, a thicker plate is required. In any case, the thickness is generally made equal to or slightly larger than that of the thickest part to be connected. For light trusses, 10 mm to 12 mm thick gusset plates are generally used. For heavier trusses, 16 mm to 22 mm thick gussets are more commonly used.

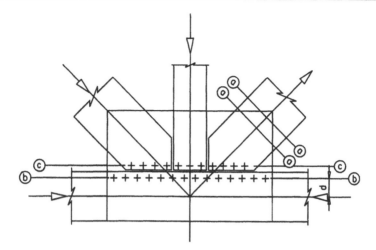

Fig. 18.13 Gusset

Typically, the thickness of the gusset plate is to be checked against the following conditions:

(a) The gusset must be capable of resisting the the maximum force from each web member across critical sections through the gusset (e.g., 'aa' in Fig. 18.13), assuming that the maximum axial force at any section is proportional to the number of rivets or bolts above the section.

(b) The gusset must resist shearing by the combined co-existant forces in all the web members across the section of the gusset through the first line of rivets or bolts connecting the gusset to the chord (line 'b-b'). For simplicity, this shear Q can be taken as the sum of the horizontal components of the maximum forces in the web members.

(c) The gusset must also withstand the bending moments due to the maximum shear force Q, the bending moment at any section being:

$$Q \times d$$

where d is the distance from the centre of intersection of joint to the section considered.

The critical section will probably be either:

(1) Section 'b-b' through the first line of rivets/bolts connecting the gusset to the chord, or

(2) Section 'c-c' approximately through the first line of rivets connecting the gusset to the web members.

For joints near the position of maximum shear in the truss, the thickness of the gusset is controlled by the above considerations. For joints near the position of maximum moment and small shear, the thickness of gussets will generally be controlled *not* by the above, but by the splice requirements. For intermediate positions either of these may be the criterion.

18.8 CONNECTIONS IN PLATE GIRDERS

18.8.1 Flange-to-Web Welds

Flange-to-web welds are located in areas of bending stresses and are designed to transfer the longitudinal shear forces between the flange and the web. Fabrication drawings usually specify fillet welds for these welds (Fig. 18.14a). In a modern fabrication shop, however, submerged arc automatic welding process is usually used to make these welds. The resultant weld is shown in Fig. 18.14b, where the two fillet welds penetrate deeply within the web and intersect each other, giving complete fusion. Some authorities prefer to specify this type of deep penetration welds in the drawing itself. For thicker web plates, a double bevel edge preparation may be required to acheve complete penetration. This detail is shown in Fig. 18.14c. For thicker plates, pre-heating of the components is desirable. In case of manual welding, low hydrogen electrodes should be used for such work.

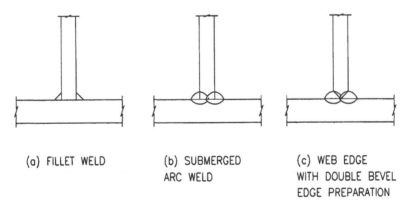

(a) FILLET WELD   (b) SUBMERGED   (c) WEB EDGE
                             ARC WELD   WITH DOUBLE BEVEL
                                                     EDGE PREPARATION

Fig. 18.14 Flange-to-web welds

18.8.2 Shop and Site Splices

Splices are provided between two adjoining units of a girder to enable them to function as an integrated unit. These splices may be either welded type or bolted type. The type to be used usually depends on whether the splice will be installed in the shop or in the site. In case of shop splices, butt welded joints are most often used. For site splices, bolted connections are generally preferred as these are easy to install and do not require particularly specialised workmanship. However, where a clean and aesthetically pleasing appearance is required by the owner, butt welded joints are preferred in site splices. Full penetration butt welded splices can offer sound connections, and if properly executed and closely inspected during the welding process at site, they can be an attractive alternative to bolted site splices.

Bolted splices

In a bolted splice, cover materials are placed on both surfaces of the flanges as well as the web, and are extended symmetrically on each side of the joint and are connected by high strength bolts. It is recommended that both surfaces of the spliced parts be provided with cover materials, appropriately proportioned as per codal provisions. If this is not practicable, the effects of eccentricity should be considered in the design of the splice. This is likely to pose complexity and, therefore, effort should be made to make the joint with cover material on both surfaces.

When two different sizes of flanges are connected, splices are normally designed according to the properties of the heavier section, since this will give greater flange forces to cater for. As regards the layout of a web splice, it is a good practice to provide a minimum of two rows of bolts on each side of the joint. Figure 18.15 shows a typical bolted splice in a plate girder.

It should be noted that the splices do not occur at a specific point, but rather cover a length of the girder which can range between 300 mm to 600 mm. Along this length, both the moment and shear will vary. This variation needs to be considered in the design.

It should also be noted that fatigue check, as per codal requirement, should be carried out for the flange and web joints.

As regards detailing of a splice connection, care must be taken to ensure that the bolts are placed in such a manner, that they can be readily tightened without interference from other bolt heads or cover materials.

Welded shop splices

For making up the required lengths of flanges and webs, shop splices are almost always done with welded butt joints. These are normally made before the girder is fitted together and welded. These shop splices need not be in a single vertical plane. They are located to suit the available plate length or

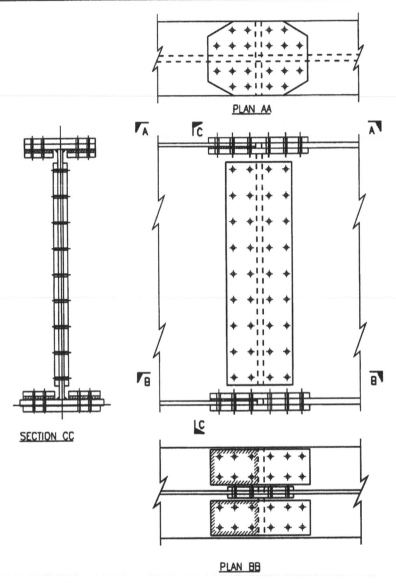

PLAN AA

SECTION CC

PLAN BB

Fig. 18.15 Typical bolted splice in a plate girder

where a transition in section is required. Semi-automatic or automatic submerged arc welding facilities can be used in plate widths in the range of 600 mm to 900 mm. In other cases, manual metal arc (MMA) process is used.

In shop splices, single-V butt joints are generally used for thinner plates. For thicker plates, however, double-V butt welds are normally used, since these plates can be easily turned over in the shop. Double-V butt welds consume less weld material compared to single-V butt welds. Apart from this economy in use of weld metal, double-V butt welds produce balanced welding, thereby minimising the chances of angular distortion.

Welded site splices

Site welding demands considerable knowledge and expertise on the part of the welders, the supervisory staff as well as the inspectors. This aspect is very important for achieving acceptable weld quality.

Except in very big bridges with long run of horizontal seam welds, manual metal arc (MMA) process of welding is normally used in most site operations. Therefore, while finalising the weld details, it is necessary to ensure that the welder's task is made as easy and straightforward as possible. There should be sufficient room for the welder's head and the mask so that he can see what he is doing. There should also be adequate access for the stick electrode to be placed at the required location. Additionally, structural detail should allow most of the welding work to be performed in the downhand position, which makes the job easier and faster. In short, simplicity, ease of access for welding, and ease of inspection should be the primary objective in the designer's approach.

Welded site splice for a plate girder can be detailed in three ways:

- Placing the two flange splices and the web splice in the same vertical plane (Fig. 18.16a)
- Placing the three splices in three different planes (Fig. 18.16b)
- Placing the two flange splices in the same plane and slightly shifting the web splice (Fig. 18.16c)

The first alternative has an advantage in that it is comparatively easier to prepare the joints and maintain proper fit up when all the splices are located in the same vertical plane. Also, the web-to-flange fillet weld can be fully done at shop unlike the other two alternatives, where some site welding, including overhead fillet welding, will be required to be done at site. Apart from this advantage, the shop fillet welds to the very end of the girder in the case of the first alternative, enables the flanges to be clampled together by means of draw cleats for temporary support during erection (Fig. 18.16a). In the case of the other two alternatives, however, the flanges are not shop welded at the ends and the temporary draw cleats are attached in the webs only (Fig. 18.16b and c).

It is sometimes argued that splices are weak points in the structure, and should preferably be staggered. In the case of butt welded splices, however,

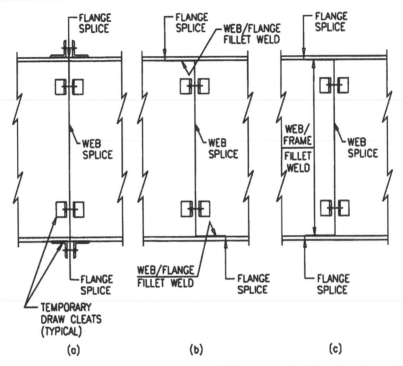

Fig. 18.16 Alternative locations of welded flange and web splices in a plate girder

this arrangement does not improve the performance of the girder in any way. Therefore, the first alternative, i.e., locating all the three splices in the same vertical plane, appears to be the most attractive solution.

In general, except for fatigue consideration, no design calculations are necessary in butt welded joints, since these joints are considered to be a continuation of the plate itself, assuming that the selected electrodes are compatible with the parent material and the workmanship is satisfactory. The resultant welded butt joint becomes at least as strong and as malleable as the parent material. Therefore, in detail drawings, a full strength butt weld is specified without calculation.

In site splicing, normally the web plates are of single-V preparation. For thicker webs, say over 12 mm, a double-V preparation may be preferred in order to balance any angular distortion. This also reduces the consumption of weld metal, compared to a butt weld with single-V preparation.

For splicing flange plates, which are normally much thicker than the web plates, either single-V or double-V preparations may be used. Single-V butt welds have a tendency for angular distortion. In this context, double-V butt welds would, no doubt, balance the distortion, but would involve overhead welding. In site operations, manual arc welding (MMA) process normally dominates and downhand welding is preferred, being easier and faster than overhead welding. A suggested compromise detail would be to have an asymmetrical double-V preparation, where two thirds of the flange thickness would be on the top and the remaining one third on the bottom. This would substantially balance the tendency of the angular distortion to some extent and also reduce the amount of overhead welding. Figure 18.17 illustrates these alternative preparations for butt welding.

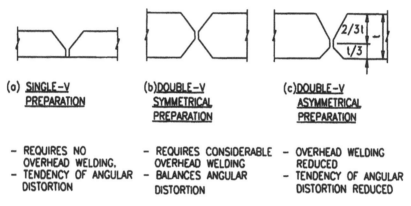

(o) __SINGLE–V PREPARATION__

- REQUIRES NO OVERHEAD WELDING.
- TENDENCY OF ANGULAR DISTORTION

(b)__DOUBLE–V SYMMETRICAL PREPARATION__

- REQUIRES CONSIDERABLE OVERHEAD WELDING
- BALANCES ANGULAR DISTORTION

(c)__DOUBLE–V ASYMMETRICAL PREPARATION__

- OVERHEAD WELDING REDUCED
- TENDENCY OF ANGULAR DISTORTION REDUCED

Fig. 18.17   Alternative preparations for butt welded flange splices

It should be noted that weld metal contracts on cooling. Thus, in a butt welded joint, shrinkage of the girder components is inevitable, and therefore, good ductility of the weld metal is vital for producing a good joint. Furthermore, shrinkage allowance in the length of flange and web plates must be provided while cutting the plates. As for example, in a girder with different thicknesses of flange plates, because of greater number of weld runs, the amount of shrinkage will be more in the thicker flange plate. Unless this aspect is considered beforehand, there may be a change in the profile of the girder after welding. Also, if the flanges are welded first, the initial web gap should be larger than the flange gaps, so that after the flanges are welded, even with their consequent shortening, the remaining web gap is sufficient for welding. Figure 18.18 illustrates the point. For general guideline, the following welding sequence, in which both flanges and webs are alternately welded to a portion of their depth, is suggested [6]:

1. Weld full width of about half to one-third of the thickness of both flanges

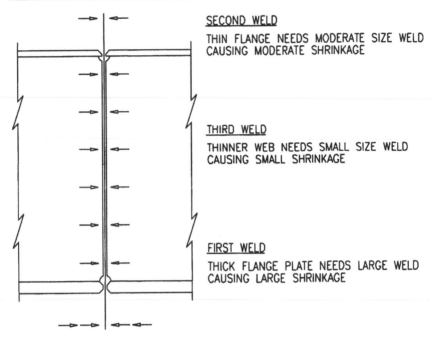

Fig. 18.18 Effect of weld shrinkage in a site welded splice [3]

2. Weld full width about half of the thickness of the web
3. Complete the welding of the flanges
4. Complete the welding of the web

For deep webs, the vertical weld length is generally divided into two or three sections and welding is done first in the top section, followed by the next, and so on. The direction of the welding for each section should start from the bottom and proceed vertically up sequentially (Fig. 18.19).

Coped holes in web at splice

It is customary to provide coped holes in the web plate to allow field welding of the butt joints of the flanges (Fig. 18.18). While this detail helps in butt welding the flange splice properly, it reduces the fatigue strength of the girder to some extent because of the notch effect of the coped holes. It is, however, generally felt that, inspite of this drawback, coped holes should be provided to ensure quality weld for the flange splices. If necessary, the holes can be plugged by weld metal after the butt weld has been completed and inspected.

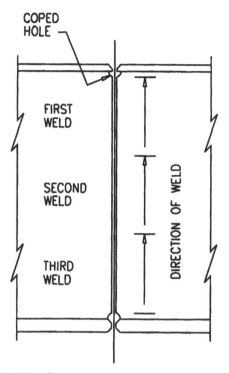

Fig. 18.19 Sequence of welding for deep webs [6]

REFERENCES

[1] Dowling, P.J., Knowles, P.R. and Owens, G.W. (eds.) 1988, *Structural Steel Design*, The Steel Construction Institute, London, UK.

[2] Pask, J.W. *Manual of Connections*, The British Constructional Steelwork Association Ltd., London, UK.

[3] Needham, F.E. 1983, *Site Connections to BS 5400 : Part 3*. In: The Structural Engineer, March, 1983, the Institution of Structural Engineers, London, UK.

[4] Owens, G.W. and Knowles, P.R. (eds.) 1994, *Steel Designers' Manual* (fifth edition), Blackwell Scientific Publications Ltd, Oxford, UK.

[5] Brokenbrough, R.L. and Merrit, F.S. (eds.) 1994, *Structural Steel Designer's Handbook*, (second edition) McGraw-Hill Inc, New York, USA.

[6] Blodget, O.W. 2002, *Design of Welded Structures*, The James F. Lincoln Arc Welding Foundation, Cleveland, USA.

[7] Ghosh, U.K. 2000, *Repair and Rehabilitation of Steel Bridges*, A.A. Balkema, Rotterdam, Netherlands.

Chapter **19**

Fabrication Procedures

19.1 INTRODUCTION

The role of the fabricator is to convert rolled steel into finished goods by adding value. He achieves this by selling workmanship and machine utilisation, which are directly related to time. Since time is the basis of cost computation, economy can be achieved by introducing standardisation by repetition of dimensions, geometry, member sizes and shapes, centres and diameters of bolts, etc. This situation is particularly true for multiple span bridge steelwork projects where this kind of rationalisation of similar components may be attempted.

Traditionally, structural fabrication work, including bridge steelwork, is qouted on 'per ton' basis. For arriving at this figure, the fabricator has to estimate costs by separating the various activities involved into categories such as, laying out, cutting, drilling, assembling, welding, etc., and allocating and pricing manhours as well as machine hours to be spent for such activities. Of course, he has to add other expenses such as costs of materials, consumables, overheads, etc., as well as profit element to arrive at the final price.

19.2 FABRICATION ACTIVITIES

Fabrication work starts only after the approved detail drawings are received at the works. Meanwhile, procurement action must have been initiated for raw materials and consumables, so that these are received at the works store in time to match the sequence of operations. A flow chart used typically for the fabrication of bridge steelwork is shown in Fig. 19.1. Major activities involved in the process are discussed in the following paragraphs.

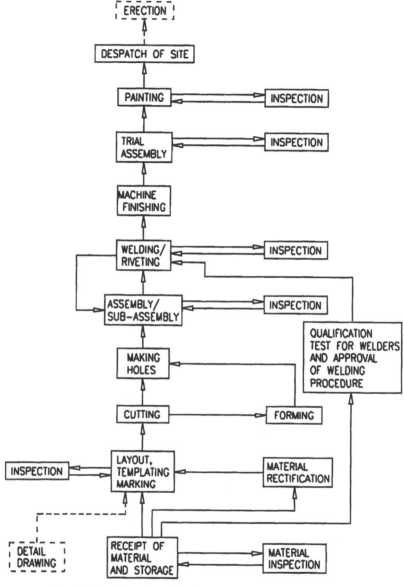

Fig. 19.1 Flow chart for fabrication of bridge steelwork

19.3 DETAIL DRAWINGS

Detail drawings are based on the design drawings prepared by the designer and are the only means by which the designer's ideas are communicated to the fabrication shop. These are very important documents for fabrication of steel bridges. Normally, preparation of detail drawings is the responsibility of the contractor entrusted with fabrication work. He may get the work done in his own drawing office or an outside detailing firm. In both the cases, the drawing office is responsible for interpreting the engineer's design drawings, calculations and specifications and produce detail drawings correctly for fabrication at workshops. Detail drawings show the controlling dimensions, member sizes, grades of steel, sizes and locations of holes for fasteners, types, sizes and lengths of welds, and all other information pertinent to the fabrication process. These drawings are generally approved by the engineer/designer before issue to the workshop for fabrication. Nowadays, detail drawings are mostly prepared by detailers using computers with appropriate software available in the market.

One of the first actions of the drawing office is to provide the material procurement department with complete lists of all materials (steel, bolts, etc.). Normally, materials are to be ordered well in advance, indicating the sequence of requirement to suit the production schedule.

Detail drawings should include 'marking plans' showing the location and orientation of each erectable piece with its distinguishing erection mark. This mark is normally hard-stamped and painted on the erectable piece, to enable the erector to identify the piece and erect it correctly.

Although the drawing office is normally responsible for the development of joints and connections, for bridge steelwork, it is a good practice for the designer to indicate on his design drawings the type of connection that he has considered in his analysis and design. This would help the detailer to produce a detail which would be compatible with the behaviour of the structure envisaged by the designer.

The despatch lists are normally prepared by the drawing office. These lists show the total number of each individual type of erectable piece, the overall dimensions and weight of each piece, and the total weight. This information is used for despatching purpose at the workshop, as well as assessing the required lifting capacity of cranes at site. These lists are also used for monitoring the sequence and quantity of production at shop and site.

19.4 MATERIAL PROCUREMENT

Fabrication shops obtain steel sections and plates either from steel producing units or from steel stockholders. Normally, there is a minimum quantity of any size and grade for the producers to accept an order. They also require some lead time to supply the material in accordance with their production schedule. Smaller quantities can be obtained from the stockholders, who normally anticipate demands of certain sections and plates and procure these in advance in bulk quantities from the producers. Consequently, materials can be made available by them within a very short time. Needless to add, steel procured from stockholders is generally costlier than that obtained directly from the producers. Some materials in small quantities may also be available from the fabricator's own stock of off-cuts and part-lengths remaining from previous orders.

In order to make sure that the material to be used for fabrication conforms to the specified quality, it is necessary to carry out inspection of such material. If available, certificates issued by the producers may be checked with the distinguishing mark on each plate/section. Alternatively, test pieces may by cut from the ends of material procured and subject them to tensile test, bend test, Charpy test, etc.

19.5 MATERIAL RECTIFICATION

Steel sections may become bent or twisted during transit from the hot rolling mills or stock yards, or during the course of fabrication. Straightening of such bars is normally carried out in the workshop in the cold state, by an electrically driven beam bender and straightener. Usually, the bar is 'over-bent' in the reverse direction of the deformation (a little more than the actual deformation), to allow for a certain amount of 'springing back'. Flattening of bent plates is done by passing the plates through an electrically driven mangle with a series of rollers, first in one direction and then in the reverse direction. Smaller plates are generally manually flattened. In order to remove waves in plates of thickness of 50 mm or over, these are first heated and then flattened in a press whilst still hot.

Sections and plates can also be straightened by local application of heat. In this method, the principle of thermal expansion and contraction is utilised by applying concentrated heat from a blow pipe to a wedge-shaped area on the member. When a bent member is heated uniformly at a particular location, the heated portion will tend to expand and this expansion will be prevented by the surrounding cooler metal. The forces hindering the expansion will cause the relatively weak heated part to expand (bulge) externally. When the area cools, the contraction causes the bend to straighten. The temperature should be restricted to approximately 700 degrees Centigrade, i.e., equivalent to dull red colour of the heated steel. Heating should proceed from the base to the apex of the wedge and the heat should penetrate evenly through the plate thickness, maintaining an even temperature. Figure 19.2 illustrates the methodology of flame straightening of bent members. Although the method appears to be quite simple, it requires sufficient field experience and practice to be successful. As a result, the method, though very effective, has not gained popularity amongst technicians.

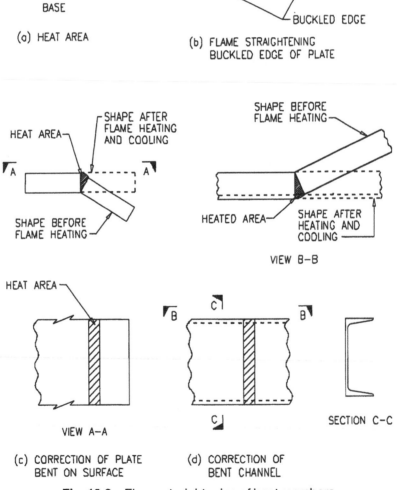

Fig. 19.2 Flame straightening of bent members

19.6 LAYOUT, TEMPLATING AND MARKING

Traditionally, full scale layouts of components are marked on templates in the template shop. For bridge fabrication, thin steel sheets are generally used for templates. Various details, such as shearing, notching and cutting lines, locations and diameters of holes, and other data are indicated on these templates. Mistakes committed in the drawings are often detected during this operation. These templates are used in fabrication shops by smiths, benders, assemblers and machine shop operators for cutting, shaping, punching, drilling, etc. Where fabrication is not repetitive in nature, markings are often made directly on the work piece instead of using templates. The markings are punched and paint impressions are normally given.

When a number of bridge spans are required to be fabricated, it may become necessary, for ease of erection, to have the open holes for site connections of all the similar members of different spans to be exactly same, so that the erectable pieces are interchangeable. This interchangeability can be successfully achieved if these holes are drilled by jig system. For this purpose, steel drilling jigs are used as templates, and hard steel bushes with initial tolerance of (–) 0.0 mm, (+) 0.1 mm are fixed centrally at the hole locations to ensure that the disposition of holes are maintained in all the similar components during the drilling operation (Fig. 19.3). The locating holes are to be drilled to zero tolerance by very skilled operators. For drilling a batch of long plates, drifts and clamps are to be used to keep the plates properly stretched and to minimise gaps in between the plates. Tolerance of these bushes needs to be checked from time to time and these must be replaced when the tolerance exceeds (–) 0.0 mm, (+) 0.4 mm. This system ensures elimination of mismatch at site to a considerable degree and as a result, reduces the chance of secondary stresses in the structure.

Now-a-days, with the introduction of computer numerically controlled (CNC) machines in fabrication industry, the use of traditional methods of layout, templating and marking on steel for fabrication, at least for large members, is being reduced considerably. These new generation machines have many advantages, such as dimensional accuracy, higher productivity, ease of assembly, and interchangeability, apart from saving in time and labour. Consequently, the work of the template maker is being curtailed progressively. However, in site fabrication, where modern CNC machines may not always be available, the traditional marking methods by templates are still very useful.

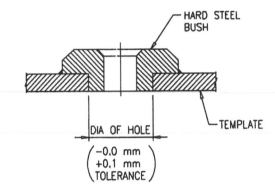

ENLARGED DETAIL OF BUSH ATTACHMENT

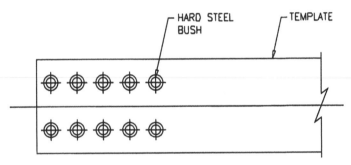

Fig. 19.3 Typical drilling jig

19.7 CUTTING

Cutting operation may fall into one or more of the following categories:

- Flame cutting
- Shearing
- Sawing

19.7.1 Flame Cutting

Oxygen flame cutting is now a general practice for splitting and cutting plates, slabs, or other sections for fabrication work. In fact, oxygen gas torch has developed more as a cutting and forming tool than as a welding tool in structural fabrication industry. Flame cutting can be done by manually held nozzles or by mechanically guided machines. However, for fabrication of bridge steelwork, only machine flame cutting is recommended.

The principle involved in flame cutting is that the torch burns a pressurised mixture of oxygen and gas (normally propane or acetylene), which passes through a ring of small holes in a cutting nozzle, to pre-heat the steel to about 1,600 degrees Fahrenheit (about 900 degrees Centigrade), at the point where the cut is to be made. The torch then releases a suitable jet of high pressure oxygen on the pre-heated part, through a separate hole situated at the centre of the nozzle, which cuts the metal in the direction of the movement of the torch at a suitable speed. Cutting speed, gas pressures and jet sizes are dependent on the cross section of the steel to be cut. Figure 19.4 illustrates a typical oxygen flame cutting head. By using the proper type of nozzle, and adopting correct speed, gas ratios and pressures, it is possible to achieve flame cut edge of extremely high quality. Consequently, flame cutting has now become an indispensable process in steel fabrication industry. For oxygen gas cutting machine, the safe limit of carbon content in the steel is taken to be 0.3% under all conditions of ambient temperature and humidity found in practice. Structural steels normally used for bridge construction fall within this category. Consequently, flame cutting process has become very popular in the fabrication of steel bridges.

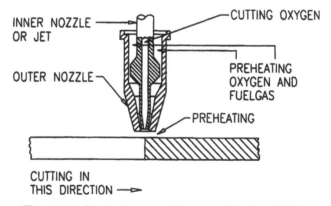

Fig. 19.4   Typical oxygen flame cutting head [1]

The process is extensively used in the straight line slitting machine which operates with two or more torches for multiple cutting of long plates of narrow widths frequently used for flanges of welded plate girders. This system of application of heat on both edges simultaneously limits distortion by imposing almost the same amount of shrinkage stresses on both the edges of the plate. For this reason, while ordering plates meant for multiple cutting in the workshop, sufficient allowance should be provided in the width, so that an extra flame cut adjacent to the rolled edge can be done.

Also in use for many years, is the small portable single torch straight line cutting machine mounted on a powered trolley which runs along a special track commonly laid on the job plate itself. The machine can also be used to give bevel edge preparation for butt welding, by setting the torch in an inclined position. A double bevel cut and root face cut can be made simultaneously with three torches suitably mounted on the carriage.

Flame cutting process is also used to produce curved shapes or intricate profiles, such as those encountered in finger-type expansion joint plates for bridge decks. For profile cutting, the nozzle is mounted on a movable horizontal arm attached to a travelling gantry, the pattern being traced by a device following the contour of a metallic template. Profile cutting can also be performed by numerically controlled (NC) machines, which, in addition, can mark locations of holes on steel job plates. Also in use are the machines operated by optical controlling head which can follow outlines of profiles drawn on papers. These machines are very useful for cutting gusset plates to proper shapes.

19.7.2 Shearing

Shearing process is used to cut certain classes of steel material to size. Guillotine-type shears are used to cut plates of upto 25 mm thickness. Long plates may be slit lengthwise by using wheel shears mounted on a carriage which moves along the length of the plate. The latter may also cut bevel edges suitable for butt welding. Although the cut is performed rapidly, considerable time may be needed for handling and proper alignment of the plate before the cutting stroke is finally made. Also, the 'shear-cut' edge for plates of above 12 mm thickness gets deformed quite often, requiring edge planing for acceptance by the inspector. Angles and flats of usually upto 12 mm thickness can be cut by guillotine principle, commonly known as cropping. This method is widely used in workshops for batch production of angle cleats, lacings and similar small items.

19.7.3 Sawing

Saws are normally used for cutting rolled sections to the required length. Broadly, there are three types of motor operated saws available to the fabricator:

- Circular cold saw
- Band saw
- Hack saw

Of these, circular saw is the most popular, and is more productive than either the band saw or the hack saw. With circular saw, the ends may be square or bevel cut. Accuracy in squareness is 0.2% of the depth of the cut, and in length is within a fraction of a millimetre. As the rotating blade is fed steadily into the material, a jet of cutting fluid is sprayed to reduce friction and heat. Automatic sawing lines are controlled by a computer program, which not only registers and cuts, but also transfers the cut material to the next workstation by means of mechanised longitudinal and transverse conveyors.

As an alternative arrangement, band saws are also used for cutting light beam or channel sections.

19.8 FORMING

Trapezoidal shaped troughs, commonly used for orthotropic steel decks, are formed by cold bending of steel plate by press work. Essentially, this is a bending process, where the outer fibres of the material are in tension and the inner fibres are in compression. For a given thickness and mechanical properties of the plate, if the internal radius decreases, the strain at the outer fibre increases, and the material may crack after a critical limit is reached. In order to avoid the risk, a minimum internal radius should be determined before the press work is commenced.

Cold bending of sections and plates can also be done with rolling machines, typically with three adjustable rollers, one placed opposite the gap between two others (Fig. 19.5). In some cases the rollers are horizontal, while in others, they are vertical. Plates of upto 175 mm thickness can be accommodated in modern rolls, in which an accuracy of tubular rolling of about 6 mm 'ovality' can be achieved on a diameter of 1.5 m.

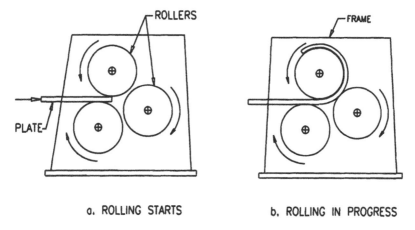

a. ROLLING STARTS   b. ROLLING IN PROGRESS

Fig. 19.5 Typical plate rolling arrangement

19.9 MAKING HOLES

In bridge steelwork, most field joints are made with bolts (or rivets). Accuracy of holing largely determines the quality of the joint. Holes are normally formed by drilling. In the traditional method, each hole is drilled individually, by a radial arm drilling machine. This method is time-consuming and is generally used in smaller fabrication shops, particularly in site fabrication units. Larger fabricators now have machines with NC systems which enable groups of holes to be drilled simultaneously along the web and both flanges of I beam sections. The more sophisticated machines incorporate sawing also to save further time.

Punching is a fast process to make holes. However, this process is generally not allowed in bridge steelwork for reasons of durability (see Chapter 25). Some codes allow sub-punching of undersized holes followed by reaming to the desired size. However, this method also is time-consuming and not suitable for fast work.

19.10 ASSEMBLY AND JOINING

At this stage different components are assembled and joined in the shop. Before welding became popular, shop connections used to be made by riveting process. Now-a-days, shop connections are mostly done by welding. Shop bolting has only limited use–mostly for trial alignment. For field joining, high strength bolts are generally preferred to welding, because of ease of operation and reliability. Field welding is mostly limited to structures where bolted joints would appear to be unattractive from aesthetic point of view.

19.11 RIVETING PROCESS

In the workshops, hydraulically operated riveting machines, commonly known as 'iron man', 'horse shoe' and 'scissors' are normally used. Varying sizes of these are in use. Some machines are heavy and can handle riveting at a distance of 2 metres form the edge of the work. Others are lighter in weight and can be used more easily, but cannot span more than about 500 mm. For small riveting work, pneumatically operated portable machines are normally used. Steel rivet cups, shaped to form the various sizes and shapes of rivet heads, are fitted into the machines as necessary.

Rivets are heated in electric, oil, gas or coke furnaces or coke burning blast fires. A rivet, already formed with one head and shank, is heated and inserted into pre-aligned oversized holes on the components to be joined, with the head in close contact with the steel, tightly pressed by means of a dolly or bucking bar with a head die, against one of the components. The shank of the rivet which protrudes through the rivet hole is shaped into a driven head by pressure from the riveting machine (Fig. 17.11). The shank should be of appropriate length to provide the exact amount of material required to swell it to fill the rivet hole and to form the second head.

19.12 WELDING PROCESSES

Common arc welding processes used in steel bridge fabrication work are:

- *Manual metal arc (MMA) welding*: This process is used where short lengths of welds are required, such as for welding small fittings.
- *Metal-active gas (MAG) welding*: This process is used where continuous welds are required.
- *Submerged arc welding (SAW)*: This is an automatic process for depositing long and high quality welds of uniform size. This process is extensively used for plate girders and box girders.
- *Stud welding*: This process is used for fixing stud shear connectors on the top surfaces of top flanges of plate girders used in composite construction.

These processes have been discussed in Chapter 15.

19.13 WELDING OPERATION

Some activities which need particular attention during fabrication, both for durability and economy, are discussed in the following paragraphs.

19.13.1 Edge Preparation

Edge preparation needs to be executed accurately. Machine flame cutting normally produces accurate edge preparation. Oversized preparations and irregular gaps must be avoided as these consume excessive amount of weld metal and consquently increase distortions, apart from increasing the cost. A larger gap will also drastically reduce the welding speed.

19.13.2 Assembling with Jig and Fixtures

Jig and fixtures are used to locate and clamp the component parts of a structure in position. In principle, a jig consists of a framework built around the work piece and to which are attached locating and fixing (or clamping) devices to ensure accurate assembly of the component parts of the work piece in their correct relative positions for tack welding, and possibly also for complete welding.

In addition to keeping the component parts in position, a jig also helps in preventing distortion of parts due to welding by holding correct alignment during welding and subsequent cooling. Thus, a jig should necessarily be strong and suffuciently rigid to ensure accurate alignment.

For quantity production, it is often economical to have two sets of jigs and fixtures to expedite the work. One assembly can be fitted up by a helper, while the welder is welding another assembly. Thus, by additional cost of one jig and one helper, the production can be doubled.

Design of jig and fixtures should be simple and inexpensive with minimum machining work. Welded construction should be adopted as far as possible. Accuracy is important, but need not be greater than required by the guiding specifications.

19.13.3 Tack Welding

Tack welding is done to hold the assembled components in correct position for final welding. Normally, a large amount of tack welding is required, and care should be taken to ensure that these tack welds are of sufficient lengths and numbers to perform the duty.

19.13.4 Positioning

The joint should preferably be placed in a position, where it is easily accessible to the welder and also the welding can be performed in the downhand position. This positioning can be done either manually, or by using mechanical equipment termed 'manipulator' or 'positioner'. The latter method is commonly used in modern fabrication shops. Manipulators are normally so designed that they can be readily adjusted to receive units of varying dimensions. In bridge work it is usual to employ manipulators with uni-axial rotation, since assemblies are generally long members and as such are rotated about their longitudinal axis only. Figure 19.6 illustrates a typical manipulator for plate girder fabrication.

The main advantages of using manipulators for welding are:

- The work can be rapidly and economically turned over and positioned so that down hand welding, instead of horizontal, vertical or overhead welding, can be performed. Studies reveal that change from vertical or

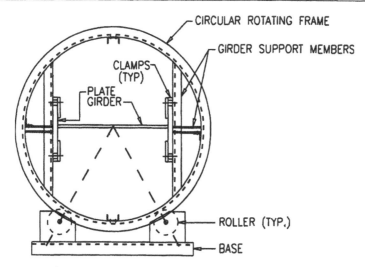

Fig. 19.6 Manipulator for plate girder fabrication

overhead position to downhand position can increase the speed to as much as 400% for some joints. Also, joints in the downhand position take advantage of the force of gravity and in this position electrodes can be used in their maximum efficiency.

- Welding procedures can be suitably selected to ensure welds with maximum strength and more accurate alignment of the finished job (i.e., reduced distortion), apart from achieving better profile at the minimum cost.
- Safety is obviously much greater, as manual handling is reduced to minimal.

19.13.5   Welding Distortions

Welding distortions arise due to the shrinkage of the molten weld metal while cooling. The amount of distortion primarily depends on the weld size, heat input of the process, number of runs, the degree of restraint of the joint and the thickness of the materials. These aspects have already been discussed in Chapter 16 and are not repeated here.

19.14   MACHINE FINISHING

It is often considered both necessary and best practice to machine the ends of compound compression chords at an edge-planing or rotary end-milling machine. Most fabrication shops for bridge steelwork are equipped with such facilities. In modern machines, optical laser beam methods are often used to position the cutting heads normal to the axis of the member to be machined. The objective is to obtain smooth square ends so that members, when butted together, will make very close contact. Machine planing process is also used where unacceptable levels of hardness on the edge of the plate caused by flame cutting has to be removed. Needless to add, members requiring machining should be fabricated with appropriate allowances in their dimensions.

19.15 CAMBER

Camber is the upward curvature built into a bridge structure during construction, so that, when subjected to a specific loading, it deflects and regains its original geometric shape. The decision to camber and how much to camber (i.e., for dead load, or dead load + partial live load, or dead load + full live load) should be taken by the engineer considering the recommendations given in the code being folowed.

The procedure for providing camber in truss bridges has been discussed in Chapter 10. The fabrication shop has to follow the dimensions of the components as given in the detail drawings.

For beams and plate girders upto 25 m spans, where only small cambers are theoretically required, the extra cost incurred for providing camber may not be justified. Thus, where aesthetics is not important, a permanent sagging profile may be allowed, and the required profile of the roadway top can be made up by adjusting the slab or the finishes.

Rolled beams are generally cambered in workshops by large presses. Beams can also be cambered by local application of heat in the same principle of straightening up by application of heat discussed earlier in this chapter. In this case the flange to be shortened is heated. As the flange is heated it will tend to expand. The expansion, however, will be prevented by the surrounding cooler materal of the unheated web and the flange will increase inelastically in thickness. Since the increase in thickness is inelastic, it will not return to its original thickness on cooling. Thus, when the flange is allowed to cool, it shortens (to return to its original volume) which produces the camber. Application of heat needs to be carefully controlled for the process to be successful. The method, though simple, requires considerable experience and practice.

For plate girders, the common practice is to cut the web to the cambered profile prior to attaching the flanges.

19.16 TRIAL ASSEMBLY

Most contracts provide that the accuracy of fabrication should be proved by a trial assembly at the works before the components are despatched to erection site. This is normally done by erecting the bridge structure complete, or in parts, as agreed with the engineer, on supports of blocks of wood or steel trestles. Distinctive erection marks, as per detail drawings, are hard stamped and painted on all the components prior to trial assembly. These are shop assembled with sufficient number of parallel drifts and bolts in order to bring, and keep the parts in place. Dimensional accuracy, squareness, alignment, camber, accuracy of the holes in the field connections of the main members, etc., are checked at this stage.

Normally, every span should be trial assembled before despatch. However, if the work is made completely interchangeable by use of steel jigs and hard steel bushes or by modern NC machines, and monitored adequately, shop erection and checking of only one span may suffice, at the discretion of the inspecting agency.

Some national standards specify that lower sized holes should be made at shops for bolted field connections and splices and that these should be reamed to the required size while the members are assembled in the shop. This requirement entails sub-punching or sub-drilling, shop assembly and reaming operations for the joints in every span. These operations are bound to add significant costs to the project. Since modern fabrication shops using NC machines can produce full sized holes located with high degree of accuracy, the procedure of making undersized holes during fabrication and reaming them to correct size after assembly appears to be unnecessary.

19.17 INSPECTION

In order to maintain a high standard of workmanship and to ensure that all work is performed in accordance with the required specifications, it is necessary to have a strict inspection regime at various stages of fabrication. In fact, inspection is an integral part of the fabrication process. The topics which mainly cover this activity, viz., inspection, testing and quality control, are discussed in the next chapter.

19.18 PAINTING

A protective priming coat is applied on the steel to provide a compatible surface for subsequent coats of paint to be applied after erection of the bridge. The type of the priming coat depends on the coating system (see Chapter 26) to be adopted for the bridge, which, in turn, depends on the exposure condition of a particular location and the desired life cycle of the coating system. For this purpose, recommendations of the governing codes supplemented by specialist literature are normally followed.

Before application of primer paint, surface of the steel must be prepared by thoroughly cleaning all loose mill scale, dirt, loose rust and other foreign matter. The traditional method of cleaning is by using manual tools such as steel scrapers, wire brushes, etc., or alternative power driven tools. For more sensitive protective systems, a comparatively sophisticated cleaning procedure is required. The common practice is to use grit or shot blasting process to achieve a clean surface. In this process, abrasive particles are projected at high speed on to the steel surface by either compressed air, or centrifugal impeller wheel. The abrasive particles are either spherical (shots) or angular (grits) . The particles impinge on the steel surface, removing mill scale and rust, and roughening the surface for paint coating. Grit blasting produces a rougher surface to provide an adequate key for metal spraying or special paints. Many authorities have produced their own standards or specifications for surface cleanliness and other aspects of surface preparation for bridge steelwork. Most of these are based on the Swedish Standard (e.g., SA 2-1/2, 3 etc.).

There are two methods of blast cleaning and applying priming coat to bridge steelwork:

- In the first method, priming coat is applied in two stages. Steel material is first blast cleaned prior to fabrication and applied with a pre-fabrication primer. After fabrication, heat affected zones and surface abrasions are made good by cleaning and applying the primer, and then the post-fabrication final priming coat is applied. This system has an advantage of applying primer in the permanently inaccessible surfaces before assembly.
- In the second method, blast cleaning is done only after fabrication, which eliminates the need for further cleaning and remedial work. In this system, the blast cabinet must be large enough to accommodate the completed components. Blast cleaning is followed by application of priming coat.

The following aspects need particular attention during the application of priming coat:

- Surfaces which transfer forces by friction (as in HSFG bolt connections) should not be painted with the priming coat.

- Shop contact surfaces should be given priming coat and should be brought together while the paint is still wet.
- Surfaces not in contact, but inaccessible after assembly, should receive the full specified protective treatment before assembly.
- Parts to be in contact with concrete should not be painted.
- Surfaces should be perfectly dried before applying priming coat.
- Priming coat should be applied immediately after the surface is prepared.

19.19 DESPATCH

On completion of final inspection, steelwork is transferred to the storage yard. Here fabricated components are stored systematically, as per their erection marks, so that despatch to site can be done in the correct sequence to suit the erection schedule. Loose components are bundled or wired to await despatch together with assembly bolts. Material is usually transported by railways for distant shipments. Local shipments are transported by trucks. The yard is normally equipped with loading facilities for each type of transportation.

Knowledge of highway and railway regulations regarding clearing dimensions, maximum permissible loads, etc., is essential, particularly for apparently oversized consignments. Overall dimensions of large despatchable units, therefore, need careful consideration. Another aspect which should also receive due attention is that framed units obviously occupy considerable space while despatching. This situation can be mitigated by intelligent disposition of existing splices and/or introduction of additional splices. Similarly, members with projecting gussets are not only susceptible to damage during transit, but are also awkward for economic and secured storage during transit. Such details should be avoided as far as possible. However, these matters relate primarily to the design concept and should receive adequate attention at that stage.

REFERENCES

[1] Farrel, J. 1954, *Construction Steelwork Shop Practice*, Iliffe & Sons Ltd., London, UK.

[2] Peshek, C. and Marshall, R.W. 1994, *Fabrication and erection*. In: Structural Steel Designer's Handbook, Brockenbrough, R.L. and Merritt, F.S. (eds.) McGraw-Hill, Inc, New York USA.

[3] Dowling, P.J., Knowles, P.R. and Owens, G.W. (eds.) 1988, *Structural Steel Design*, The Steel Construction Institute, London UK.

[4] Tardoff, D. 1985, *Steel Bridges*, The British Constructional Steelwork Association Ltd, London, UK.

[5] Taggart, R. 1986, *Structural steelwork fabrication*, In: The Structural Engineer, August, 1986, Institution of Structural Engineers, London, UK.

[6] Palmer, M.F. and Fowler, R.J. 1958, *Modern methods of fabricating steelwork*. In: The Structural Engineer, Institution of Structural Engineers, London, UK, November, 1958.

[7] *Procedure Handbook of Arc Welding Design and Practice*, 1959, The Lincoln Electric Co, Cleveland, USA.

Chapter 20

Quality Control in Fabrication

20.1 INTRODUCTION

Quality control at the fabrication shop is of utmost importance for ensuring that the completed bridge structure behaves in the manner envisaged at the design stage and thereby achieves the desired durability. For this purpose it is necessary to ensure that the fabricated steelwork strictly complies with the approved detail drawings. To achieve this, a strict inspection regime in accordance with an agreed Quality Assurance Plan (QAP) is normally enforced. This plan needs to be tailor-made for the type of structural system adopted for a particular bridge, and should specially address the areas which may be sensitive to imperfections and consequently need special care during fabrication. The frequency and intensity of inspection may need to be higher in such cases.

The concept of QAP rests mainly on the activities of two agencies, viz., the fabricator's in-house quality control department and the owner's inspection agency. For attaining the desired level of quality in fabrication, the fabricator follows an accepted quality control programme. The owner's inspection agency, in turn, is reponsible for checking and accepting (or rejecting) the items offered by the fabricator. To make the excercise a success, it is necessary to have properly skilled persons to be associated with this process. Training of personnel on a continual basis is thus essential.

QAP envisages quality control at every stage of the production process. Typically, detail inspections take place at the end of each stage of fabrication process. The results of such inspections are recorded properly and communicated to the shop floor personnel, who take corrective measures, if necessary. This system ensures that initial errors are not repeated, particularly in case of repetitive items.

Final inspection is done after completion of fabrication and trial assembly, when emphasis is generally given on overall dimensions, open holes for site connections and similar items to ensure proper alignment during site erection. Framed elements, such as latticed girders, are self-checking to a certain extent, as correct fitting of the members during trial assembly often proves that the structure has been fabricated correctly.

In this chapter the relevant aspects which need to be considered to achieve quality fabrication are discussed.

20.2 DOCUMENTATION

Proper documentation of stage inspection activities is an important step towards achieving high level of quality of the final product. This ensures that defective or unsatisfactory items or sub-assemblies cannot be used in the structure, unless these are rectified to the satisfaction of the inspectors. Some of the items/activities, for which maintenance of records is recommended, are:

- Steel sections/plates, consumables *vis-a-vis* their certifications
- Templates, jigs and fixtures
- Sub-assemblies/components: cutting, fit-up and alignment
- Welding procedure including the parameters adopted in the job
- Qualification tests for welders and welding operators
- Non-destructive and destructive inspections
- Rejections of items/sub-assemblies including repair methods adopted
- Manufacturing deviations allowed in the fabricated items such as plate girders, open-web girders including open holes

20.3 STORAGE OF MATERIALS

On receipt of steel materials at the storage yard, they should be carefully unloaded, examined for defects, checked with documents, and stacked specificationwise and sizewise, above the ground on platforms, skids, or other suitable supports, to avoid contact with water or ground moisture. Electrodes should also be stored specificationwise and kept in dry, warm condition. Bolts, nuts, washers and other fasteners should be stored on racks above the ground. Paint should be stored under cover in air-tight containers.

20.4   INSPECTION OF MATERIALS

To ensure quality assurance of the final product, it is necessary to carry out inspection of raw materials such as, structural plates and sections, welding consumables, fasteners, gas, bought out items, etc., before they enter the shop floor.

Structural plates and sections are generally checked for their physical and chemical properties by verifying the distinctive castmark/heat mark numbers die stamped on the steel, with the corresponding numbers mentioned in the certificates issued by the rolling mills. Over and above this, independent sample tests are often carried out to confirm the accuracy of mill test results. For this purpose, test pieces are cut out from the end of the material to be used in fabrication, and subjected to various tests. Such test pieces should be cut and processed in the presence of the inspectors and duly documented.

Since lamellar tearing is generally associated with welded joints, materials meant for welded construction should particularly be checked against presence of lamination. Tests for lamination should also be carried out for materials in which tension stresses are transmitted through the thickness, or where lamination could affect the buckling behaviour of the member under compression. Lamination can be detected by non-destructive testing (NDT) method, such as ultrasonic examination, discussed later in this chapter.

Dimensional tolerances, i.e., permissible variations from theoretical dimensions, on sections and plates from hot-rolling mills, sometimes present certain problems during fabrication. Also, materials may become bent or twisted during transit from the mills to the fabrication shops. These need to be properly checked and rectified, as necessary, before being used in the job.

20.5   INSPECTION OF LAYOUT, TEMPLATING AND MARKING

It is expedient to carry out, on a routine basis, dimensional checks of all full scale layouts, templates and markings prepared in the template shop. Mistakes detected at this stage should be rectified before these are issued to the fabrication shop. Drilling jigs with hard steel bushes, meant for multiple use, should be checked from time to time during fabrication process and must be replaced when the deviation exceeds the allowable limit.

20.6 INSPECTION AFTER ASSEMBLY

Inspection is carried out after the different components are assembled and joined (but before riveting or welding is undertaken), to ensure that the fit-up has been done as per orientation and measurements shown in the drawing.

20.7 FABRICATION TOLERANCES

Most standard codes allow a certain amount of deviation from the required dimensions of fabricated parts. The limits of such deviations are clearly spelt out in the specifications for workmanship. These deviations are normally taken care of in the factor of safety while computing the member capacity, and hence are in conformity with the corresponding design specifications. Imperfections beyond these limits, however, need to be rectified before further processing of the deficient workpiece.

20.8 INSPECTION OF RIVETED CONNECTIONS

Rivets and riveted connections should be inspected and tested in accordance with the requirement of the governing code. Driven rivets are checked with a rivet testing hammer. When struck sharply on the head with a rivet testing hammer, the rivet should be free from any movement or vibration. Loose rivets and rivets with cracked, badly formed, or deficient heads or with heads which are unduly eccentric with shanks, should be cut out and replaced. The firmness of the joint is typically checked by 0.2 mm feeler gauge, which should not go inside under the rivet head by more than 3 mm.

20.9 QUALIFICATION TESTS FOR WELDERS

Qualification tests for welders and welding operators are conducted to determine their ability to produce welds of acceptable quality with the welding processes, materials and the procedures that are to be used in the production. Normally standard codes lay down the procedures for such tests. It must, however, be accepted that the welders and operators are bound to pay special attention and effort for the qualification test welds. They may not do so under every production condition. Therefore, complete reliance should not be placed on these tests. Quality of the welds on the job must also be checked during and after welding.

20.10 APPROVAL OF WELDING PROCEDURES

Before taking up welded fabrication work, examination of the proposed welding procedure is carried out to substantiate that satisfactory weld is achievable with this procedure. For this purpose, a specimen of welded joint of adequate length of same cross section and material, as will be used in the job, is produced with the electrode, wire, flux, current, arc voltage and speed of travel that are proposed to be used in actual construction. This joint is then subjected to the required tests for acceptance of the welding procedure by the inspecting agency.

Normally, the welding procedure document includes the following records in respect of each type of welded joint:

- Detailed description of how the weld is to be made.
- Sketches showing weld joint detail and welding parameters (electrode/wire type and size, welding current, arc voltage, speed of travel, etc.) for each run.
- Results of testing. The tests required for acceptance of the welding procedure are laid down in most national codes and these are generally followed in the workshops.

Having established the acceptable welding procedure, the actual job should be welded strictly in accordance with it.

20.11 RUN-ON AND RUN-OFF PIECES

For submerged arc welding, it is a common practice to tack one run-on and one run-off piece at each end of the joint to be welded. These pieces are to be of the same material and thickness as the parent plates and prepared in identical manner. The welds should be so deposited that the run-on and run-off pieces form a part of the main weld. These may be tested to determine the quality of the weld.

20.12 INSPECTION OF WELDED JOINTS

In order to ensure that the fabricated structures perform their intended functions satisfactorily, the welded joints are to be properly inspected. The inspection methods commonly employed can be broadly divided into two categories, viz., non-destructive and destructive inspections.

20.12.1 Non-destructive Inspection

In non-destructive inspection, the defects and discontinuities in the welds are detected without impairing the quality and usefulness of the material. The various methods most frequently used are :

- Visual inspection
- Liquid penetrant inspection
- Magnetic particle inspection
- Radiographic inspection
- Ultrasonic inspection

Visual inspection

Visual inspection is the simplest and the most widely used inspection method. It is fast and economical, and does not require special equipment. It can furnish vital information, provided the inspector's power of observation is reasonably good. The method requires good lighting arrangement. Equipment needed are: inspection gauges, magnifying glasses, mirrors, scales, callipers, etc. Normally, visual inspection is carried out in the following three stages of fabrication:

(a) Inspection prior to welding: Proper inspection at this stage could eliminate conditions that may lead to serious weld defects. The items inspected include:

- *Parent metal*: All plates and sections should be inspected before welding is undertaken. Verification of the quality is carried out by reference to the relevant test certificates. If considered necessary, spot checks on the chemical composition and physical properties may be made by cutting test specimens from the parent metal. Freedom from harmful defects, such as cracks, surface flaws, laminations and rough or imperfect edges should be verified. Dimensions of components should also be checked.
- *Welding consumables:* These should be checked for quality.
- *Welding equipment:* This should be checked for workablity in accordance with the approved welding procedures.
- *Welders' qualification:* Welding should be permitted only by welders and welding operators who are qualified in accordance with the standard procedure.

- *Assembled parts:* After the parts are assembled in position for welding, these should be checked for root gap, edge preparation, dimensions, finish, fit-up including back strips, alignment, cleanliness, etc., that might impair the quality of the welded joint.
- *Welding procedure:* Before any welding on actual job is allowed, the inspector should verify that the procedure to be adopted for the particular job is in accordance with the approved welding procedure.

(b) Inspection during welding: While welding is in progress, visual inspection is carried out to ensure that the approved procedure is followed throughout. The items to be checked include maintenance of proper current, arc voltage and speed of travel, sequence of welding, cleaning of slag after each run in multi-run welding, electrode spattering, etc. Care should be taken particularly during the early stages of the work when teething problems generally arise.

Consumables should be inspected periodically to ensure that the specified quality is maintained consistently.

Also, the weld run which will become inaccessible after subsequent welds should be thoroughly checked.

(c) Inspection after welding: In order to carry out visual inspection, the weld surface should be thoroughly cleaned of oxide layers and adherent slags. Often, chipping hammers are used to remove slag. This practice should be avoided, since the hammer marks are likely to obscure the evidence of fine cracks. Brushing with stiff wire brush or grit blasting may be adopted for this purpose.

The items to be inspected include:

- *Weld profile:* The finished weld profile should conform with the size and contour recommended in the standard specifications.
- *Surface defects* such as unfilled craters, lack of fusion, cracks, porosity, slag inclusion.
- *Weld appearance* such as irregular ripple marks, peening marks, surface roughness.
- *Disposition of welds vis-a-vis* detail drawings
- *Dimensional accuracy* of the fabricated items (distortions, warpage, etc.) *vis-a-vis* allowable tolerance limits.

Liquid penetrant inspection

Liquid penetrant inspection is carried out to detect the surface defects or discontinuities such as cracks, porosities, etc. Commonly, a dye of red colour is used for this inspection. First, the weld surface is cleaned and dried. Then the penetrant is applied by spraying to form a film over the surface. The penetrant seeps into (and is drawn into) any surface cracks or other defects by capillary action. After the penetration time (about 20 minutes), the excess penetrant on the surface is cleaned off using a solvent. A developer (like a chalk powder dry or suspended in a liquid) of contrasting colour with high absorbant property is then applied. If any cracks or surface defects are present, the penetrant bleeds out of the defects and appears as a stain on the chalk surface. The surface is then examined using a magnifying glass. The surface is to be properly cleaned after inspection.

Liquid penetrant inspection can be carried out either by using dye penetrant or fluorescent penetrant. In the former case, vivid red dye is used which contrasts sharply with the white background of the developer. The fluorescent penetrant, on the other hand, is more prominently visible with a brilliant glow under ultra-violet light and consequently more sensitive compared to dye penetrant. This method requires a dark room and a source of ultra-violet light.

Magnetic particle inspection

This method of inspection is used to detect surface and sub-surface discontinuities, such as cracks, porosities, slag inclusions, fusion deficiencies, undercuts, incomplete penetrations, gas pockets, etc. The method is quite useful for locating *fine* sub-surface defects and can be gainfully used where radiographic or ultrasonic methods are not available or not practicable to apply because of shape or location of weldment.

In this method, the magnetic properties of steel are used by setting up a magnetic field within the piece to be tested and dusting the test area with fine iron powder or spraying the area with liquid detection medium carrying magnetic iron powder. If there is a crack, the magnetic lines are disturbed and a small north-south pole area at the crack zone is created and the outline of the crack becomes clearly visible. A surface crack is indicated by a sharp line of magnetic particles following the crack outline, and a sub-surface defect by an indistinct collection of the magnetic particles on the surface above the discontinuity.

There are a few techniques for creating the magnetic field in the test area. In the first method, the poles of a strong permanent magnet or an electro-magnet are placed on either side of the area to be tested. In the second method, electric current is passed through the job itself. In the third method, magnetic field is produced by induction by winding a coil around the test area and passing a current through the coil.

Radiographic inspection

This method is used for detecting both surface and sub-surface defects in welding, such as cracks, inclusions, porosities, fusion deficiencies, incomplete penetrations, undercuts, etc. The system essentially consists of passing x-rays or gamma rays through the weld being tested and creating an image on a photo-sensitive film. The film is placed in a cassette or holder behind the weld to be examined. The x-rays or gamma rays pass through the weld and expose the film. If there are voids in the weld, less radiation will be absorbed by the steel and more radiation would pass through that area to the film. Thus, these defects will appear on the film as areas darker than the image produced by the surrounding areas of sound material. Similarly, inclusions of low density, such as slag will appear as dark areas, whereas inclusions of high density, such as tungsten will appear as light areas.

This method has an advantage of providing a permanent record of every test carried out. However, it requires specialised knowledge in selecting the angles of exposure, as well as in interpreting the results recorded on films. The system also requires access from both sides of the test area, with the radiation source placed on one side and the film on the other side. Another point needs special mention here, viz., the safety aspect posed by radiation hazards associated with the system. Certain statutory safety precautions are necessary in the transporting and storage of the radio-active isotopes used in the system, as also in radiographing the weld so as to avoid radiation exposure to human body.

Ultrasonic inspection

This ia a highly sensitive method for detection of surface and sub-surface flaws in the weld as well as in the parent metal. It is widely used as an important tool for quality control in steel fabrication industry. It involves introduction of a high frequency sound beam into the area to be tested by means of an ultrasonic transducer. The sound beam travels in a straight line through the steel until a discontinuity is met, when it reflects back to the transducer. This produces a voltage impulse, which appears on the cathode ray tube oscilloscope. This system has an advantage, in that, it requires access from only one side of the material, unlike radiography, where access from both sides is required. However, the method does not provide any permanent test record as in the case of radiography. Also, the method requires a well-trained and experienced operator who has to correctly interpret the pulse-echo patterns appearing on the screen.

20.12.2 Destructive Inspection

Destructive inspection of welding is conducted for assessing the quality of welding and is commonly used for approval of welding procedures, qualification tests for welders and quality control during production process. Broadly, these can be divided into three categories:

- Chemical analysis
- Metallographic testing
- Mechanical testing

Chemical analysis

The solidified weld metal is essentially a mixture of the molten parent metal and steel from the electrode. During welding process, many elements get evaporated or lost in the process. Chemical analysis is carried out to ascertain the resultant chemical composition of the weld metal and compare it with the parent metal. It is not essential that the compositions of the weld metal and the parent metal should be identical. However, these should be compatible.

Chemical analysis is particularly important if a bridge structure is located in a corrosive environment, where the weld metal may be more susceptible to corrosion than the parent metal.

Metallographic testing

As the welded joint cools, the metallurgical structure of the steel in the fusion zone and the heat affected zone (HAZ) suffers certain changes. Metallographic testing (macro-etch and micro-etch test) is conducted to examine the soundness of the final weld in respect of distribution of non-metallic inclusions, the metallurgical structure, depth of penetration and identification of the number of weld passes, quality of fusion, etc. A section of the weld joint is cut and the specimen is prepared by using standard metallographic practice for examination. The macro-etch testing is conducted by etching the specimen to bring out the macro structure and examining by naked (unaided) eye or at low magnification of the order of 30 times. In micro-etch testing, the specimen is polished, etched, and examined under high magnification (generally, more than 50 times) using an optical or electron microscope.

Mechanical testing

Mechanical testing is done to determine the mechanical properties of weld, such as tensile strength, ductility, etc. Specimens are prepared by machining in accordance with the appropriate specification. When specimens are cut by thermal process, the material at the edges is likely to be affected by heat and needs to be discarded while preparing the test piece.

There are a number of tests recommended by various authorities. A discussion on some of the common tests follows.

(a) Longitudinal tensile testing: In this testing, the specimen is prepared from the weld metal only (Fig. 20.1), and ruptured under tension load to determine the breaking strength, yield strength, percentage elongation, percentage reduction in area, as also type and location of flaws on the fractured surface.

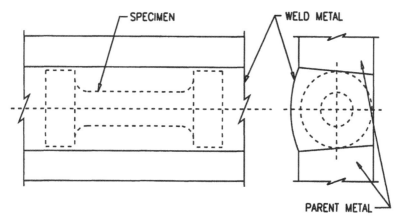

Fig. 20.1 Preparation of specimen for longitudinal tensile test of weld metal

(b) Transverse tensile strength test: This test is carried out to determine the tensile strength of a butt welded joint in the direction transverse to the weldment. The specimen (Fig. 20.2) is prepared in accordance with the code and subjected to tensile test. The remarks made above for longitudinal tensile testing of weld metal apply for this test also.

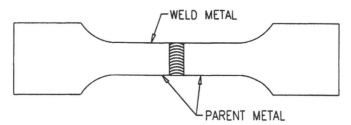

Fig. 20.2 Transverse tensile test specimen

(c) Hardness testing: Hardness measurement is relevant in welded fabrication work for ascertaining whether there has been any significant change in the property of steel by the heat evolved during welding operation. The commonly used hardness testing methods are:

- Brinell method
- Rockwell method
- Vickers method

Brinell method can be used for hardness survey over a relatively large area, such as face of weld and base metal. Rockwell and Vickers methods, on the other hand, can be used for hardness measurement over a small zone like weld cross section, HAZ, and individual weld bead.

(d) Guided bend test: This test is for ascertaining soundness and ductility of a butt welded joint. It requires inexpensive equipment and is widely used in fabrication industry. Essentially, it consists of bending the joint 180 degrees around a pin of specified diameter (around 40 mm). The test piece of specified size is bent in such a direction that the weld tends to pull away from the two edges of the joint. The outside surface elongates due to tension, while the inside surface shortens due to compression and the test piece bends without failure.

(e) Nick-break test: In this test, two plates are butt welded and the test specimen is made by cutting out a section of specified dimensions and nicking it on each side with a suitable instrument (e.g., saw). When the specimen is struck with sufficient force, it will break at the nicks. The exposed fracture surface is examined for internal defects, such as porosity, slag inclusion, degree of fusion, fracture type, ductility, etc.

(f) Free bend test: This is a simple test which indicates the ductility of weld metal in a butt welded joint.

(g) Fillet weld break test: This test is carried out to determine the soundness of fillet welds. A test piece of fillet welded joint of specified detail (Fig. 20.3) is prepared and is subjected to force in a press or a testing machine, or by blow of hammer till it fractures. The fractured surface is then examined for porosity, slag inclusions, root fusion, etc.

(h) Shear strength tests for fillet weld: These tests indicate shear strength of fillet welded joints. Tests are carried out separately for longitudinal and transverse welds by preparing fillet welded test pieces to specified details and subjecting them to tensile load to break or fracture the joints. These tests

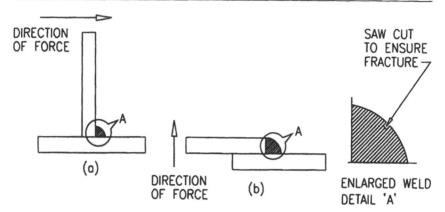

DIRECTION OF FORCE

SAW CUT TO ENSURE FRACTURE

A

(a)

DIRECTION OF FORCE

(b)

A

ENLARGED WELD DETAIL 'A'

Fig. 20.3 Fillet weld break test specimens

are simple and inexpensive and are frequently used in workshops as periodic check on the quality of work.

(i) Notched bar impact test: It has been noticed that weldments which are ductile under uniform static loading at normal temperature often show marked reduction in their ductility under suddenly applied loading (impact) at low temperatures. In general, impact strength of the weld metal is affected by heat, chemical composition and presence of certain embrittling elements in the weld metal.

Notched bar impact test is carried out to ascertain the impact strength of the weld metal. Typically, a V-notched bar test piece of specified dimension is used for the test. The test is done by breaking the test piece by a blow from a swinging hammer and the energy required for doing so is determined. Two types of swinging hammer tests are available, viz., Charpy and Izod. In both types, the plane of the swing of the hammer is vertical. In the Charpy test, the test specimen is supported at both ends placed on a split anvil and broken by a blow opposite the notch. In the Izod test, the bar is held vertically in a vise, with the notch placed just above the jaws, and broken by a blow applied on the side of the notch. Figure 20.4 shows a schematic arrangement of these tests.

The swinging hammer is slowed down when it strikes the test piece which absorbs some energy. The difference of the height to which the hammer rises after striking the test piece and the height from which it started is measured. From this difference and the effective weight of the hammer, the energy absorbed by the test piece is determined.

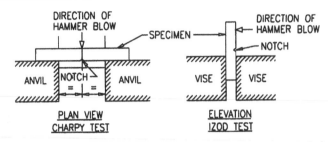

Fig. 20.4 Schematic arrangement of impact test

20.13 INSPECTION OF TRIAL ASSEMBLY

Inspection of the trial assembly is the final inspection, prior to application of primer paint and despatch of components to site for erection. Normally, a joint inspection of the fabricator, and the inspecting agency is carried out, preferably with the participation of the customer's representative in charge of the contract.

The items normally confirmed during an inspection of trial assembly include:

(a) Dimensional accuracy

- Overall length and length between centres of bearings; also length of individual panels.
- Height of trusses, girders, etc.
- Distance between centres of trusses, girders, etc.
- Squareness (in plan) of the entire span, as well as of each panel.
- Verticality of trusses, girders, etc.
- Fabrication camber at no load condition.
- Deflection due to dead weight of the span.

(b) Connections

- Condition of drilling (alignment of holes, accuracy of diameter, condition around hole, etc.).
- Joining surfaces (gaps, irregularities).
- Edge preparations of field welding (accuracy of shape, size, gap, etc.).

(c) Overall appearance

(d) Inspection of shear connectors

(e) Inspection of other items, such as, expansion joints, bearings, handrails, water drainage system, etc.

20.14 CONCLUSION

The quality assurance plan and the inspection regimes described in the foregoing sections, particularly for welded fabrication, are, no doubt, expensive and will add to the cost of the job, directly in themselves, and indirectly by time delays due to their disruptive influence on production. However, because of the dynamic nature of bridge loadings, which induce fatigue conditions and the possibility that the structures may be subjected to brittle fracture conditions through low ambient temperature due to their location, a good case is made for their use in modern steel bridges. In all cases, however, the importance of the joint under consideration should dictate the type of inspection and the acceptance criteria to be adopted. The controlling specifications generally provide the guidelines.

REFERENCES

[1] *Bridge Inspection Guide*, 1984, Her Majesty's Stationery Office, London, UK.

[2] Gourd, L.M. 1995, *Principles of Welding Technology*, Edward Arnold, London, UK.

[3] *Steel Construction Guidebook*, 1983, The Kozai Club, Tokyo, Japan.

[4] *Procedure Handbook of Arc Welding Design and Practice*, 1959, The Lincoln Electric Co, Cleveland, Ohio, USA.

Chapter 21

Erection

21.1 INTRODUCTION

Erection of the steel superstructure is the culmination of the process that started in the design office. During this process, fabricated components which were despatched from the workshops are received at the site of erection, stored, sub-assembled, and finally erected in their desired locations. The process primarily comprises of a number of routine activities. These include:

- Receiving structural components, handling and storage
- Rectification, if necessary
- Sub-assembly
- Transportation to erection front
- Erection
- Alignment and levelling
- Painting

No doubt, close inspection is to be carried out at various stages of erection. A typical flow chart of these activities for erection of a common type of steel bridge is shown in Fig. 21.1.

The responsibility of the erector is to ensure that the erected bridge steelwork conforms to the quality of workmanship required by the governing specifications. He has also to see that the erection is done within the time limit specified in the contract. From the contractor's point of view, the erection has also to be done within the estimated cost limit. Any overrun in time or cost jeopardises the good work done in an otherwise successful erection work. Thus, the erector's responsiblity is numerous and varied.

In this chapter, some aspects of the erection process which contribute to the successful discharge of the erector's responsibility are briefly discussed.

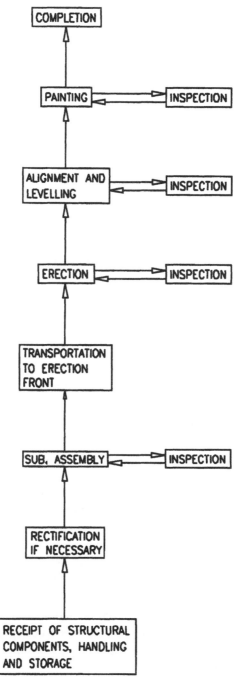

Fig. 21.1 Typical flow chart for erection of a steel bridge

21.2 THE SITE

In a bridge, the location and configuration of the area to be spanned, as well as the ground condition, largely determine the span and the structural system to be adopted. Similarly, these same site conditions have profound influence on the method of erection for the chosen bridge structure as well. In fact, quite often, the erection method, which is suitable for a particular site, may, indeed, dictate the choice of the span and the structural system for the bridge.

21.3 PLANT AND EQUIPMENT

Contractors normally have a central plant yard where plant and equipment are stored after they are returned from one job. These are repaired and tested on a regular basis and are kept ready to be sent out on to the next job. It is an economic necessity that existing plant and equipment are utilised as far as possible for different bridge erection jobs. Mobile cranes, light weight erection cranes to suit standard repetitive spans are examples of such equipment. Special 'job specific' erection devices can be manufactured only for substantially really big jobs where the high cost can be written off against that particular job.

The type of crane to be used for erection in a particular bridge is generally dictated by the weights of the components to be lifted, and the corresponding capacity of the crane at each relevant radius. In order to effect the maximum utilisation of the cranage capacity, often two erection cranes are used in tandem, one for lifting heavy materials only, such as chords, verticals, diagonals of trusses, etc., while a follow-up crane is employed to erect lighter components, such as stringers, bracings, etc.

Where railway connection can be arranged to the site prior to the commencement of erection activity, transportation of plant and equipment does not present much problem. Otherwise, these have to be transported by road, in which case the condition of the road and the distance of the site from the plant yard or the nearest rail-head need careful study. A truck-mounted mobile crane can be driven on the roads, but a crane mounted on caterpillar tracks would need a special trailer-truck to move it from one location to another. The erection cranes would also have to be transported in pieces, to be assembled at site. The curves to be negotiated for carrying such equipment on road to a site situated in a mountainous region may also present problems.

If the location of the worksite is far away from the contractor's central plant yard, the possibility of hiring erection plants may be considered. In case, however, the worksite is far too remote and inaccessible for transportation of heavy plant and equipment, the erection scheme may need to be tailored to use conventional and lightweight equipment only.

It is a good practice to inspect all equipment, machinery, tools, tackles, etc., prior to actual commencement of erection, to ensure that these are in serviceable condition, and are well upto the capacity for which these are required. It is also necessary to carry out frequent visual inspection of critical items during actual erection process to detect any distress or damage which may have developed after the initial inspection.

21.4 CENTRE-LINES AND LEVELS

It is always easier to place bridge components onto their correct positions and levels initially, than it is to pull the whole structure into position afer it has been erected. It is, therefore, expedient to carry out a thorough survey of the foundations before starting the erection. The points that need particular attention in this respect are :

- Centre-lines of foundations—in both longitudinal and transverse directions—should be clearly marked on the top of the abutments and piers, to ensure accurate positioning of the steelwork after erection.
- Positions of the anchor bolts *vis-a-vis* the centre-lines should be checked, and rectified, if necessary.
- Levels of the top of foundations should be checked, and rectified, if necessary.

21.5 FOUNDATIONS

Design of permanent substructures and foundations of a bridge is normally carried out considering the effects of the loadings which can occur during service condition. However, the temporary loadings which will be applied to these structures during the various stages of erection normally do not receive the proper attention they deserve, although these may very well exceed the permanent design loads. This aspect needs to be considered in the design.

Also, foundations of temporary supports should receive adequate attention. Where loads are significant on these foundations, or where the effects of their settlement might cause instability of the partially or fully completed structure, it is advisable to investigate the ground conditions by carrying out test borings in the location and ensure adequacy in the design and construction.

It is common to have steel brackets or extensions temporarily fixed to the permanent abutments and piers. Adequacy of such brackets or attachments, paticularly their connection details with the permanent structure, should be given careful consideration, and incorporated in the original design.

21.6 STOCKYARD

The storage area or stockyard for fabricated structures received from the workshop should preferably be located near the site of work.

The size and shape of the storage area is naturally governed by the available land. Ideally, the stacking area should be on a levelled and firm ground, and be large enough to *always* store sufficient quantity of steelwork, itemwise and without damage or distortion, to cope wih the requirement of the erection front. Also, adequate road access—and railway connection where required—must be provided to give all-weather service. Axle loads of the vehicles nomally in use in a bridge site are heavy. Therefore, the access road, as well as the roads inside the storage area, must be maintained properly.

The stockyard will have to be provided with adequate handling equipment, such as mobile cranes, gantry cranes, derricks, chain pulley blocks, winches, etc. For a long and busy stockyard, a Goliath crane is often suitable as this can cover the large area comfortably.

The stockyard should have adequate lighting arrangement. It should be properly planned with platforms, skids and other supports on which fabricated materials can be stored above the ground to avoid contact with earth surface or water. For long members, such as chords, diagonals, etc., skids should be placed near enough to prevent damage due to deflection. Materials should be stacked methodically, with their identification marks visible, so that they can be transported out whenever required without resorting to double handling. The area should, therefore, be easily accessible to handling equipment.

Proper control in the storage and timely distribution of fasteners like rivets, bolts, etc., is crucial for the smooth operation of erection work. Correct numbers of correct sizes of fasteners required for particular connections or areas are often bagged and kept ready separately, to be sent out to the erection front, in order to save time and also to cut down the wastage.

Similarly, proper storage of welding electrodes and their systematic issue contribute substantially to the smooth progress of the erection work.

Cables and items with threaded parts, such as anchor bolts, need special care. These should be stored in a dry shed and should be so placed that they are clear of the ground. Regular inspection of such items would ensure that they remain in satisfactory condition. Machined and threaded items should be kept properly greased.

21.7 ASSEMBLY YARD

In order to achieve economy, it is necessary to reduce the number of manhours on the erection front by maximum use of pre-fabrication and sub-assembly. Since the size in which the fabricated components are received at site depends on the limits imposed by roadway or railway authorities, it is a common practice that these components are sub-assembled and bolted (or riveted) to larger transportable and erectable sizes in an assembly yard, and then sent to erection front for hoisting. The assembly yard is normally located on the ground near the site. As many numbers of permanent bolts or rivets as possible should be fixed in the assembly yard. Bolting or riveting in the yard is easier to perform and is more economical than doing it aloft on the bridge. Care should be taken in the yard to see that all loose material to make a complete joint are properly cleaned and sent to the proper place and in sequence, so that little time is lost in the erection front in locating these small parts.

The assembly yard should preferably be located adjoining the stockyard, and close to the bridge, and should be served by adequate crane facility for handling and assembling materials. The yard should also be connected by a suitable roadway or rail track system for hauling assembled materials to the bridge. Adequate facility for loading assembled pieces on transport trolleys or bogies in the yard and also unloading these pieces at the bridge site are to be arranged.

The assembly yard is usually equipped with a plant repair shop, having facilities to repair minor defects and also to maintain the plant and machinery being used in the assembly yard and the bridge.

21.8 SITE WELD

Except for very big bridges with long runs of horizontal seam welds, the manual metal arc process is usually used in most bridge sites. This process demands that the details should be such that adequate access can be obtained for the electrode as also for the welder to be able to see the work. This means that there should be adequate space for the welder to manoeuvre without unreasonable twist of his body.

Since downhand welding is easier and consequently, faster than overhead welding, structural details should allow most welds to be made in the downhand position. The aim should be to ensure that the site welding activity is easy and straightforward, and therefore, faster and more economical, without compromising on quality. The welding engineer in charge for the site welding, therefore, should study the details in the drawings well in advance of the actual work, and determine the appropriate strategy to be adopted in the site.

One aspect that needs special attention is the care with which the electrodes are to be stored at site in order to achieve quality weldments. Electrodes should be kept dry, particularly where bare wire process is used. Use of rusty electrodes can induce hydrogen cracking. Furthermore, electrodes should be kept free of oil, since hydrocarbons may also be a source of hydrogen. Stick electrodes have advantage in this respect, as these can be dried in an oven, and the flux covering also provides protection against rusting to a considerable extent.

All weld metal contracts on cooling. In a butt welded joint in a plate girder with asymmetrical flanges, the longitudinal shrinkage will be greater in the thicker flange, because of the larger cross section of the weld metal in this flange compared with the thinner one. This may lead to change in the profile of the girder. It may be necessary to arrange to have the components pre-set, in order to be able to compensate for the effects of shrinkage and distortion as the welded area cools. For plate girder splices, it is a common practice to weld the flanges first, keeping the gap between the webs sufficiently large, so that, after the flanges are welded, the web gap is reduced to acceptable dimensions [3].

Setting of site butt weld joints also needs careful inspection. The root gaps must not be too small, nor too large, since in both cases, the quality of the resultant weldment may suffer. The corrective measures for inaccurately fabricated butt weld joints are generally expensive. It is, therefore, imperative that this aspect should be considered during inspection at the fabrication shop itself. There should, thus, be proper interaction between the site welding engineer and the fabrication unit, in order to sort out such problems.

It is a common practice to weld temporary 'draw cleats' at the ends of plate girder pieces to be welded at site (Fig. 21.2). Bolts are used to draw the pieces together, to achieve appropriate fit-up and secure them in position

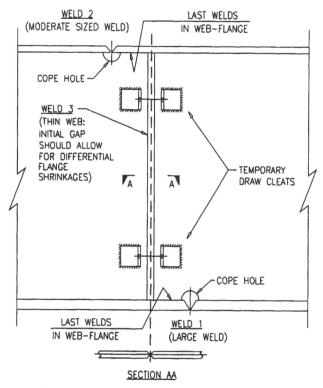

Fig. 21.2 Typical detail of plate girder splice for site welding [3]

during the welding process. After the welding has been completed, the cleats are removed by gas cutting, and dressed properly, avoiding damage to the main structure.

As in the case of shop welding, site welding also demands a high level of knowledge and expertise of the members of the participating team, viz., the welders, operators, supervisors and inspectors—perhaps, in a greater degree, since the work is carried out away from controlled environment of an organised workshop. The non-destructive inspection regimes, described in connection with quality control in welded fabrication (see Chapter 20) are equally applicable for site welded joints also.

One of the main risks that has to be countered in a site welded connection is the effect of moisture. It is necessary to protect all parts adjacent to the proposed welds for a reasonable time before the commencement of the welding and keep the area absolutely dry till the welding is completed. It has been observed that even if the wet area is dried by flame heating, moisture is held by capillary action in micro-fissures in the job, which is likely to cause hydrogen cracking. It is, therefore, prudent to play safe and cover the area well in advance to avoid the risk.

21.9 SITE BOLTED/RIVETED CONNECTIONS

Compared to site welded joints, bolted or riveted joints are easier to perform in the site, and need much less supervision and inspection. These are, therefore, mostly used in common bridges. The procedures of riveting and bolting have been discussed earlier in relevant chapters. Some of the other aspects which need particular attention during making site connections are discussed in this section.

All steelwork intended to be bolted or riveted together should be in contact over the whole surface. Before the parts are assembled, it should be ensured that all burrs are removed and no surface deffects exist. The contact surfaces are then applied with the specified primer coat of paint and assembled when the paint is still wet. The connection is secured together by service bolts or permanent bolts along with parallel barrel drifts (Fig. 21.3) . Each member must be accurately fitted and securely held in its correct position in the girder, before final bolt tightening or riveting is done. Achievement of this would entail careful selection of holes for drifting and careful application of forces, so as to obtain the accurate position of the members in the joint. Drifts are used for drawing light members into positions—but their use in heavy members should be restricted to only securing them in their correct position. Drifting should be done with great care so that the holes are not distorted or enlarged. To help overcome the

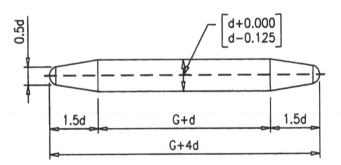

d = NOMINAL DIAMETER OF HOLE

G = GRIP = THE COMBINED THICKNESS THROUGH WHICH THE DRIFT HAS TO PASS

Fig. 21.3 Typical parallel barrel drift

closing forces when fairing joints, split drift device shown in Fig. 21.4 is often used. These drifts are very useful. For example, it has been observed in actual practice that drifts designed for 24 mm diameter holes can successfully close holes even when the out-of-fair dimension is as much as 5 mm [4].

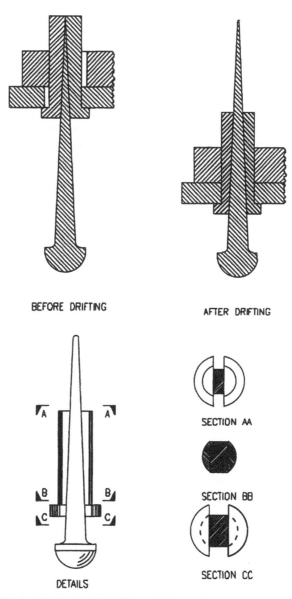

BEFORE DRIFTING   AFTER DRIFTING

SECTION AA

SECTION BB

SECTION CC

DETAILS

Fig. 21.4 Typical split drift for fairing holes in gussets [4]

In view of the strict quality control system normally enforced during the fabrication process of bridge girders in the workshop, it is unlikely that any serious mismatch in the holes should occur. Should error still persist, a moderate amount of reaming of the mismatched holes may have to be resorted to. This should be done in consultation with the inspection agency. However, the effect of reaming on the soundness of the joint must be examined.

In cases where joints have to withstand substantial erection forces, such as in the cantilever method of erection, provision should be made in the joints to withstand such forces prior to final bolting or riveting. This is normally done by inserting sufficient numbers of temporary parallel barrel drifts and turned and fitted bolts of the size of the holes to provide slip-free connections.Service bolts may be used in the balance holes to hold the joint together. The drifts, along with the turned and fitted bolts are replaced, one by one, by permanent bolts or rivets. The number and disposition of the drifts and bolts should be finalised for each joint beforehand.

21.10 BRACINGS

In general, the erected steelwork, particularly the partially erected steelwork, should be securely bolted or otherwise fastened and properly braced. This is necessary to take care of the stresses from the erection equipment, and the loads carried during erection, including the lateral forces and wind/seismic loads according to local conditions. There have been instances of bridge failures due to inadequate fastening or bracing system during the erection process (see Chapter 2).

21.11 TOLERANCES

The main areas where deviations from theoretical dimensions may affect the performance of a bridge structure are :

- Geometry of the structure in the vertical plane
- Geometry of the structure in the horizontal plane
- Matching of components at connections and splices
- Distortions and deformations during fabrication and erection

Codes of practice provide guidelines of tolerances upto which deviations in the various areas can be accepted. The best way of achieving the specified tolerances is to monitor the operations from the very beginning. It should start from the design stage, through the fabrication stage, and finally end with the erection stage. It is easier to do anything correctly in the first instance, than to rectify a wrongly executed work at a later stage.

21.12 ALIGNMENT AND LEVELLING

Aligning and levelling is the final activity in the erection process of all types of bridges. Jacking operation is an essential activity in almost all methods of erection for common types of bridges. This is done with the help of hydraulic jacks.

Jacks are portable tools for lifting heavy loads, and are very useful during the erection process. They are particularly used during the final operation of levelling and placing the erected bridge structure to its desired position. Traversing bases are often used under the jacks for moving the erected bridge in the horizontal direction. Synchronised hydraulic jacks, which are normally used for the final operation, require specialised skill and experience. When these jacks are used (in two or more numbers), care should be taken to ensure that each jack is taking its share of the load. Adequate number of suitable packs of varying thicknesses should be available during the jacking operation. Packings, piled in stacks, are put under the girder and their heights are made to closely follow the vertical movement of the jacks (up or down), so that, should the jacks fail, the load is allowed to be transferred to the packings without falling through any large distance. Normally, the gap between the top of the packings and the underside of the girder should be in the region of 20 mm to 25 mm. Each upward or downward movement of the jack should also be well within the lift specified for the particular jack being used. Furthermore, the expected 'give' in the packings when the load is transferred to them should be considered while selecting the overall height of the packings. For this operation, it is advisable for the crew to study and clearly understand the sequence of the activities involved. Perhaps, a practice operation would give an idea about the time which will be required to complete the job, as also the hazards which should be guarded against.

REFERENCES

[1] Arch, W.H. 1989, *Steelwork erection*. In: Civil Engineer's Reference Book, Blake, L.S. (ed.), Butterworth & Co (Publishers) Ltd., London, UK.

[2] Peshak, C. and Marshall, R.W. 1994, *Fabrication and erection*. In: Structural Steel Designer's Handbook, Brockenbrough, R.L. and Merritt, F.S. (eds.), McGraw-Hill Inc., New York, USA.

[3] Needham, F.H. 1983, *Site connections to BS 5400: Part 3*. In: The Structural Engineer, March, 1983, Institution of Structural Engineers, London, UK.

[4] Turley, S., Savarkar, S.G., Williams, J. and Tweed, R.J.C. 1960, *Design, fabrication and erection of Ganga bridge*. In: Proceedings of the Institution of Civil Engineers, March, 1960, London, UK.

[5] Barron, T. 1956, *Erection of Constructional Steelwork*, Iliffe & Sons Ltd., London, UK.

Chapter 22

Erection Equipment, Tools and Tackles

22.1 INTRODUCTION

Equipment commonly required in erection sites of steel bridges may be broadly classified into two categories:

- Equipment required for loading and unloading in the storage and sub-assembly yards. These can be stationary cranes (such as steel derrick, shear legs, guy derrick crane, scotch derrick crane), mobile cranes or Goliath crane.
- Equipment used in the erection front for erection of structural components. These are placed either on the ground from where they operate, or on the erected portion of the bridge itself and are moved forward as the erection progresses.

In this chapter the salient features of some of these equipment are discussed in general terms. Selection of the type of equipment for a particular site will, however, depend on a number of inter-related conditions, such as economics, capability and capacity of the equipment, layout, topography, and condition of the site, quantity and type of materials to be handled, time period allowed for completion of the project, etc.

22.2 STEEL DERRICK

Steel derrick is generally a guyed latticed steel pole with a pulley block hung from a point near to the top of the pole. The leading rope from this pulley is taken through a snatch block at the foot of the derrick to a winch placed some distance away. The stabilising guys are anchored at a distance in the ground by means of suitable anchorages.

A steel derrick is usually fabricated from angle sections or tubes laced together with flat, angle or tube sections. The length of the derrick may reach 25 m or more, made up of one tapered top piece and one tapered bottom piece along with a number of identical parallel intermediate pieces of suitable length for convenience of transportation. With this system of construction the derrick can be built up to any desired length.

While lifting a load, the axis of the derrick should not be moved more than a few degrees out of the vertical, as there is a danger of the derrick slipping out from the initial stable position. In order to safeguard from this movement, and also to resist the pull from the lead line, the foot of the derrick is to be securely anchored by wire ropes by means of a suitable anchorage system.

22.3 SHEAR LEGS

Shear legs offer a simple and inexpensive method of lifting heavy loads. This equipment is generally used for loading and unloading bulky materials. It is a modified version of the derrick, and consists of two poles both tied together at the top ends, with the bottoms opened out to approximately 20 degrees angle. The arrangement is held in an inclined plane by two guy ropes, one placed in the front and the other at the back. The lifting device is hung from the intersection of the poles. Chain block is a useful accessory to be used as the lifting device. Winch with a snatch block may also be used, particularly for lifting heavy loads.

22.4 GUY DERRICK CRANE

This is a simple and inexpensive form of static crane consisting of a latticed guyed mast fitted with a jib. The mast stands vertically, and is stabilised by five or more guys, and rests along with the jib on a foot block similar to that used in a steel derrick. The guys connected to the top of the mast are to be securely anchored in the ground. The mast is free to revolve with the bearings of the spreader plate at top and foot block at the bottom. The jib is made slightly shorter in length than the mast height, so that it can rotate through 360 degrees without fouling the guy ropes when raised in the near vertical position. Thus the crane allows both changing of radius (luffing) and slewing, which makes it a very versatile lifting equipment. The standard jib length is 25 m with a capacity of 5 t (tonnes).

22.5   SCOTCH DERRICK CRANE

Scotch derrick crane has been widely used in bridge sites, both in the stock-yard and in erection work. However, with continued development of Goliath and other types of cranes, it is gradually losing its advantage for heavy lifting over long and high reaches.

Essentially, this equipment consists of a slewing mast and a luffing jib. Stability is achieved by using fabricated latticed guy and stay members. Two guys are fixed at the top of the mast at 45 degrees angle with the horizontal. The lower ends of these guys are connected to the ends of the horizontal stays which are fixed to the base of the mast forming an angle of 90 degrees in plan. A horizontal latticed diagonal member is fixed between the ends of the guys and the horizontal stays, forming a triangle in plan, which completes the triangulation of the system in three planes between the mast, the guys and the stays. The jib attached to the base of the mast may be rotated through 270 degrees between the latticed guys and is capable of hoisting, slewing and luffing. Standard models range in capacity from 3 t to 50 t with jib lengths varying from 15 m to 35 m.

22.6   MOBILE CRANE

Mobile cranes are very useful equipment for bridge sites, both in the storage yards, as well as in the erection fronts for certain types of bridges. These cranes can be classified into the following three groups:

- Self-propelled cranes
- Lorry-mounted cranes
- Crawler-mounted cranes

22.6.1   Self-propelled Crane

This type of crane is generally of low lifting capacity (upto about 10 t), and is fitted with tyres. It is extremely mobile, but requires a hard level surface from which to operate. It is generally fitted with jack outriggers built into the chassis for stabilising it during lifting operation.

Normally, the small capacity crane of this variety has a fixed jib length. A crane with higher capacity can have sectional latticed jib which allows the jib to be extended, if necessary. Alternatively, it may be fitted with telescopic boom which is very convenient for working at different radii. A crane with telescopic boom can be prepared for use within a very short period after arrival at site and is consequently very popular with construction agencies.

22.6.2 Lorry-mounted Crane

This type of mobile crane is mounted on a specially designed lorry or truck. The lorry is driven on the road, from one site to another, from a conventional cabin, similar to that of a normal lorry. For operation of the crane engine and controls, a separate cabin is generally provided. A lorry-mounted crane has a capacity upto 20 t in the free standing position. The capacity can be increased by using jack outriggers built into the chassis. The crane has a folding lattice jib or a telescopic boom to enable it to operate at different working radii. Like self-propelled crane, lorry-mounted crane fitted with telescopic boom is very popular. As in the case of self-propelled crane fitted with tyres, lorry-mounted crane also needs level ground for operating efficiently.

22.6.3 Crawler-mounted Crane

A crawler-mounted mobile crane is very useful in a site with uneven terrain and poor ground conditions, where rubber tyred mobile crane cannot work easily and effectively. Since the weight of the crane is spread over a large bearing area under wide and long tracks, it is comparatively stable in soft and poor ground. The crane is powered by diesel engine and consists of three sections, viz., base, superstructure and jib. The base is mounted on crawler tracks and supports the superstructure carrying cabin, engine, winches, kentledge, jib and other mechanical parts. The superstructure revolves on a large turn-table mounted on the base frame to give 360 degrees slewing. The jib is generally of lattice construction with provision for additional construction and may have a fly jib to obtain different radii.

The main disadvantage of this type of crane is that it is very slow in travelling and, therefore, needs a special low loading truck to transport it from one site to another. It is thus not a practical option when the need is for only a short period of time.

22.7 LOCOMOTIVE CRANE

Locomotive crane is self-propelling, and travels on a standard gauge rail track and can be used in bridge sites. This type of crane requires only one operator and has the advantage of lifting capability for the entire circular area covered by the swing of the boom. Often, extensions are added to the top of the boom to give greater working radius with, of course, reduced lifting capacity. The weakest lifting position is at right angles to the track. For heavy side lifts, outrigger beams can be pulled from the sides of the carriage and packed up from the ground, thereby increasing the lifting capacity without overturning the crane. The lifting capacity may be upto about 50 t and the jib may be 20 m long or above.

22.8 GOLIATH CRANE

Goliath crane is very useful for stockyards of major bridge sites, where a large quantity of steelwork has to be handled.

Essentially, it is a rail mounted mobile crane consisting of a horizontal transverse girder which carries a trolley with hoist gears travelling across at the axis of the girder. The girder is supported on two rail mounted latticed 'A frames'capable of moving along the direction of the rails on powered bogies. The driver's cabin is normally situated in the level of the trolley which gives the driver an excellent all round aerial view of the site. The system allows the hook three-way movements, viz., vertical, longitudinal and transverse directions, making the Goliath crane very useful in a large stockyard.

22.9 ERECTION CRANE

The equipment to be used in the 'front' for erection of bridge components depends on various factors such as location of the bridge, its type and typical details, site conditions, the method of erection, etc. Thus, for a bridge on dry ground where height is reasonably within limit (e.g., fly-over, road over-bridge) a mobile crane of suitable capacity and jib length may be used. Where time is not the criterion, a steel derrick may also be used. For cantilever erection of an open web bridge across a waterway or gorge, a custom-made lightweight erection crane of steel or aluminium alloy may be used. In such a case, the erection crane is normally placed on the top chord level and it travels forward as it erects bridge components panel by panel. The travelling erection crane should have the provision of slewing the boom through 180 degrees so that it can pick up erectable steelwork from trolleys travelling on the bottom deck of the already erected back panel and bring it to the front by the slewing arrangement for erection in the forward panel. Scotch derrick cranes have often been used for erection purpose in many bridges in the past.

22.10 WINCH

Winch is a very useful equipment for any erection site. It is a mechanism which can be used along with wire ropes and sheaves for lifting or pulling heavy loads, often in conjunction with steel derrick or crane, which forms its hoisting mechanism. Winch may be hand-driven (i.e., manually operated) or power-driven.

22.10.1 Hand-driven Winch

Hand-driven winch is limited to short lifts, slow speeds and intermittent hoisting. It consists of a wire rope drum geared by one or more reductions to hand cranks and is usually provided with a band brake, and also a ratchet and pawl arrangement for holding the load and preventing the drum from running backwards under load. The drum and gears, along with the other components, are mounted on a suitable steel frame fixed on a base structure. Hand-driven winch is seldom used these days.

22.10.2 Power-driven Winch

Power-driven winch is a very useful tool and is available in a variety of designs and capacities. As it requires only one man to operate, it is normally economical with labour. Essentially, it consists of prime mover, drive, rope barrel, mechanical or power brake and friction clutch. A suitable bed plate serves to locate and hold the components which are mounted on it. The power source may be petrol, diesel, electric, steam or compressed air. Where electricity is not available, diesel engine is normally recommended as it is reasonably reliable and economical to operate.

Because of its basic simplicity, very often maintenance of the power-driven winch is neglected. It is, however, imperative that maintenace is carried out at regular intervals. Also, an experienced operator should be engaged to avoid accident.

22.11 ROPE

Rope is an essential item in any erection work. Rope can be either soft rope or wire rope.

22.11.1 Soft Rope

Soft rope is made by twisting together vegetable or other fibres. It is generally made from abaca, sisal or hemp and is known to the trade as manila, sisal or hemp rope. Rope made from synthetic fibres, such as nylon, is now in the market and is also used in some special cases. Although, wire rope has largely replaced soft rope for lifting and hauling work, soft rope is still in use for lighter work in bridge sites.

The capacity of soft rope is generally designated by the makers. However, it deteriorates if used improperly or not maintained adequately. Wet soft rope should be properly dried before storing. Also, it should not be left on the ground to dry on its own as this is likely to cause damage to the rope and consequently reduce its capacity.

22.11.2   Wire Rope

Wire rope is built up from strands of wires laid together generally around a fibre core saturated with a lubricant. Wire rope is usually designated by the number of strands and wires in each strand in the manner '6×9 construction', '6×37 construction' etc. The first numeral denotes the number of strands in the rope and the second numeral is the number of wires in a strand. Thus, the term '6×19 construction' denotes that the wire rope is made up of 6 strands and that the strands are made up of 19 wires.

There are a variety of different possible constructions of wire ropes available in the market. Selection of the most suitable rope depends on many factors. A flexible rope is comparatively easy to work with. This characteristic depends on the number of wires used in a particular construction. The more the number of wires, the more flexible will be the resultant rope. The breaking load depends on the quality of steel from which the wires have been drawn. Apart from the breaking load, selection of wire rope is influenced by two other criteria, viz., resistance to bend fatigue and resistance to abrasion. Smaller the size of the wires in a a rope, greater will be its resistance to bend fatigue. On the other hand, larger the size of the wires in a rope, greater will be its resistance to abrasion. Thus, a suitable balance between these two factors is required to be worked out for selection of the right wire size. In bridge erection work '6×37 construction' wire rope is mostly used. The large number of small sized wires makes it comparatively flexible and resistant to bend fatigue and at the same time resistant to wearing.

Wear of Wire Rope

Repeated bending and straightening operation of a wire rope as it passes over sheaves of pulley blocks or winch drums is one of the main causes of its wear. When a rope is bent, the wires farthest from the centre are subjected to an additional tensile force due to the load. Similarly, the wires nearest to the centre of the rope are compressed, thereby reducing the tension due to the load. Even when there is no load in the rope, the wires farthest from the centre are subjected to tensile force and the wires nearest to the centre are subjected to compressive force. This continual change in the pattern of stresses in the extreme wires in a rope, produces fatigue condition which reduces its strength considerably. Thus, with smaller diameter pulley sheave, the rope will be bent more while passing over it and consequently greater will be the effect of fatigue on the rope. This effect is magnified when a rope passes over a series of pulleys which bend it first in one way, and then the other way. In order to minimise this effect, some national standards recommend the minimum size of pulleys and drums to be used with each size of wire rope.

Wire rope may also be damaged due to incorrect shape of the groove of the pulley sheave. If the groove is too wide for a particular rope in use, the rope tends to flatten as it passes over the sheave. Conversely, if the groove is too narrow, the rope is gripped too tightly, and is consequently damaged. This aspect should be checked while selecting pully blocks. Also, with continuous use, the bottom of the sheave also gets worn out. Pulleys should, therefore, be inspected regularly for signs of wear.

Wire rope is generally used outdoors in an erection site and consequently is liable to damage due to atmospheric corrosion. Frequent inspection followed by cleaning and oiling is essential to counter this problem.

Quite often wire ropes get damaged during storage or when handling. They should be stored in dry sheds and kept in spools placed on supports above the ground to avoid damage from dampness. The storage area should be free from fumes, chemicals, and local heat. Wire ropes should be handled properly during uncoiling operation. Instead of pulling the rope, the spool itself should be rotated on a small turntable; otherwise the rope may suffer kinks, thereby seriously affecting its durability. When a turntable is not available, an alternative method would be to roll the coil on the ground with its axis in the horizontal position so that the rope rests in a straight line on the ground as the coil moves forward.

22.12   WIRE ROPE SLING

In bridge erection sites, wire rope sling is widely used in preference to chain sling for moving fabricated steel components. It is lighter, and comparatively more flexible, and consequently, is easier to handle. The rope used for a sling should be of similar standard as that for crane work. A two-legged sling is very handy for slinging steelwork components which are required to be placed horizontally.

The lifting capacity of a two-legged sling varies with the included angle between the legs; it decreases very rapidly as the angle increases. This point must be kept in mind while using a two-legged sling, and the included angle kept to the minimum as far as possible.

22.13   JACK

Jack is a portable tool used for lifting heavy loads through short distances. It is a very useful and essential tool for any bridge site. There are three types of jacks in the common use : screw jack, ratchet jack and hydraulic jack. Owing to the large reduction in the gear system, the lifting speed of all the varieties of jacks is rather slow. The capacity range is limited by the weight of the jack, which must necessarily be portable.

While using a jack for erection purpose, particular attention should be paid to ensure that the base area is large enough to avoid accidental overturning, or crushing of the concrete of the pier or abutment underneath. Also, care should be taken to insert thin pieces of timber packings between the top of the jack and underside of the steel material to be lifted to avoid the risk of slipping between the two metal surfaces.

22.13.1   Screw Jack

Screw jack is a simple tool used for all classes of light work. It consists of a ram with heavy screw which is threaded into a 'nut' built in the top portion of the jack base. The ram is turned by a bar passing through holes in its side and is lifted by the screwing action with the 'nut' of the jack base. The ram head is fitted with a ball and socket arrangement at the top bearing plate so that the ram can be turned without disturbing the bearing plate. Screw jacks are available in standard sizes ranging from 5 t to about 30 t capacity.

22.13.2   Ratchet Jack

Like screw jack, ratchet jack (or rack and lever jack) is a simple tool for use in light work. It consists of a cast-steel or malleable-iron base and post, to which is pivoted a lever and pawl. A rack-toothed bar passes through the hollow post and is raised by ratchet arrangement actuated by the lever.

22.13.3   Hydraulic Jack

Hydraulic jack is generally used for heavy lifts and is very commonly used in bridge erection sites. It is manufactured in a great variety of forms according to service required. The fluid used in the jack is generally oil. The jack is available either, with the pump unit attached with the jack unit, or separate.

In the latter case the two units are connected by flexible piping. This arrangement makes the jack very adaptable in confined spaces. For jacking operation in bridges, two separate jacking units (for two girders) with a common pump unit are often used to ensure synchronised (equal) lifts in the two girders. The pump is generally operated by a lever.

The jack should be fitted with a screw collar as a safety precaution against failure; alternatively, some form of packing should be used instead.

Hydraulic jacks upto 50 t capacity are readily available. Bridge jacks with lifting capacity of as high as 600 t or even more can also be built in special cases.

22.13.4 Traversing Jack

Traversing jack is a very convenient tool for aligning a bridge structure, where, apart from lifting the load vertically, horizontal shifting (traversing) is also necessary. It is essentially a normal jack mounted on a special carriage and base, which permits the load to be lifted and moved some distance horizontally before being lowered.

22.14 CHAIN PULLEY BLOCK

There are many varieties of chain pulley blocks. The most commonly used one is of worm gear type. It is light in weight and easy to operate. It consists of a hanging hook on the casing to hang from a fixed position such as top of shear legs. It has a worm over which an endless operating chain is placed and a worm wheel on which a hoisting chain with a hook for slinging a load is put. The gearing provides easy operation, but slow movement of load. Chain pulley block can also be conveniently used for pulling two components together during erection work.

22.15 PULLEY BLOCK

Pulley block is generally used by the erector where the capacity of hauling or lifting is required to be increased. This tackle is available as one-, two-, three- or four-sheave block. Generally the block is used in pairs, such as, 'three- and two-block', 'four- and three-block', etc. The single sheave is used for light hauling or lifting purposes.

A pulley block may be used either for soft rope (such as manila or hemp rope) or for steel wire rope. The shape in the groove in the sheave depends on the manner in which the particular type of rope strains under the load. Thus, the shape of the groove for a soft rope is different from that for a wire rope. The erector would do well to remember that a block intended for a soft rope should never be used for wire rope, lest this might flatten and damage the rope.

22.16 SNATCH BLOCK

Snatch block is a single sheave block which is used for altering the direction of the lead from the winch. The most common use of a snatch block is at the foot of a derrick for transferring the horizontal lead from the winch to a vertical direction towards the block fixed at the head of the derrick. A typical snatch block has a hinged arrangement on one side so that the rope can be easily lifted out or into position of the sheave. The hinged portion can be closed and locked by a pin which prevents accidental release.

22.17 UNION SCREW

Union screw is used to pull two components closer, or to strain wires. Basically, it consists of three parts, viz., two threaded steel links connected to a central threaded bar. The central bar can be rotated by inserting a suitable toggle into a central hole. The two halves of the central bar have opposite-handed threads so that, as the bar is rotated, the two steel links are moved wider apart or closer together, as required.

22.18 IMPACT WRENCH

Where a large number of bolts are involved, tightening of these bolts by manual effort may prove to be a long drawn out process. Use of impact wrench in such a case will be very helpful as it will cut down the time of operation drastically. Nowadays, impact wrench has become an essential tool for use in any site. Many types of impact wrenches are available in the market. Essentially, this wrench has a special die fitted to the shaft of the machine which grips the nut. When the control is operated, the die runs the nut down until it becomes tight. The action of the machine can be reversed so that the nut can also be loosened and removed.

22.19 DRILLING EQUIPMENT

Modifications to fabricated steelwork is often required to be carried out at site. Therefore, a portable drilling equipment should always be available in any erection site. In larger sites, miniature workshops are generally set up with a special bench fitted with a permanent drilling stand.

Drilling equipment may be of different types, viz., hand drilling gear, electric or compressed air driven units, etc. Hand drilling gear with a ratchet brace is usually very useful when the work is awkwardly placed. However, hand drilling is a slow process. For normal drilling operation, power-operated machines should be arranged, particularly when a large number of holes are required to be drilled. Power drills are available in various models for both electricity or compressed air. Selection of the type of a drilling equipment depends largely on the type of power available, as also on the amount and type of drilling anticipated in a contract.

22.20 GRINDING MACHINE

Grinding machine is primarily used for cleaning welding work or edges of plates cut by an oxy-acetylene torch. An abrasive wheel of 150 mm or 200 mm diameter is generally fitted at the tip of the spindle. In some cases the wheel may be fitted with wire brushes; when the spindle is revolved, the wires fly out due to centrifugal force and form a very hard brushing surface. Nylon wheels are also used for special grinding requirement. Grinding machine is available with straight handle or 'pistol grip' handle. The former is commonly used in bridge work. To avoid injury to eyes from flying pieces of metal, the operator should wear goggles while using the grinding machine.

22.21 MISCELLANEOUS SMALL TOOLS AND TACKLES

Drift: Drift is used in riveted or bolted multi-plate connections. When connections are formed, service bolts are inserted first, followed by drifts to hold the plates in line. Service bolts are then tightened. Drifts are generally made from high tensile steel. Figure 21.3 shows details of a typical parallel barrel drift commonly used in bridge work.

Shackle: Shackle is very useful for connecting slings to steelwork components. Two types of shackles are generally used, viz., Dee shackle and Bow shackle. Shackle pin used for bridge work is generally screwed at one end and has an eye on the other end, so that it can be tightened or slackened with a toggle spanner.

Bulldog grip: Bulldog grip is used to connect two wire ropes alongside each other. Generally, a minimum of three numbers of bulldog grips are used in any connection. This connection is quite reliable and has proved to be very useful in bridge sites.

C-clamp: A typical C-clamp is normally used to hold two pieces of steelwork together when no holes are available for bolting, such as to hold components together preparatory to welding.

Spanner and wrench: Spanners and wrenches are indispensable tools in any erection site. These are available in a number of varieties. The most commonly used types are straight and cranked podger spanners. A set of double ended spanners of different sizes should prove to be useful in bridge sites. Also, Stillson wrenches of different sizes should be available for tightening hand railings, holding down bolts, etc.

Hammer: Hammers generally used in erection sites are 1 kg ball pein hammer, 3 kg to 4 kg flogging hammer, and 7 kg to 8 kg sledge hammer.

Pinch bar: Pinch bar is essentially a steel bar of about 40 mm diameter and 1.25 m length which is tapered in one side (handle side) while the other end is made heavier and shaped into a heel and a toe. The pinch bar is used as a lever for lifting heavy loads with the heel acting as a fulcrum about which the toe and handle can be moved to obtain mechanical advantage.

22.22 BOGIE

A two wheeled bogie consists of a small platform of about 1,000 mm × 300 mm, set over a short axle and fitted with two rubber wheels of about 250 mm diameter. It is a very useful item in any erection site for transporting steelwork.

A four-wheeled bogie is a similarly constructed bogie consisting of a platform over two pairs of steel wheels placed at a distance corresponding to the gauge of a railway track. The wheels are shaped to the profile of railway wheels and can travel on pre-laid rail tracks. It is very useful for transporting materials by means of manual effort or, often by a power-driven vehicle.

REFERENCES

[1] Harris, F. 1989, *Modern Construction Equipment and Methods,* Longman Scientific & Technical, Harlow, UK.

[2] Barron, T. 1956, *Erection of Constructional Steelwork,* Iliffe & Sons Ltd., London, UK.

[3] Chudley, R. 1987, *Construction Technology, Vol. 4,* Longman Scientific & Technical, Harlow, UK.

Erection Methods

23.1 INTRODUCTION

A variety of methods can be used for erection of steel bridges, and the choice is influenced by a number of factors. Some of the important ones are:

- Site conditions
- Structural system of the bridge
- Weight of erectable pieces
- Crane/equipment available
- Quality of skilled workmen available
- Requirement of traffic flow underneath the proposed bridge
- Safety requirements

Salient features of some of the methods used for common types of bridges are described in this chapter.

23.2 ERECTION WITHOUT TEMPORARY SUPPORT

Simply supported plate girder bridges are often erected in single length pieces directly on abutments or piers. Commonly, simple hoisting equipment, like crawler crane, or truck-mounted crane is used for the purpose. A typical example of this method is illustrated in Fig. 23.1. This is a simple method and can be used routinely by an experienced erector without any appreciable problem. Perhaps, the only problem that an erector is likely to come across is the lateral instability of the top flange during lifting or before fixing the permanent bracings of the plate girder. If required, temporary support for the compression flange should be provided. Another alternative would be to assemble the two adjacent plate girders on the ground along with their lateral and cross bracings and erect the assembly in one piece.

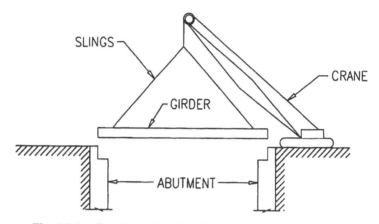

Fig. 23.1 Erection of girder without temporary support

This is possible only where the weight of the assembly is within the capacity of the crane available at the work site.

Figure 23.2 illustrates a simple lifting method for small span bridges, where crane is not available [3]. Individual trusses are assembled on the ground, and lifted by tackles hung from special frames placed on pier tops. The frames consist of light members which are pulled up and assembled manually on the pier top. Two sets are used, one at the top of each pier. The lifting tackles are hung from small trolleys, which can move transversely, towards the centre of the piers for placing the truss on the bearings, after it has been lifted to pier cap level. While one truss is being lifted, the other truss acts as a counterweight on the other side of the pier.

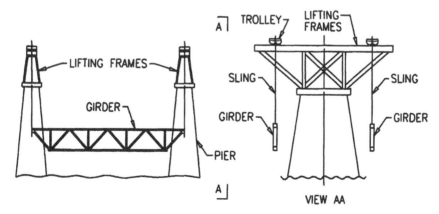

Fig. 23.2 Erection of girder by lifting frame

23.3 ERECTION ON FLASEWORK

Provided the site conditions allow, erection of spans on falsework is a very common method. In this method, temporary supports, typically steel trestles are put up at regular intervals, and the girders are assembled on these supports with the help of suitable hoisting equipment. The method of erection of a simple through truss bridge is described below, and illustrated in Fig. 23.3.

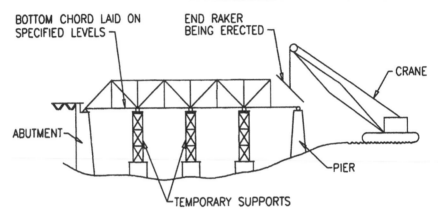

Fig. 23.3 Erection of truss bridge on temporary supports

First, suitable foundations are made at locations of node points. Steel trestles are then erected on these foundations to required heights so as to attain the levels defined in the camber diagram. Normally camber jacks are used for fine tuning these specified levels. It is essential that the camber is accurately maintained throughout the process of erection. Constant checks are, therefore, to be made on the levels at these points to ensure that no sinking has occurred. Whenever trestles become considerably tall, their stability for vertical as well as horizontal loads due to wind should be checked, and bracings provided, if necessary. The bottom chords along with the bottom lateral bracings are then placed on the camber jacks, carefully levelled and checked for straightness. Next the web members (verticals and diagonals) and the the deck members (cross girders and stringers) are erected, followed by top chords and top lateral bracings. The joints are made progressively. It is important to check that the span has been correctly aligned and all the joints have been properly formed by riveting or bolting before the trestles are removed and the loads are transferred to the end bearings.

The main disadvantage of this method is that it is labour intensive and rather slow.

23.4 LAUNCHING

23.4.1 General Procedure

Where erection by falsework is not practicable due to presence of adverse river bed conditions or high piers, launching method can be used to erect both plate girder and truss bridges. Typically, the span to be launched is erected on the approach embankment and then pulled or pushed forward across the gap. The launching process primarily comprises of the following operations:

- Pulling or pushing the span
- Restraining
- Rolling mechanism
- Controlling the operations

Pulling or Pushing the Girder

Pulling arrangement is a primary feature in all launching methods. Typically, the arrangement consists of a winch and wire rope with snatch block arrangement through pulley system for reducing the operating force as also chances of sudden jerk. The winch is normally placed on the forward pier or abutment and pulls the span by the wire rope connected to the tip of the girder.

Launching is also done by pushing the span from behind by applying jacking system. The mechanism consists of specially designed hydraulic jacks, which are placed horizontally between a permanent structure (fixed to abutment or pier) and a suitable cross member of the structure to be moved. As the pressure in the jack is increased, the ram pushes the steel girder forward to a specified distance. The process is repeated by re-adjusting the gears [4].

Restraining

It is a common practice to restrain the span and release it gradually to avoid any sudden forward movement. This can be conveniently achieved by fixing a pulley block to the rear of the span and running a wire rope to a winch securely tied back to an anchorage. Care should be taken to see that there is no loose winding of the restraining rope on the drum of the winch, as this may result in slips and jerks when being pulled out under the full load.

Rolling Mechanism

A traditional rolling mechanism is shown in Fig. 23.4. The arrangement employs short pieces of rails chamfered at each end on which roller bars carry the bottom flange of the moving girder. As the girder moves forward, rollers which come clear from the forward end are picked up by workmen and fed in again behind. An alternative gear used for rolling is shown in Fig. 23.5. Rollers with suitable bearings are housed in frames fixed on a flat base, so that the rollers remain in their own locations instead of moving forward while the girder is moving.

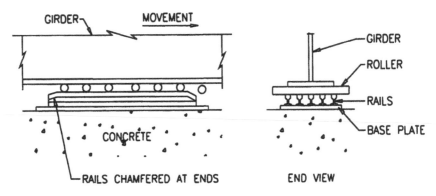

Fig. 23.4 Traditional rolling mechanism

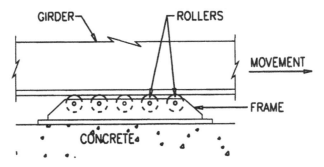

Fig. 23.5 Alternative rolling mechanism

Modern practice is to dispense with the rollers and use sliding devices instead. A sliding device essentially consists of a polished or chromed stainless steel sheet sliding on a low-friction material such as PTFE (poly-tetra fluoro ethylene) which has an extremely low coefficient of friction. During launching, the PTFE panels are inserted between the contact surfaces of the polished stainless steel sheet and the underside of the bridge girder. Elastomeric bearings are often used additionally to improve the distribution of reaction loads along the length of the sliding device [4].

Rolling of truss bridges needs certain special considerations. The members of the truss are primarily designed for axial forces and are somewhat weak in flexure between two nodal supports. Thus, when the structure is launched, the support reaction is moved along the lower chord of the truss, resulting in a high concentrated force acting outside the node, and thereby causing bending in the member. The flexural strength of the bottom chord should be checked for this condition. To avoid this situation, temporary beams with sufficient flexural strength are often clamped under the bottom chords, with arrangement to transfer the reaction to the truss system through the node points only. Often, stringers are used for this purpose. Figure 23.6 illustrates the arrangement.

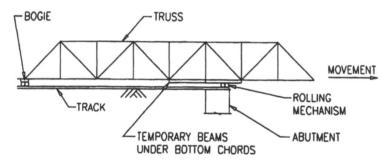

Fig. 23.6   Launching of truss bridge with temporary beams under bottom chords

Bending moment on the bottom chords can also be avoided if rollers are fixed under the bottom chords *at the node points*, and the truss structure is supported on these rollers and moved on suitable steel beams placed on the abutments and piers. Figure 23.7 illustrates the arrangement. A typical rolling mechanism used for the purpose is also shown in Fig. 23.8.

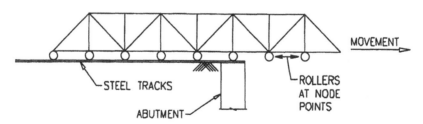

Fig. 23.7   Launching of truss bridge with rollers fixed at bottom chord node points

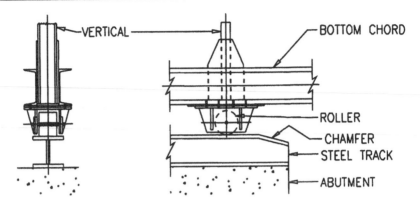

Fig. 23.8 Typical rolling mechanism for fixing at bottom chord node points

Controlling the operations

Launching of bridge spans is always fraught with elements of risks and need careful control during the process. The method should best be carried out with experienced and well trained workmen only. The factors which need particular attention and control during launching are:

(a) Deviation of pier tops: Piers may be subjected to significant horizontal forces during launching, resulting in appreciable deviations at the tops, particularly in case of tall and slender piers. To contol this effect, these piers are often stabilised at the top by steel cables anchored to the piers/abutments. This aspect needs to be considered at the design stage itself, while finalising the erection scheme.

(b) Deviation of superstructure in plan: Any slight error in the alignment of the bridge when it is assembled on the approach embankment may cause this deviation. Also, as the bridge is being moved forward, it should be ensured that both the sides of the bridge are advancing at the same speed. Errors tend to increase during movement unless corrected immediately, and become more difficult to rectify as the launching operation advances.

(c) Wind velocity: The launching scheme should clearly specify the maximum wind velocity in which the operation could be carried out. Constant watch on the wind speed is, therefore, essential. In case the actual wind velocity exceeds the specified limit, the operation must be stopped forthwith, and all structural elements secured appropriately.

(d) Cantilever-end deflection: This is an important aspect in any launching operation. Normally, the anticipated deflection at the tip is calculated in advance, and catered for in the design so that the tip reaches the forward pier

at a level higher than the top of the pier. The operation at site demands careful surveying and continuous check of the deflection.

23.4.2 Launching Methods

The various methods commonly used in launching bridge spans are briefly outlined in the following paragraphs.

Launching with derrick

The most common launching method, illustrated in Fig. 23.9, is the derrick method. This is mostly used to bridge small gaps. A small derrick (or 'A-frame') is first erected on the forward pier. The span to be launched is erected on the approach embankment and is then pulled forward across the gap by means of wire rope tackle from the top of the derrick and lowered in position on bearings.

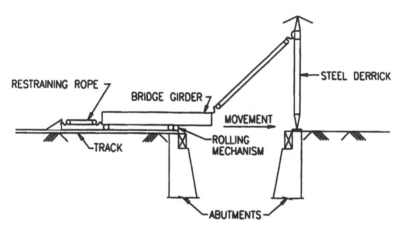

Fig. 23.9 Launching with derrick

Launching with moving support

When situations allow, launching can be done by using a moving support for the leading end. A typical arrangement is shown in Fig. 23.10. The truss bridge is first erected on the approach embankment with the rear end supported by a pair of bogies placed on rail tracks laid on the embankment, and the forward penultimate node resting on trestles, erected adjoining the abutment. These trestles are mounted on bogies running on temporary tracks laid on the ground in the same alignment as the bridge. The bogies are hauled along this track with the aid of winches, until the forward end of the span

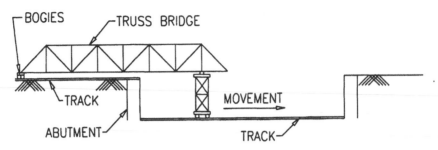

Fig. 23.10 Launching with moving support

reaches the abutment. The span is then lowered on bearings by jacking down.

Launching on falsework

In situations where erection of falsework is practicable, launching a bridge girder on falsework can be resorted to. In this method, a temporary launching platform consisting of steel beam deck system with intermediate trestles is erected to support the bridge span during launching. A rail track, on which bogies can move, is laid on the embankment and is extended over the temporary deck upto the next abutment. The bridge steelwork is assembled on the approach embankment and placed on four bogies in the locations of the bearings. The entire assembly is then rolled on the launching platform rails by means of cables and pulleys connected to a winch, until the forward node reaches the next abutment. The bridge is then lowered on to bearings. A typical arrangement is shown in Fig. 23.11.

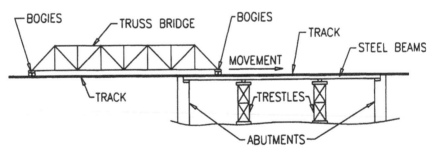

Fig. 23.11 Launching on falsework

End launching

In this method, usually used for bridging larger gaps, a temporary light steel or aluminium extension piece or nose is attached at the forward end of the assembled span, so that the entire assembly, resting on a pair of bogies at the rear and rolling mechanism on the abutment, can be hauled across from one end to reach the other end, without overturning or deflecting excessively. A kentledge is often placed at the rear of the span to prevent overturning. Figure 23.12 illustrates the arrangement. The nose is connected to the span by temporary connections (usually high strength bolts) and is designed to withstand the forces due to the reaction from the forward roller support on which it rests upon reaching the other end. The temporary connection of the nose with the span also needs to be adequately strong to withstand the forces arising at different stages of the launching process. The success of this method depends largely on the proper design of the rolling mechanism, and the pulling arrangement.

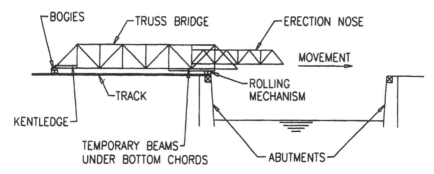

Fig. 23.12   End launching with erection nose

Figure 23.13 illustrates an end launching method in which a bridge span is fixed at the rear end, instead of fixing a nose at the forward end of the forward span. The span at the rear acts as a kentledge and prevents over-turning. The two spans are connected by providing suitable link-pin ar-

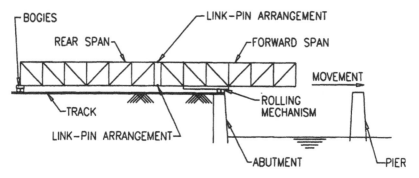

Fig. 23.13   End launching with a span fixed at the rear

rangements between the spans at the top and the bottom levels. This method is very convenient when multiple-spans are required to be erected, in which case, one of the job spans can be used as kentledge. Figure 23.14 shows how this method was used in launching a four span (4 × 32 m) truss bridge. The first two spans were linked with special pinned connections at the top and bottom chords. The subsequent spans were connected by pin connections at the ends of the bottom chords only. Rolling mechanisms were located at the first abutment and the piers. Since the bottom chords were not strong enough to withstand the bending due to vertical reaction when the spans were moving over the piers, beams were temporarily clamped to the underside of the chords to transfer the rolling loads to the panel points [3].

Often, a combination of the above systems are utilised, such as, fixing a girder at the rear, and also attaching a nose in the front. Figure 23.15 illustrates such a system which was used in launching a road bridge in India in 2002.

Where sufficiently long embankment behind the abutment is not available, bridge girders are oftened launched by adopting a principle commonly known as *incremental launching method*. In this method, the front part of the girder with a nose attached is assembled on the embankment to the maximum length possible, and moved forward to the extent it is safe against overturning. Further length is then assembled at the rear and launching is proceeded again. The process is repeated until the nose reaches the forward support. In some cases, it may be expedient to use an intermediate support by way of a temporary trestle. The intermediate support helps in reducing the overturning moment of the cantilevered portion of the span, and also the deflection at the tip. The top of the trestle should have suitable rolling arrangement, so that the span can move easily past this point. Figure 23.16 illustrates the method.

Figure 23.17 illustrates a simple method of end launching which was used for a multiple plate girder railway bridge [3]. Two spans were assembled and linked together on the approach abutment over bogies on tracks and launched out. When the forward end of the leading girder reached the first pier top, it was supported on a special supporting frame placed on the pier top, and then disconnected from the rear (second) span, and jacked down to its final position on bearings. The third girder was then moved forward and linked to the rear of the second girder. These two girders were then moved forward after laying necessary track on the erected first girder and adjusting the positions of the launching bogies. Launching was carried out until the forward end of the second girder reached the second

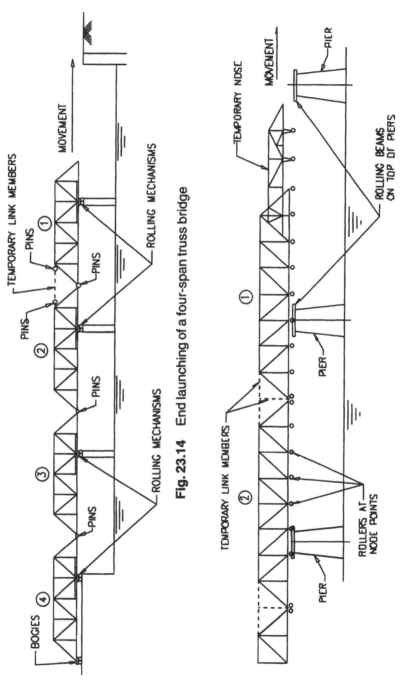

Fig. 23.14 End launching of a four-span truss bridge

Fig. 23.15 End launching with a combination of nose in front and a span fixed at the rear

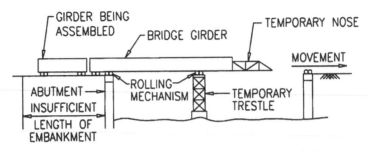

Fig. 23.16 Incremental launching of bridge girder

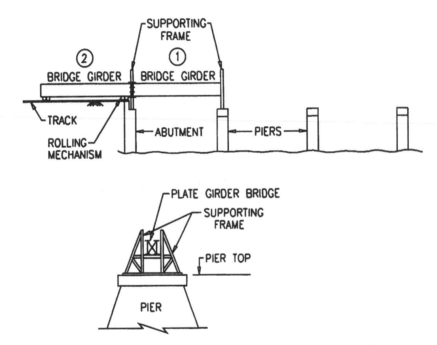

Fig. 23.17 End launching of a multiple plate girder bridge

pier top. The process was repeated for all subsequent plate girder units, except the last one. The final span in a bridge such as this, has to be launched by other method such as by using derrick, described earlier.

Erection by using launching girder

This method is specially useful for long multispan bridges where a large number of simply supported span girders predominate. A launching girder is a specially designed truss structure, which is assembled on the embankment and launched by rolling across the gap with adequate kentledge at the rear, to reach the forward support without overturning. It incorporates a movable trolley support system from which the permanent girders are hung and moved across to bridge the gap and then lowered on the bearings. Figure 23.18 shows a commonly adopted procedure. In the next stage, the launching girder is rolled forward along tracks laid on the permanent girders to reach the next pier for erection of the next girder set. Design of the launching girder should take care of the various stages of its operation, viz., first as a cantilever, supporting its own weight during its launching, and then as a simply supported span to carry the moving permanent girders.

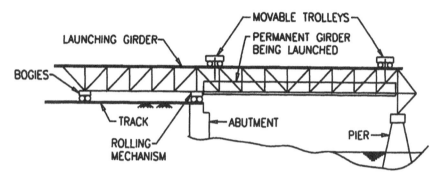

Fig. 23.18 Erection by using launching girder

23.5 CANTILEVER METHOD

23.5.1 General Procedure

Cantilever method is a most useful method of erection for almost all types truss bridges. It has the very great advantage that the erection work can be carried out despite seasonal river conditions, heights, and other site conditions, where support from underneath by falsework is not practicable. The method provides a fast track solution where a number of similar spans are involved, since the repetitive nature of the erection work enables cantilevering time to be reduced as more and more spans are erected. One other advantage of this method is that it can be used in all seasons and the pace of erection is not affected during rainy season. Also, compared to launching method, this method involves less hazardous site activities.

The method envisages erection of one span by traditional method on temporary falsework and then, by utilising this span as anchor span (counterbalancing weight), the second span is cantilevered. Normally a light-weight erection crane with slewing arrangement is placed at a suitable forward node point of the top chord of the anchor span. From this position, members of the forward span are erected in sequence, so as to first form a triangulated configuration, cantilevering out from the anchor span. This triangulated configuration is connected to the anchor span by temporary erection members. These comprise a link-pin arrangement at the top chord level and a buffer system at the bottom chord level. The former transmits the tensile forces developed at the top level of the cantilevered second span to the anchor span. Similarly, the compressive forces at the bottom level are transmitted to the anchor span through the buffer slab/block system located at the bottom chord level. Figure 23.19 illustrates the basic principle. To ensure pure hinge connections, the tension link is connected by a large diameter pin designed to resist the entire force by shear, and the compression connection is made through a rocker face, like that in a bearing. This allows the gradual change in direction with progress in cantilever to be accommodated smoothly, without causing any secondary stresses in the joints.

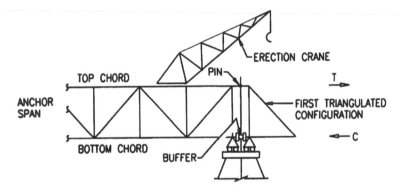

Fig. 23.19   Cantilever method of erection

After the connection between the anchor span and the triangulated configuration of the cantilevered span is made, erection of the rest of the members of the first panel is completed. The crane is then moved forward to the first panel point of the next span to erect the elements of the second panel. Following this procedure, erection of the second span is continued until the next pier is reached. The temporary connecting members are then removed. The process is repeated for erection of the subsequent spans.

23.5.2   Link and Buffer Arrangements

Link and buffer arrangements between the anchor span and the cantilevered span vary, depending on the type of the span being erected. The following paragraphs briefly describe the systems in respect of some common types of spans:

Trusses with end rakers

The arrangement is illustrated in Fig. 23.20. The mechanism of transmitting the forces consists of:

Tension at the top

- Two sets of top extension pieces, one for the anchor span, and the other for the cantilever span are attached to the extreme top nodes of the spans by means of turned and fitted bolts, using existing open holes of the joints. These extension pieces have holes for link pins for connection with the link members.
- Two sets of link members (consisting of two pieces each, connected by a central pin) are connected to the extension pieces attached to the two spans by means of two sets of pins.
- A set of two temporary props support the links until the cantilevered triangular configuration is completed and the link and buffer arrangements are operative.

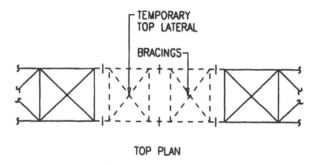

TOP PLAN

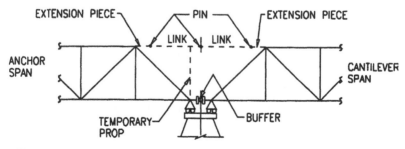

Fig. 23.20   Link and buffer arrangement for trusses with end rakers

- Top lateral bracings between the link members are provided to transmit the lateral forces on the cantileverd span to the anchor span.

Compression at the bottom

- Buffer arrangement comprises two sets of buffer slabs fixed at the ends of the bottom chords of each span with a set of buffer blocks with rocker faces placed in between. Alternatively, pin connections may also be adopted.

Trusses with end verticals

In this case the arrangement as shown in Fig. 23.21 comprises:

Tension at the top: Two sets of top link pieces, one for the anchor span, and the other for the cantilever span are attached to the ends of the top chords of the two spans by means of turned and fitted bolts, using the existing open holes of the joints. These two sets of link pieces are connected by two pins. Where, due to lack of space between the two spans, two sets of link pieces cannot be used, the tension is transmitted through 'single piece' link members fixed between the top ends of the two spans by means of bolts.

Compression at the bottom: A similar arrangement as in the case of spans with end rakers may be used.

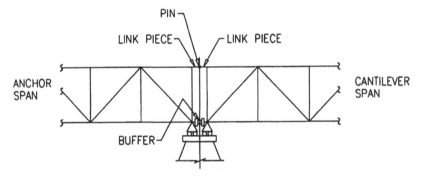

Fig. 23.21 Link and buffer arrangement for trusses with end verticals

Underslung trusses

In this case, the arrangement is shown in Fig. 23.22.

Tension at the top: A similar arrangement as in the case of trusses with end verticals may be used.

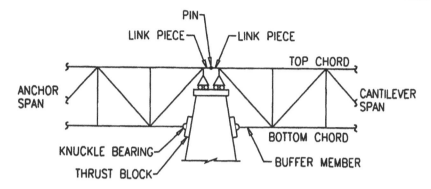

Fig. 23.22 Link and buffer arrangement for underslung trusses

Compression at the bottom

- Two sets of horizontal buffer members are connected to the last and the first nodes of the bottom chords of the anchor and cantilever spans respectively.
- Two sets of thrust grillage blocks are fixed on the sides of the pier to transmit the compression forces in the bottom chord level of the forward span to the bottom chord level of the anchor span through the concrete pier.
- Two sets of knuckle bearings are placed between the buffer members and the thrust blocks to transmit the forces from the former to the latter as well as to allow any rotational movement of the buffer member *vis-a-vis* the thrust blocks.

23.5.3   Analysis and Design

In cantilever method of erection, it is quite common to have reversal of stresses in a number of members in the frame during erection conditions. Therefore, the structure needs to be analysed for erection loads at each stage and erection forces in every member ascertained. The erection loads should include not only the weights of the members in the structure in the particular erection stage, but also loads from erection cranes, tools, tackles, etc. Wind load should also be added to these loads.

A member which has been designed as a compression member for service condition, but subjected to tensile force during erection, is required to be checked for its adequacy of net sectional area. If necessary, additional temporary material has to be fixed to make up the deficiency. Similarly, when a member, designed for tension in the service condition is subjected to substantial compressive force during erection, it may be necessary to provide additional bracings to the member to reduce its effective length with a view to increasing its capacity in compression. Alternatively, temporary steel material may have to be added. Sometimes, it becomes expedient to make the member heavier, and avoid the high costs for fixing and removing the additional temporary steelwork.

Also, the maximum deflection at the tip of the span being cantilevered needs to be calculated in advance. If this is too large for the span to reach on the top of the forward pier, the length of the link is shortened in advance to offset this effect, thereby enabling the span to be at a suitable level in the final stage of erection.

For longer spans, special receiving brackets cantilevering from the forward pier are often erected in advance. The brackets are utilised generally to minimise the erection stresses by shortening the effective length of the cantilever span, and particularly for resisting the effects of high wind of the cantilevered girders.

23.6 ERECTION BY SUSPENDED CABLES

This method is normally employed for erection of bridges across deep gorges or in locations which call for suspension bridge to be used as the structural system. In case of suspension bridges, the permanent cable system is utilised for erection of the deck system, while in case of other forms of bridges, a temporary cable system is slung over two temporary towers and anchored at ends. The girder pieces are then erected by hanging from the cables, and connected together to form the girder system. The girders are placed simultaneously from both ends, such that the cable profile remains mostly symmetrical. The size and weight of individual erectable units are usually kept within managable limits. Erection is done with the help of equipment mounted either on the cable itself, or on the girder being progressively erected. The girder pieces are joined with loose connections until the last unit is in position, as significant displacements of the main cable and of the towers take place while loading is progressively increased. Figure 23.23 shows a typical arrangement of this type of construction method.

The cable ends are usually anchored by adopting one of the following types of anchorages:

- Rock anchorage
- Gravity anchorage

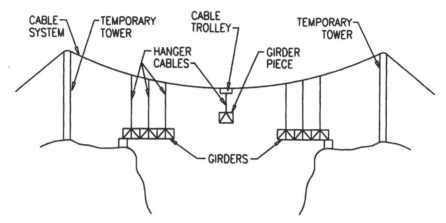

Fig. 23.23   Erection by suspended cables

Rock anchorage is done by drilling into the rock and inserting large bolt type anchorages which are grouted. The cables are attached to these grouted bolt type anchorages. In gravity anchorage system, the anchor bolts are located in concrete anchorage block to which splayed strands of the cable are attached. The cable forces are resisted by a combination of overburden, deadweight of the concrete block, and bearing friction.

23.7   FLOATATION METHOD

In the past, many bridges have been erected by using this method. The method is suitable for bridges where the water is reasonably calm and has sufficient depth throughout the span. The method also requires a suitable area close to the site for locating a yard where assembly work can be handled with reasonable ease. In this method, the girders are first assembled in the yard, and then transferred on to trestles mounted on two flat barges. The barges are towed to the site and so positioned that the span is over its final location. The span is then lowered on to its seatings by using jacks. Alternatively, water-tight ballast tanks of the barges may be slowly filled with water to lower the span. The rise and fall of the river water level due to high and low tide conditions respectively, can also be utilised for lifting, positioning, and lowering the span.

The method requires meticulous planning, and supervision, particularly during floating out from the assembly area and placing the span on supports at the desired location. The stability of the completed assembly at different stages of operation needs careful study. Needless to add, the effects of tides and variation in the levels of the river water need special attention while formulating the sequence of operations. Figure 23.24 illustrates a typical example of erection of a simple truss bridge by using this method.

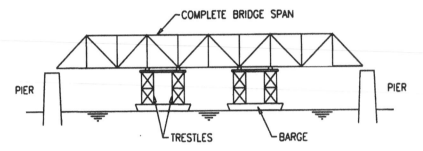

Fig. 23.24 Erection of truss bridge by floatation method

23.8 COMBINATION OF DIFFERENT METHODS

For some bridges, site conditions may be such that no single method of erection is suitable. In such cases, a combination of two or more methods may have to be adopted. Figure 23.25 shows an example of erection of a truss bridge partly on falsework and partly by cantilever method. In this case, since the entire deadload on the cantilevered portion of the bridge structure

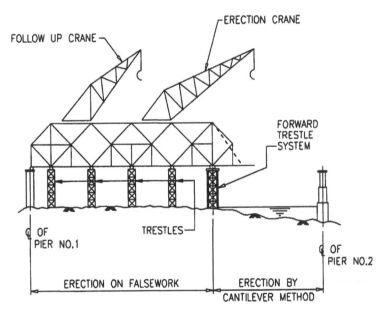

Fig. 23.25 Erection of truss partly on falsework and partly by cantilever method

is supported at the forward trestle system, it is necessary to design this trestle system and its foundation to cater for this load.

23.9   ERECTION OF ARCH BRIDGES

The most common method for erection of steel deck type arch bridges is to use falsework to support the steel rib segments. Where it is not possible to use falsework underneath the structure, the arch can be erected by means of tieback cables.

For tied arch bridge, the ties along with the deck can be erected on temporary falsework, while the arch ribs, bracings and hangers can be erected from the deck, with the falsework underneath still in place. Alternatively, all the components may be erected by means of tieback cables, without using any falsework. Where conditions permit, floatation method can be used for erection of steel arch bridges, following the sequences already discussed earlier in this chapter. Arch ribs can also be erected by first erecting half of the span on each abutment in the vertical position (like a column), and then rotating the pieces down to meet at the centre between the abutments. Figure 23.26 illustrates some of the above schemes.

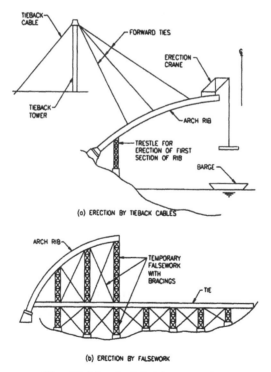

Fig. 23.26   Arch bridge erection

23.10 ERECTION OF CABLE SUPPORTED BRIDGES

For suspension bridges, the towers are erected first, followed by the main cables. Once the main cables are erected, the deck and stiffening truss sections can be erected with hanger supports from the main cables.

Until 1960s, suspension cables used to be layed by the aerial spinning method (AS method). This method of spinning the parallel wire cables was invented by John Roebling and was used for the first time in Niagara Falls bridge in 1855. In this method, an endless hauling rope is first erected, and a set of grooved spinning wheels is hung from it. Wires are supplied in reels at one end of the bridge and the ends of the wires are looped over the wheels and anchored back to the abutment. These loops are carried forward over the span on the spinning wheels (hung from the hauling rope), until they arrive at the anchorage on the other side. There, the loops are detached from the spinning wheels and placed around semicircular strand shoes connected to the anchorage. The spinning wheels return empty to the originating anchorage to pick up the second set of loops. The operation is repeated as the wheels shuttle back and forth across the span, till the spinning is completed. A system of counterweights is used to keep the wires in constant tension as they are spun. When the spinning is complete, the wires are bound together with clamps into a single cable. Figure 23.27 illustrates the spinning system.

Around 1960s, a method of erection of prefabricated parallel wire cables was developed (PS method). This method was first used in the New Port

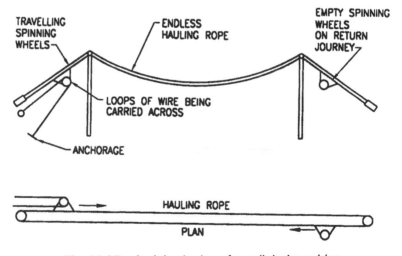

Fig. 23.27 Aerial spinning of parallel wire cables

bridge in USA. Nowadays, prestretched structural locked coil strands and ropes are also used for cable supported bridges. These systems have reduced the site work significantly.

Erection of the deck and stiffening trusses of suspension bridges is carried out in one of the two following sequences:

- Start erection from the centre of the main span
- Start erection from the towers

In either case, erection should be carried out systematically, and in sequence, in order to keep the deformation of the main cables uniform and symmetrical, as far as possible. The deck section joints are normally left unconnected until the erection of the stiffening truss is completed. Also, it is necessary to calculate beforehand the cable tension, shape and deformation of the tower during different stages of erection of the stiffening truss, to ensure that the deformations are within the permissible limits.

Cable stayed bridges are normally erected by cantilevering into the main span from the abutments. As an example, a typical three span cable stayed bridge is considered (Fig. 23.28). The side span girders are first erected on falsework (Fig. 23.28a). The first girder in the central span is then cantilevered out from the tower pier, and connected to cable No. 3 (Fig. 23.28b) emanating from the tower. The corresponding cable (No. 2) is also connected simultaneously to the corresponding girder in the side span. Cable No. 1 is then attached to the first girder in the side span (Fig. 23.28c). The second girder in the central span is then cantilevered out and connected to cable No. 4. This procedure is repeated on the other side of the crossing. If site conditions permit, the entire cable stayed bridge can also be erected on falsework.

23.11 CONCLUDING REMARKS

In the limited space available, it is not possible to go into the finer details of, the various methods of erection coverd in the preceding sections. Attempt has been made to focus only on the distinctive features of these methods, which can be utilised for selecting and developing erection schemes for common types of bridges. Nevertheless, the engineer has to use his ingenuity, weigh considerations of economy and speed, and of course, of safety, to arrive at an acceptable solution. A good engineer should, in any case, never take a known risk, and should always endeavour to follow codal provisions. However, one must accept that, despite taking all precautions, failures, nevertheless, still occur—often, contributed by human errors. Therefore,

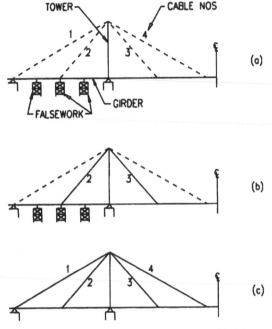

Fig. 23.28 Erection of cable stayed bridge

experience and skill of the staff and workmen must be given due importance as they play an important role in the success of any bridge erection project.

REFERENCES

[1] Harris, F. 1989, *Modern Construction Equipment and Methods*, Longman Scientific and Technical, Harlow, UK.

[2] Barron, T. 1956, *Erection of Constructional Steelwork*, Iliffe & Sons Ltd., London, UK.

[3] Williams, J. 1964, *Steel bridge girder erection*. In: The Annual, 1963-64, The Institution of Engineers (India), Bengal Centre, Culcutta, India.

[4] Zhuravov, L.N., Chemerinsky, O.I. and Seliverstov, V.A. 1996, *Launching steel bridges in Russia*. In: Structural Engineering International, Journal of the International Association for Bridge and Structural Engineering, Zurich, Switzerland, August 1996.

[5] Ghosh, U.K. and Ghoshal, A. 2001, *Erection of open web steel bridges by cantilever method*. In: The Bridge & Structural Engineer, Journal of the Indian National Group of the International Association for Bridge and Structural Engineering, New Delhi, India, February-March, 2001.

[6] Francis, A.J. 1989, *Introducing Structures*, Ellis Horwood Ltd., Chichesten, UK.

[7] Podolny, W. (Jr.). 1994, *Cable-suspended bridges*. In: Structural Steel Designer's Handbook, Brockenbrough, R.L. and Merritt, F.S. (eds.) McGraw-Hill Inc, New York, USA.

Chapter 24

Safety during Erection

24.1 INTRODUCTION

Safety during erection work can be broadly described as the freedom from the risk of injury caused by accidents. Accidents primarily cause terrible human tragedies, leading to physical injuries and sometimes to death. They occur mostly, but not confined only, to employees on the site. For example, local public may get affected by accidents. Similarly, design engineers, on visit to site for inspection may also be exposed to the hazards of erection work.

While human tragedies form the primary outcome of accidents, they also affect the project indirectly in many respects. Some of these effects include:

- Loss of rhythm in the working environment, resulting in reduced work rate until normal morale is restored
- Occasional disruption in work while investigations are carried out by different agencies
- Damage to already erected portion of work
- Damage to plant and equipment
- Loss of time while debris is cleared and damaged work and plant/equipment are repaired or replaced
- Legal and allied expenses

The above effects may involve substantial economic impact, which has to be balanced against the costs for introducing safety norms in the work site. Studies reveal that the introduction of safety norms is always well worth the investment, since the risk and cost of serious accidents are very high. It is, therefore, an economic necessity also, to ensure accident-free operation during erection.

24.2 COMMON ACCIDENTS

The basic factor that influences safety in any activity is the ability to foresee the hazards and take pre-emptive action in order to eliminate them. Inability to do so, is the primary cause of most accidents. Some of the common accidents in a bridge site are:

- Fall of people from one level to another, or people falling at the same level
- Stepping on, or hitting against stationary or moving objects
- Striking against sharp, rough and projecting objects
- Hit by flying, falling, sliding, or moving objects, including a collapsing structure or part thereof
- Caught in, or between objects, while loading or unloading material
- Fires or explosions
- Caught in a moving machinery, contact with electric current, etc.

24.3 PREVENTION OF ACCIDENTS

Safe systems of operation and safe working places are the two most important factors which largely contribute to the prevention of accidents in any bridge construction site.

Many authorities, such as Railways and Public Works Departments, have their own safety rules and regulations. Most of these rules, however, are meant for their departmental use for normal day-to-day working. Also, most countries have statutory requirements applicable for construction work in general. These may not address the specific requirements of bridge erection work. Some such requirements are therefore discussed in the following paragraphs.

24.4 SAFE SYSTEMS OF OPERATION

It is the duty of the supervisor to ensure that the workers are not only skilled and physically fit, but are also trained in safe methods in using tools and tackles for the particular job they are assigned to perform. Training should enable them to distinguish a safe practice from an unsafe one. This safety-consciousness would urge them to use safety accessories such as helmets, safety goggles, safety boots, etc., as a matter of habit. The authorites should, on their part, ensure that the safety accessories supplied to the workmen are of good quality and conform to the required standard. In this section, some of the safe systems of work are briefly described.

24.4.1 Safety Belts

Workmen required to work at a height, where provision of safe working platforms or suspended cages (discussed later in this chapter) are impracticable, should wear safety belts attached to a reliable structure above the working position.

24.4.2 Temporary Construction Screens

In cases where roadway or railway movements are allowed during construction work, the necessity of protecting the traffic becomes important. Temporary construction screens are to be erected to isolate the construction works from the areas being used for vehicular or pedestrian movements. Whenever overhead works are carried out, movable overhead screens (e.g., wire netting) should be erected to catch falling items such as tools, loose bolts/nuts, debris, etc., falling from a height. Pedestrian enclosure with overhead cover gives better protection to pedestrians than that given by simple barricade without overhead cover. Red indication lamps should be provided so as to be visible at night.

24.4.3 Warning Signs and Lighting Arrangement

Easily visible warning signs indicating hazards (e.g., 'Men at work', etc.) should be displayed at all strategic locations. Proper lighting should be provided for good visibility for the movement of men and vehicles, and also for the work.

24.4.4 Electrical Connections

Electrical cables for the construction works should be identified and segregated so as to avoid accidental contact during construction work. It is imperative that the supply is cut off wherever work has to be done across a cable line. Proper earthing of all electrically-operated equipment and hand tools must be ensured. Also, all electrical connections should be periodically checked by a competent electrical supervisor. Whenever electrical faults are noticed, they must be immediately attended to.

24.4.5 Temporary Track and Roadway

All temporary ramps or steelwork supporting track and roadway for carrying materials, equipment, etc., should be properly designed, and constructed under a competent supervisor. These should be inspected at regular intervals for safety.

24.4.6 Transportation of Material

Accidents do occur while transporting material and equipment from one place to another, and also during loading and unloading. Steel members should be so placed on the carrier that they will not shift, slide, or overturn during transit. Also, the loading should be so arranged that there will be room for placing the unloading slings safely, and that the materials would not trap any workman as a part of the load is removed at the time of unloading. The unloading equipment should have the reach and capacity to handle the loads being lifted off the carrier, and moved to the point where the material is to be unloaded.

24.4.7 Containers for Small Materials

Workmen using small material such as bolts, drift pins, etc., should be provided with appropriate containers with handles of sufficient strength and attachment to carry the loaded containers. This arrangement would reduce the danger of such small materials rolling or being kicked, and falling on persons working below, passing pedestrians or vehicles. It also helps a man to avoid stepping inadvertently on such a material and losing his balance and resulting in an accident.

24.4.8 Prevention of Slipping

Workmen should be advised to clean their shoes of mud, grease, oil, snow, ice or other slippery material before going on to a steel or planked floor, or climbing a ladder, so as to avoid slipping. They should also not drag slippery materials on the floor or ladders, causing others to slip and get injured.

24.4.9 Safety against Fire

In order to guard against spreading of accidental fire, sufficient number of fire extinguishers should be kept ready at all strategic locations, such as cranes, compressors, etc. Extreme care should be taken in storing inflammable materials like petrol, kerosene and gases. It is a good practice to have the telephone number of the nearest fire station displayed at prominent places so that in case of emergency, the fire brigade can be contacted without any delay.

24.4.10 Welding and Gas Cutting

Essentially, there is hardly any great danger associated with either electric arc welding or gas cutting operations, provided appropriate precautions are taken when the equipment is used. In any welding operation, the work should be earthed properly. Unless proper precautions are taken, welding work should not be undertaken while it is snowing or raining, since there is a chance of the electric current travelling back through the operator and causing shock. The cables from the welding equipment should be placed in such a way that they are not run over by traffic.

Helmets, shields, and goggles fitted with suitable filter lenses should always be used by the welder to protect his eyes from harmful effects of the rays emitted during welding process. People working nearby should also take precautionary measures to guard against the rays.

Special care should be taken to provide suitable blanketing screens to check spreading of sparks while undertaking welding operations near combustible materials. Welding should best be avoided where inflammable liquids and gases are stored.

Risk of fire should always be considered when the oxy-acetylene cutting torch is used. As a precaution, fire extinguishers should be kept nearby. When gas cylinders are in use, the valve key should always be placed in position. Valves should be closed when the torches are not in use. Care should be taken to store the gas cylinders away from any source of heat.

24.4.11 Riveting

Riveting operation is a fairly safe site activity, and only a few sensible precautions are necessary to avoid any serious accident.

Care should be taken while handling rivets, so that they do not fall, strike or cause any injury to people working in the vicinity or below. Snaps and plungers of pneumatic riveting hammers should be secured to prevent the snap from dropping out of place. The nozzle of the hammer should be inspected frequently, and the wire attachment renewed when worn out.

The rivet heating equipment (oven) should be as near as possible to the place of work. Water should be on hand for quenching the fire during riveting operation and to prevent fire when working near inflammable materials.

24.4.12 Lifting Gear

Hooks, rings, and shackles, along with chains and wire slings are normally covered by the term lifting gear. Accidents associated with lifting gear may occur either due to defective components or their unsafe and negligent use.

It is very important that all components of the lifting gear are tested for their safe working load and marked before being allowed to be used in the work. The safe working load should never be exceeded.

Wire ropes should be uncoiled properly, and lubricated regularly. The wires at the open ends should be tied together.

Wire ropes with broken strands should not be used in erection work. Apart from reducing the lifting capacity of the rope, broken strands are likely to cause injury to the fingers of workmen. Kinks in hoisting ropes should also be avoided.

Knotted chains or wires should not be used as the practice causes severe strain on the material. Chains should not be joined by bolting or wiring. A chain in which the links are locked, stretched, or do not move freely, should not be used. The chain should be free of kinks and twists. Proper eye splices should be used to attach the chain hooks. Shackles should be fitted with proper pins.

It should be ensured that appropriate size of pulley-blocks are used so as to allow free play of the wire rope in the sheave grooves and to protect it from sharp bends under load.

While using multi-legged slings, each sling or leg should be loaded evenly. They should be of sufficient length to avoid a wide angle between the legs. Also, they should not slide while erecting diagonal members. Slings should be protected around sharp corners by providing suitable packing pieces. Pipes cut longitudinally into two pieces are often used for this purpose. Slings for erecting the steelwork should be light and flexible enough to grip the pieces, and at the same time, strong enough to carry the load.

Inadequate or improperly installed wire rope clamps may cause accidents. They should be checked immediately after the rope has been stressed, as the rope, specially if new, tends to stretch under the applied load, causing it to shrink in diameter. The clamps must then be tightened again to take care of the new condition and prevent the rope from slipping. Frequent inspection of clamps is necessary to avoid accident due to such slipping.

As a general rule, all lifting tackle components should be periodically inspected for their condition and replaced as necessary.

24.4.13   Cranes

Studies reveal that most of the crane accidents occur due to operational errors, rather than inadequate design or inferior quality in the manufacture of cranes. It is, therefore, necessary that the crane operators and the workmen engaged in slinging should be trained in the safe use of cranes.

The main risks associated with operations of cranes and other lifting appliances in bridge erection sites are:

- Overturning or structural failure of the crane
- Dropping of the suspended load
- Entrapment of people
- Contacting electric current
- Incorrect erection and dismantling procedure
- Improper signalling

Implication of these risks and their remedial measures are briefly discussed in the following paragraphs.

Overturning or structural failure of the crane

Overturning or structural failure in a crane is generally caused by overloading. It is, therefore, necessary that the safe working load corresponding to its associated radius is never exceeded. The weight of the load to be lifted must be determined beforehand, and the corresponding radius for safe working load ascertained, and then the erection planned accordingly.

Cranes should always be on level and firm ground. Differential settlement of crane supports (e.g., for mobile cranes) may cause overturning. The ground condition should, therefore, be examined and improved, if necessary, before locating cranes for lifting the load. Where outriggers are provided, these must always be fully extended, and jacked down to suitable plates and packings in order to distribute the load.

The path for carrying the load throughout any lift should be carefully examined beforehand to avoid any obstruction, or proximity to excavation. Also, the route should be on reasonably level ground with adequate bearing capacity.

Using a crane to drag a load may cause considerable increase in the effective load due to the frictional force incurred. This may result in overturning also. This practice must be discouraged.

Erection cranes placed on the bridge must be secured properly with the bridge structure at all stopping points. The tip of the jib should be tied to the bridge structure at the end of work everyday.

Dropping of the suspended load

To avoid this situation it is essential that the load is slung correctly, and the safety catch of the hook is set properly. The weight and the centre of gravity of the load should be ascertained to obtain the correct position for lifting. Slinging should be so done that the hoist rope is vertically above the centre of gravity of the load. This would arrest tilting or swinging of the load, which can cause unexpected and dangerous shifting or sliding of the material and can trap men. Workmen should keep clear of any load being lifted, and should keep out from under a load that has been picked. The load should be carefully and smoothly manoeuvred during its travel. Normally, a tail rope is attached to the load and held by a workman from a distance, to control its swinging tendency while on the move.

Wire ropes used in the crane should be maintained properly. Ropes with broken or corroded wires should be discarded and replaced to avoid accidents.

Entrapment of people

Generally, the slingers are the likely victims of such hazards, when they become trapped between the hook and the sling as the crane takes the load, causing injuries to their fingers, hands, or feet. The crane driver must wait till the slinger moves away from the load and gives appropriate signal for hoisting. Similar risk is present at the time of grounding the load, and the slinger must be very careful during this operation also.

The other hazard that is encountered during operations of cranes is the entrapment of people between the moving suspended load or the crane superstructure and an adjacent structure, plant, or stored material. Care must be taken to avoid such accidents.

Contacting electric current

Movement of mobile cranes should be planned in such a way that the boom of the crane does not come in contact or closely approach a live overhead power line, to avoid the danger of electrocution of the crew stationed on the crane, or the workman steadying the load by means of a tail rope.

Incorrect erection and dismantling procedure

There are many instances of accidents occurring during erection and dismantling of cranes, such as workmen being struck by falling tools or crane parts, trapped between crane components, or falling from a height. To avoid such accidents, the erection and dismantling of cranes should be done only by experienced workmen under competent supervisory staff, and strictly in accordance with the procedure recommended in the manufacturer's manual.

Improper signalling

A proper signalling system should be developed for the use of workmen. Hand signals should be adopted instead of oral signals. Only one man, specially assigned to this task, should give the signals.

24.5 SAFE WORKING PLACES

In this section safety aspects of some of the common working places where the worker is required to work during erection of a bridge are briefly discussed.

24.5.1 Scaffolding and Gangways

Temporary scaffolding and gangways serve as the most common way of providing a platform for carrying out work at a height during erection of a bridge. The main hazards related to these structures are:

- Workmen falling from a platform
- Material falling from a platform and hitting people under it
- Collapse of scaffold, causing workmen to fall from the platform, or hitting people below, or damaging the platform steelwork

It is, therefore, necessary to give adequate attention to the design and construction of these temporary structures and employ skilled workmen to carry out the work under a competent supervisor.

Materials used for these temporary structures must conform to approved standards. When completed, these structures must be inspected and approved before these are used for the first time. Periodic inspection should also be carried out to ensure the safety of the structure throughout the erection process. Where timber pieces are used for flooring, they should be in good condition, without serious splits, cracks, excessive knots or other hazardous defects, and rigidly fixed at both ends. They should preferably be unpainted, as paint can hide hazardous conditions.

Access to gangways is provided by suitable ladders or stairways. Handrailings of appropriate height as per standards is necessary on the gangways and stairs. Also, toe boards must be fitted to the former. Ladders should be properly stayed and tied in order to prevent undue sway during use, particularly during high winds. For tall ladders, provision for intermediate landing places should be made. Ladders and stairways should be regularly inspected to identify defects. For important bridges, permanent stairs/ladders and platforms are often erected along with the main steelwork.

Temporary platforms or suspended cages for bolting, riveting or welding should have adequate space so that workmen can manoeuvre comfortably with their tools on these platforms.

24.5.2 Falsework

Falsework is a temporary structure for supporting permanent members of a bridge during its erection. After the bridge is completed and is capable to span the gap by its own strength, the falsework is removed. Pre-fabricated steel trestles are commonly used as falsework in the erection of steel bridges. Falsework should be designed in accordance with recognised design standards, and clearly shown on the erection drawings to reflect the designer's intent. Construction at site should also strictly follow the drawings so as to avoid any mishap.

Failure of falsework may be due to one or more of the following causes, which should be guarded against:

Inaccurate assessment of loading

Apart from the self-weight of the falsework, the design should take into account all imposed loads, such as:

- Self-weight of the permanent steelwork to be supported
- Dead loads from plant/equipment, stored materials
- Impact while placing permanent fabricated components by crane on to the supporting structure
- Live load of workmen
- Wind load and seismic load on falsework and the completed part of the bridge

Inadequate foundations

For small bridges it is common to found falsework structures on railway timber structures, or steel grillage placed on the ground. It is essential to ensure that the underlying ground is adequately consolidated, and that any voids are filled with plain cement concrete, as necessary. Where the underlying ground is not sufficiently consolidated, concrete foundation should be constructed at the specified depth and area. Care should be taken to avoid differential settlement of the foundations.

Sub-standard quality of material and workmanship

Since falsework is a temporary structure, there is a tendency to play down on the quality of the material used and the workmanship in construction. Also, existing trestles, which have been used previously many times over on work of similar nature, are often utilised. Rough handling, poor maintenance (resulting in corrosion and other damages), abuse by overloading, etc., may have reduced the capacity of these existing trestles. Such falsework should be inspected before use in work, and rejected, if found unsuitable. This excercise would reduce the risk of collapse considerably.

Lack of stability

Lack of lateral stability is one of the main causes of collapse of falsework structure during erection. The trestles supporting the permanent steelwork should be provided with adequate lateral bracing system to resist the lateral forces coming on to the erected structure. It may also be necessary to have the falsework tied to completed parts of the permanent structure, and/or firmly anchored to suitable foundation on the ground.

Instability due to uneven track may occur in the launching methods of erection. The designer must be very particular about this aspect. As for example, in case of launching with moving support (Chapter 23), where the falsework is moved on bogies, the designer must take into account the additional forces that may be imposed on the structure, particularly in overcoming uneven track or ground conditions, which may subject the structure to torsion or bending.

Buckling

It is important that the webs of the steel beams are checked for their ability to resist buckling at all load transfer points. It is a good practice to provide web stiffeners at all such points to avoid the possibility of collapse.

Traffic collision

In busy areas, particularly in the urban environment, vehicular traffic is often allowed to pass through adjoining falsework structures. In such cases, suitable fendering system must be provided to prevent vehicles colliding with the falsework.

Inadequate construction

The completed falsework structure should be checked prior to loading by a competent supervisor to ensure that it has been constructed strictly in accordance with the drawings and specifications provided by the design office. Also, periodic checking must be done to ensure that no part has suffered any distress. Appropriate action must be taken at the first sign of any distress.

Overloading

While lowering permanent steelwork on the falsework structure, care should be taken to ensure that no appreciable impact load is imposed on the latter and thereby avoid overloading it. Also, only those plant and equipment which are specified in the erection drawing should be allowed on the falsework structure.

24.6 BRIDGE STRUCTURE

Collapse of the bridge structure during the process of erection is nearly always sudden and spectacular. Sadly, it offers little opportunity for the escape of persons working in the proximity. Chapter 2 describes case studies of a few bridge failures during erection. The primary cause of most failures during erection may be attributed to failure to recognise the types and quantities of the temporary loads which an incomplete structure has to resist, or the structure's inherent lack of stability until it is completely erected. It is, therefore, necessary for the engineer to eliminate the possible causes of failure, or at least reduce their effects to safe proportions. A few points deserve mention in this context:

24.6.1 Erection Scheme

The erection scheme is generally prepared in the design office. The importance of safety in the erection condition should, therefore, be fully explained to the designers involved in this task. The erection scheme should take into account all the forces which have to be resisted by the individual members and joints at every stage of the erection process. The scheme drawing should be clear and complete in every respect. It should *inter alia* indicate the sequence of all operations, as also provision of temporary bracings at every stage, to make the erected portion of the structure safe and stable throughout the erection process.

The erection engineer at site should critically examine the erection scheme for safety and satisfy himself that the scheme is workable and safe, or else seek modifications in the scheme. The work at site should also be checked at every stage as it progresses, so as to ensure that the agreed erection scheme is followed scrupulously.

24.6.2 Making Joints

Special care should be taken to ensure that all the open holes in the joints are made up with parallel barrel drifts and temporary bolts, and fully tightened up as soon as the joints are assembled. The number and disposition of the drifts and bolts should be finalised for each joint beforehand in consultation with the designer. The drifts and temporary bolts are to be replaced, *one by one only*, by permanent bolts or rivets.

24.6.3 Lifting Operations

The erector should never indulge in dangerous and awkward lifting operations.

24.7 CONCLUDING REMARKS

The importance of safety during erection at site cannot be overemphasised. It is both a human and an economic necessity, as has been discusssed earlier in this chapter. Since erection of steel bridge structure is one of the most hazardous construction operations, the fundamental requirements for prevention of accidents should be recognised and made effective. Everyone involved in erection work should be made to understand that safety drive should necessarily be a co-operative rather than a coercive operation, in the interest of safety of men, equipment and the structure as a whole, and that this must be the topmost priority in any bridge erection work.

REFERENCES

[1] Davies, V.J. and Thomas, K. 1996, *Construction Safety Handbook*, Thomas Telford Publishing, London, UK.

[2] Short, W.D. 1962, *Accidents on construction work with special reference to failures during erection or demolition*. In: The Structural Engineer, The Institution of Structural Engineers, London, UK, February, 1962.

[3] Barron, T. 1956, *Erection of Constructional Steelwork*, Iliffe & Sons Ltd., London, UK.

[4] IS : 7205 - 1974, *Safety Code for Erection of Structural Steelwork*, Bureau of Indian Standards, New Delhi, India.

Chapter **25**

Durability

25.1 INTRODUCTION

The technology of modern steel making was developed in the second half of the 19th century. Since then, vast improvements in the production technology have taken place. New generation of steel such as weldable, corrosion resistant and self healing (weathering) steels with high ultimate strength are available now. This aspect of steel, with its added advantage of having almost identical strengths in tension, compression and shear, as also of improved maintenance systems have made steel one of the most versatile and durable construction materials for bridges of today.

Useful life of a steel bridge is generally considered to be between 100 to 120 years. However, in order to achieve this level of durability, it is necessary that the bridge structure is designed, constructed and maintained properly. These aspects which are relevant for attaining reasonable durabilty of steel bridge structures are discussed in general terms in the following paragraphs.

25.2 CRITERIA FOR DURABILITY

Factors which impair durability may be classified into two broad categories. The first category comprises those factors which are natural in character. These include atmospheric corrosion, earthquake, flood, etc. The second category includes those which are basically man-made. Stress corrosion, fatigue, material characteristic of steel, accident, pollution, etc., come under this category (Fig. 25.1).

While it may not be possible to control all the factors that may have unfavourable influence on durability, such as service loadings, traffic volumes, budget limitations or geographic changes, it should certainly be possible to regulate the design, construction and maintenance procedures, so as to achieve maximum service life of a bridge structure.

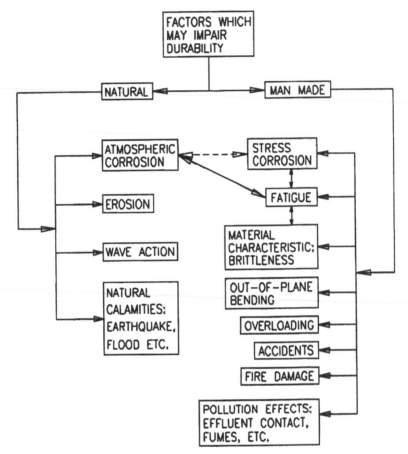

Fig. 25.1 Factors which may impair durability

In this context, it should be particularly noted that design stage has the maximum opportunity to influence quality or durability and cost of any project . This opportunity reduces significantly in the succeeding stages, viz., construction stage and service stage. Figure 25.2 illustrates this aspect. Thus, for maximising the durability of a structure, effort should particularly be made during design and construction stages.

25.3 DESIGN STAGE

Quality of design and detailing of a steel bridge depends largely on whether the factors which may impair its durability have been considered or not. Some of these factors need particular attention of the designer.

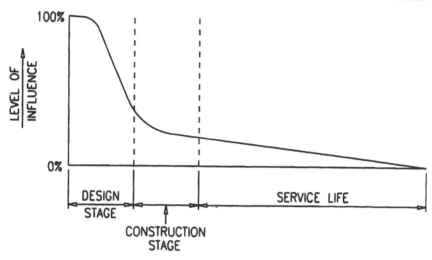

Fig. 25.2 Opportunity to influence durability and cost [1]

25.3.1 Structural System and Material

During conceptual design stage, the location of a bridge, its size, the type of its structural system as also its material, are normally determined. While the location and the size of the bridge are largely governed by site and other related considerations, the choice of the structural system and material may have an important effect on the durability of the structure. No doubt, economic consideration shapes the designer's thoughts in a very significant way. However, the lowest 'initial cost' solution may not prove to be the ideal solution with the lowest life cycle cost (see Chapter 7). Thus, the selected bridge type may well be the one with a higher initial cost, but with material requiring little maintenance (e.g., weathering steel).

One of the primary objectives of a steel bridge is to ensure that all structural loads are appropriately transferred through a series of components of the structural (frame) system to the foundations. A redundant bridge structure with multiple load paths provides alternative path for transmission of forces in case of failure of one element. A non-redundant structure, on the other hand, does not have multiple load carrying system and failure of a single element may cause collapse of the structure. In such a case, premature deterioration of a single component would require immediate replacement or rehabilitation of the damaged component. The designer is, thus, tempted to select a redundant structural system to avoid the hazards and additional expenditure due to immediate replacement or rehabilitation of the single damaged component. Designing for redundancy, on the other hand, may increase the initial project cost. This aspect also needs to be considered. Selection of proper arrangement of the structural system is, therefore, one of the important tasks of the designer.

Also, well defined load paths are required to be identified so that the loads can travel in a straight forward and un-hindered manner. This aspect needs special attention, as otherwise unintended forces may be induced in some critical areas of the structure causing it to suffer premature distress. In this context, adequate bracing system should be provided for transferring lateral loads at different levels to the foundations through well-defined load paths.

25.3.2 Loadings

With the introduction of codified approach of design and computer methods of analysis, structural design process has decidedly leaped forward in the recent past. However, correctness of the loads to be taken for analysis and design still remains as important as ever. In particular, the estimated self-loads of the elements considered in the analysis must be checked on completion of the design process. In case of significant deviation, the analysis and design may have to be done afresh.

In many bridge structures, forces in members may increase or may even be reversed during erection. Reversal of forces happens particularly in cantilever method of bridge erection. These erection forces should be considered and catered for during design stage itself. This is often done by strengthening certain affected members by fixing additional steel materials temporarily (during erection stage).

25.3.3 Analysis and Design

It is a common practice to make an initial estimate of the properties of components and of the rigidities of joints in order to calculate the forces in different elements. During subsequent stages of design and detailing, these values may change significantly. In such a case, the analysis should be repeated with the revised values, so that the analysis conditions closely relate to the actual ones. Unless this is ensured, the bridge structure, when completed, and in use, may not behave in the manner considered in the design–a condition which may impair its durability.

In the design stage, effects of local forces at critical locations should also be checked to avoid overstressing or failure of local elements during service life stage.

25.3.4 Fatigue

When a steel member is subjected to fluctuation of stresses, particularly in railway bridges, the ultimate strength of the steel is reduced considerably as compared to that for a static load applied gradually. This phenomenon is termed 'fatigue'. The characteristics of fatigue as well as the measures to be considered in the design stage to avoid its detrimental effects have been discussed in Chapter 4. The reader is advised to study these aspects carefully and ensure that the design becomes 'fatigue resistant' as far as possible.

25.3.5 Brittleness

Brittleness in steel is characterised by fracture at a relatively low tensile stress and sudden failure with little plastic deformation. The key factors that affect brittleness, and also the aspects which need to be considered to avoid brittle fracture, have been discussed in Chapter 4, to which the reader is advised to refer for guidance on this subject.

25.3.6 Welded Design

Nowadays, the elements comprising the despatchable components are, in most cases, joined by welding process. However, welding process is a sophisticated one and a designer will do well to comprehend its distinctive features in order to produce a satisfactory and durable welded design. Some of these features are discussed hereunder:

Characteristics of welds

During welding process, metals from each steel component and the electrode unite in a pool of molten metal, which, when solidified, forms a solid bond between the components. As molten metal cools, the heat flows into the adjoining parent metal, causing metallurgical changes in its structure upto a certain distance. These metallurgical changes in the structure depend particularly on its rate of cooling and carbon content.

Rate of cooling

Rapid cooling usually makes the weld brittle. The cooling rate depends on:

- Type of joint
- Thickness of steel

- Heat input in the joint
- Temperature of steel prior to welding. In thick components, pre-heating reduces the cooling rate and helps to reduce brittleness in the weld

Carbon content

Carbon is the primary strengthening element in steel. However, excessive presence of carbon increases the risk of brittle fracture. Other admixtures (alloys) are added during steel making process for improving the properties of the product while keeping the carbon level low. Carbon Equivalent formula is widely used as an empirical guide relating to composition and cracking tendency. The higher the carbon equivalent of steel, the harder and more brittle would be the heat affected zone (HAZ) near the weld and more susceptible would it be to cracking.

Characteristics of weld have been discussed in more detail in Chapter 16, to which the reader may refer for further information.

25.3.7 Corrosion Control

Design of a steel bridge structure may be perfectly satisfactory from load consideration or from aesthetic point of view, but it may include some features that accelerate corrosion or influence the cost of controlling it to a considerable extent. Therefore, corrosion control should begin at the design stage itself. If the designer and detailer develop an awareness of corrosion, many potential problems can be avoided at this stage.

Built-in water traps must be avoided while detailing bridge components. Figure 25.3a shows some examples of such unsatisfactory details. These details should be replaced by corresponding recommended details shown in Fig. 25.3b. Similarly, when hollow tubes are used, these should be hermetically sealed at the ends by welding suitable plates.

Proper access for inspection and eventual maintenance of all components must be considered while finalising structural details. If, for any reason, it is not practicable, then the designer should consider taking appropriate actions such as:

- Providing 'corrosion allowance' to the member by adding more steel to allow for future anticipated loss by corrosion.
- Using a comparatively more corrosion resistant steel.

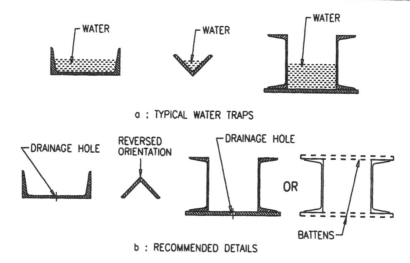

a : TYPICAL WATER TRAPS

b : RECOMMENDED DETAILS

Fig. 25.3 Features contributing to corrosion

25.4 DETAILING STAGE

Detail drawings are the primary means of communication between the designer and the fabricator. Clarity without ambiguity is, therefore, an important aspect to be remembered while producing detail drawings. Detailing must also ensure that the fabricated structure behaves in the manner envisaged at the design stage. Otherwise, durability of the structure may suffer. In general, details should be developed considering the following aspects:

- Design assumptions
- Convenience of fabrication
- Erection scheme
- Post-construction maintenance

There should, therefore, be a close and continuous interaction between all concerned parties, viz., the designer, the detailer, the fabricator, the erecter, as well as the maintenance personnel. Following are some of the critical areas which need particular attention during detailing.

25.4.1 Secondary Stress in Truss Bridges

In truss bridges, care should be taken to avoid secondary stresses in the joints by having the centre-lines of the converging members meet at a single point. This situation becomes quite challenging in case of bridge girders where sizes of members may be quite diverse and the angles between adjacent members are too small. This aspect should be considered while finalising the form of the truss configuration. Figure 25.4 illustrates the point.

25.4.2 Concentration of Stress

Sharp and re-entrant angles at notches of members, sudden reduction in cross sectional area and similar details cause concentration of stress and are

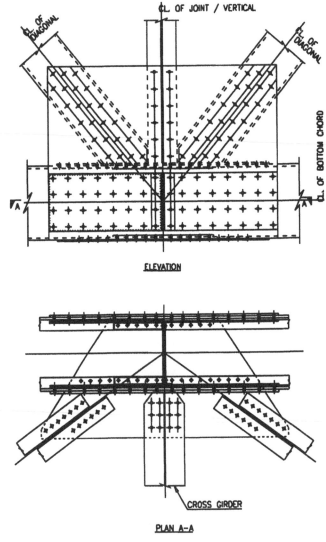

Fig. 25.4 Joint detail for bottom chord

likely to initiate cracks due to fluctuating loads leading to fatigue situation. These should be avoided. Figure 25.5 illustrates a few of such unsatisfactory details which should be replaced by corresponding recommended details.

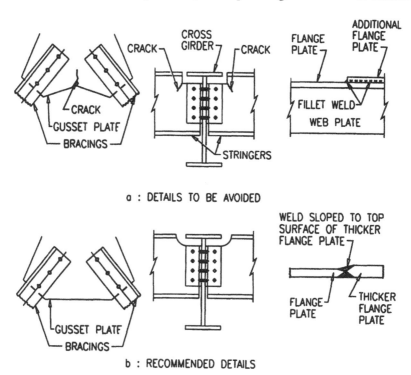

Fig. 25.5 Concentration of stress

25.4.3 Eccentricity in Connections

Behaviour of connections of cross girders to truss joints needs special consideration, as these may induce moments due to eccentricity and unwelcome deformation stresses in the joint. Proper stiffening arrangement to cater for local forces is to be provided in such cases.

25.4.4 Distortion in Welded Joints

A good weld design is represented by deposition of right amount of the right weld material in the right place. As welded joints are prone to distortions, excess welds should always be avoided. Furthermore, site joints sensitive to welding distortions should be carefully detailed (see Chapter 16).

25.4.5 Lamellar Tearing

Lamellar tearing is associated with highly restrained welded joints with heavy members under high stress condition. It is characterised by separation of parent metal caused by 'through thickness strains' induced by weld metal shrinkage (see Chapter 4).

Examples of some details susceptible to lamellar tearing are shown in Fig. 25.6a. Corresponding improved details are shown in Fig. 25.6b.

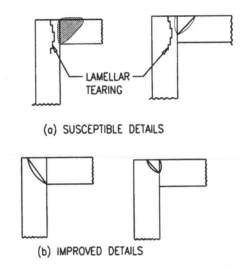

(a) SUSCEPTIBLE DETAILS

(b) IMPROVED DETAILS

Fig. 25.6 Susceptible and improved details to avoid lamellar tearing

25.4.6 Provision for Jacking

In-built provision should be incorporated in the design for jacking up the span during service life for periodic maintenance of bearings. Also, adequate space should be provided for inspection of bearings and their replacement, should such necessity arise in the future.

25.4.7 Camber

The other important aspect in the detailing of major bridges is the understanding of the need for introduction of camber and the methods for its incorporation in the detail drawings. Camber is generally provided in major bridges to offset the effects of secondary stresses when the girder is loaded. Mis-interpretation of this aspect may lead to introduction of unwelcome secondary stresses in members beyond their respective design capacities. Chapter 10 may be referred to in this connection.

25.5 CONSTRUCTION STAGE

Durability of a well-designed steel structure can be assured only if it is constructed (i.e., fabricated and erected) in conformity with the design philosophy and the basic assumptions *vis-a-vis* the material quality and sectional dimensions made by the designer. Any deviation from the basic design considerations, as reflected in the detail drawings, may introduce additional stresses in the structure and affect its strength and durability. It is in this context that quality assurance during construction of a steel bridge assumes utmost importance.

The procedure used in selection of contractors play an important role in the ultimate quality of the structure and its durability. In general, contractors are selected by 'lowest bid' method and each bidder naturally tries to minimise the time and the cost of the project by strictly sticking to the specifications incorporated in the bid documents (and no more). Thus, it becomes extremely important that the specifications clearly describe the type of structure, appropriate materials to be used and the procedures that have proven to be durable in other similar circumstances. The bidding documents must also closely indicate, without ambiguity, the criteria that will be used for acceptance or otherwise of the work produced by the contractor. Furthermore, the bidder's level of experience must also be considered in the selection process.

Construction stage of a steel bridge structure comprises two broad activities, viz., fabrication and erection. Some of the topics which may affect durability of the structure during these activities are discussed in this section.

25.5.1 Fabrication

Layout of fabrication area

Essentially, the fabricator's work starts from procurement of raw materials and ends with the despatch of fabricated items to site for erection. In the process the material has to pass through a complex sequence of many operations. A thoughtful sequential layout of the production area would largely eliminate unnecessary handling of the materials and would contribute considerably towards the quality of the product. At the same time, the layout should be flexible enough to provide minor adjustments in the sequence to cater for the requirement of individual items. For example, a top chord component may need end-milling after sub-assembly and thereafter has to be returned to the assembly line again for further assembling and welding.

Quality assurance plan

With a view to ensuring fabrication to be done in accordance with the drawings, inspection and checking at different stages of fabrication are carried out on the basis of an agreed Quality Assurance Plan (QAP). The various aspects of QAP have been discussed in Chapter 20, which should be referred to in this context.

However, when fabrication of bridge components is carried out in yards adjacent to the work site, it is not always possible to ensure sufficiently controlled environment as in an organised workshop. In such a situation, quality of workmanship is likely to suffer. It, therefore, becomes all the more important that the QAP is rigidly followed in such an uncontrolled work environment. Fabricators should be motivated to appreciate the usefulness of QAP in the workshops as well as at site.

Fabrication tolerances

It is not practicable to fabricate structural steelwork to exact dimensions and some degree of imperfection is bound to take place during fabrication process. Most specifications spell out the limits of such fabrication tolerances. These are normally allowed for in the adopted factor of safety while computing the capacity of a member. However, imperfections beyond these limits in case of some components may prove to be detrimental for the durability of the structure, whereas for some others these may not be so critical. Thus, it is the former cases, which need special attention during fabrication. Approriate perception of the implication of such imperfections on the durability of the structure is very important. As for example, imperfections on the straightness of compression flange of flexural members such as cross girders and stringers may impair the strength of the girders due to lateral torsional buckling. These may cause overall bow in the girder, and in turn, may generate twisting moments at the supports. It is, therefore, necessary for the inspecting agency to ensure that such excessive deviations in *critical areas* do not escape the attention of its personnel. Generally, high stressed locations are sensitive to imperfections and should be identified as critical areas.

Interchangeability

For multiple span bridges, it becomes convenient and sometimes necessary, to have all similar components of different spans to be absolutely identical and interchangeable, particularly, in the dimensions of the open holes for site connections. This can be successfully achieved if the holes are drilled by jig system (see Chapter 19). This system ensures elimination of mismatch at site to a considerable degree, and thereby reduces the possibility of development of secondary stresses in the structure.

Making holes

Punched holes may be critical for the durability of bridge structures subjected to fluctuation of stresses, since cold working on structural steel can cause embrittlement and cracking. Furthermore, punching may produce short radial cracks in the holes which may initiate brittle fractures at the hole when the member is loaded. Therefore, holes should be formed either by drilling or by punching undersized holes followed by reaming to the desired size (see Chapter 19).

Shop assembly and camber check

For bridge structures, it is necessary to assemble the component parts in the fabricating shop before these are despatched to site for erection. During this operation, the overall dimensions of the structure, as also alignment, squareness, camber, etc., should be checked and confirmed. Checking of camber in a truss bridge is particularly important as this is provided to eliminate secondary stresses when the span is loaded and deflection occurs (see Chapter 10). It should also be noted that inadequate or inaccurate camber may induce secondary stresses in the members, thereby adversely affecting the strength and durability of the structure. Shop assembly also ensures that the open holes provided in various components at location of joints are within tolerable limits.

Welding

Welding is a comparatively modern process of joining together metal parts. Factors which influence the design of durable welded structures have been discussed earlier in this chapter. Also, various aspects of welding process and welded joints are discussed in Chapters 15 and 16, to which the reader is advised to refer in this connection.

25.5.2 Erection

Erection of steel bridges is a highly specialised operation involving many disciplines and requires active communication and co-participation amongst the members involved. This situation calls for a high degree of coordination and meticulous planning in all stages of operation. Some of the aspects which particularly contribute to the durability of the bridge during this stage of operation are discussed in the following paragraphs.

Method of erection

It has been discussed earlier in this chapter that the method to be adopted for erection should be considered in the design stage itself. Normally, selection of the erection method is governed by the structural system adopted, site conditions, availability of equipment, quality of skilled labour, etc. However, regardless of the method adopted, the foremost aspects that should occupy the attention of the erection crew are:

- Stability of the structure during incomplete stages of erection.
- Ability of the structure to perform its intended functions satisfactorily when completed.

Bracings

In general, permanent bracings are provided to cater for lateral forces in a steel bridge. During erection stage, bracings (if required, temporary bracings) must be provided to take care of the stability of the incomplete erected structure and also of the stresses from the effects of erection equipments. Adequate provision to cater for lateral forces, such as wind loads should be made according to local conditions.

Erection tolerances

In a bridge project, setting out of the centre-lines of foundations and holding down bolts is often done at a stage when the site conditions are rather disorderly and the environment appears to be incompatible to accuracy. However, inaccuracies in marking the centre-lines or in fixing the levels, if allowed to pass uncorrected at this stage, are likely to cause immense problems afterwards, by way of misfit in connections, and also misalignment of the structure leading to secondary stresses in the members. It is, therefore, imperative that accuracy within tolerable limits is achieved from the very beginning of the site work, including the setting out operation.

Making joints

Durability of a steel bridge structure largely depends on the quality of the joints made at site.

In bolted connections, all parts intended to be bolted together should be in contact over the whole surfaces. These surfaces should be thoroughly cleaned and painted with the specified paint and the matching plates/ sections secured together, while the paint is still wet, by service bolts along with parallel barrel drifts. During assembly after erection, a joint should be made by filling at least 50% of the holes by bolts and drifts in the ratio of 4:1. Drifting should be done carefully so as not to distort the holes. Unfaired holes must not be enlarged by drifting to admit bolts. In such cases, the holes should be enlarged by reaming. However, the effect of the reaming on the soundness of the structure must be examined first. Under no circumstances, should the holes be formed by gas cutting process.

For welded connections, the components should be securely held in position before the welding is commenced, so as to ensure alignment, squareness, camber, etc.

Grouting

Grouting under the bearings should be done after the final alignment of all the spans in a bridge is over.

25.6 SERVICE LIFE STAGE

25.6.1 Maintenance Programme

Various aspects of post-construction maintenance programme are discussed in Chapter 27. It must be emphasised that this post-construction maintenance of steel bridges is essential for ensuring their durability. However, in many cases, this awareness is unfortunately lacking. As a result, regular maintenance is often neglected. Budget constraint and inadequacy of trained personnel are the two primary reasons for delaying of maintenance. This may lead to a situation when major repairwork has to be undertaken at a very high cost, which could have been avoided had the maintenace work been done at the appropriate time, at only a fraction of the repairing cost. It, therefore, becomes imperative for the bridge owner to establish a maintenance programme for the bridge and allocate the required fund for this purpose. However, allocation of the fund is not enough. The owner should also ensure that maintenance is carried out in accordance with the established programme.

25.6.2 Painting

Like maintenance, quite often, painting is deferred for want of funds. The bridge owner should understand that deferring periodical re-painting involving nominal surface preparation, would only lead to extensive surface cleaning, priming and top coating, the cost of which would be much higher than the periodical re-painting cost. Apart from this sense of false economy, deferring of routine re-painting, also exposes the 'high cost asset' to atmospheric corrosion, thereby shortening its service life. In case the cost of re-painting of the entire bridge is beyond the permissible budget, the next best solution would be to re-paint only the more vulnerable areas of the bridge, such as the joints, and the areas where the existing paint has failed prematurely. This maintenance strategy is a better option than 'no maintenance' and can retain the durability of the bridge at a much lower cost.

25.6.3 Access for Inspection and Maintenance

The necessity of inspection and maintenance has an influence over the design of a bridge. Bridges which are located at a low height above the ground are generally accessible for inspection and maintenance by simple methods (ladders, scaffoldings, platform-on-vehicles, bucket snoopers, etc.). For bridges over waterways, railways, etc., it may be necessary to provide custom-made permanent travelling gantries suspended from the structure and capable of travelling along the length of the bridge.

For box girders, access inside the decks should be provided by keeping openings with sufficient dimensions to enable inspection personnel and equipment to pass through. These openings should normally be inaccessible to the general public.

25.6.4 Debris

Debris on bridge members, particularly at joints, from pigeons nests, wind blown dust, leaking expansion joints, etc., retain moisture and may cause corrosion if allowed to remain. These should be cleaned at regular intervals. Some bridge owners have their bridges washed at least once a year as a part of maintenance strategy. This very simple action is bound to extend the service life of bridges.

REFERENCES

[1] Gordon, S. 1989, *Durability of highway bridges*, Paper presented in the Symposium on Durability of Structures, Lisbon, organised by IABSE, Zurich, Switzerland.

[2] Ghosh, U.K. 2000, *Quality assurance and training for design and detailing of durable steel structures*, Paper presented in the Seminar on Durability of Structures, organised by Indian National Group of IABSE, Aurangabad, India.

[3] Ghoshal, A. 2000, *Quality assurance and training for fabrication and erection of durable steel structures*, Paper presented in the Seminar on Durability of Structures, organised by Indian National Group of IABSE, Aurangabad, India.

Chapter 26

Protection against Corrosion

26.1 INTRODUCTION

The erected steel superstructure of a bridge needs protection against corrosion to enhance its durability. It is, therefore, necessary to study the various aspects related to corrosion process and the methods for its control and prevention. In this chapter, discussions will focus primarily on atmospheric corrosion and various protective coating systems.

26.2 ATMOSPHERIC CORROSION

26.2.1 Corrosion Process

Atmospheric corrosion is the deterioration of steel caused by its reaction with elements present in the surrounding environment. It is an electro-chemical process of flow of electricity and chemical changes in steel occurring in stages. Initially, the process starts locally by formation of anodes (steel substrate), and cathodes (surface mill scales), similar to battery cells. In the next stage, when the surface is moist, the cells start functioning in the presence of the moisture film and acidic substances present in the atmosphere, which together act as an electrolyte media. In combination with oxygen from air, hydrated ferric oxide or red rust is formed. In this connection two important points are to be particularly noted:

- For corrosion of steel to take place, both water (moisture) as well as oxygen (air) are required. In the absence of either, corrosion cannot occur.
- All corrosion occurs at the steel substrate (anode) only, and not at the mill scales (cathode).

As time passes, accumulation of the rust in the first area causes the corrosion process to be somewhat stifled, resulting in the formation of a new anodic zone, thereby spreading the corrosion area. If the original anodic zone is not stifled, the corrosion process continues deep into the steel resulting in pitting or 'pock marks'.

Formation of rust and pitting reduces the area of the member. Consequently, stress in the member increases and the member becomes vulnerable to stress corrosion and fatigue damage (see Chapter 4).

26.2.2 Corrosion Rate

Corrosion rate depends primarily on two factors, viz., period of dampness, and level of pollutant in the atmosphere.

Period of dampness

Corrosion rate is more if the period of dampness is more. Therefore, corrosion rate can be reduced if the period of dampness is reduced. Accumulation of debris on the bridge structures increases the period of dampness and consequently, the corrosion rate. Similarly, ponding of water on the bridge members increases the rate of corrosion.

Level of pollutant

Presence of sulphates and chlorides in the atmosphere makes steel susceptible to corrosion. These chemicals increase the corrosion rate by reacting with the steel to produce soluble iron salts. Thus, the type and amount of pollutants in the surrounding environment have considerable impact on the corrosion rate. Sulphur dioxide gas present in industrial environment reacts with moisture in the atmosphere to form sulphurous and sulphuric acids, which accelerate corrosion. Chlorides are found in marine environments and can increase corrosion rate in steel bridges situated near the coastal regions.

26.3 PROTECTIVE COATING SYSTEMS

26.3.1 Historical Background

Application of coatings to metals has been a practice for centuries. These coatings, originally based on naturally occurring oils, bitumens and subsequently paints, provided reasonably satisfactory protection to many structures, including bridges, many of which were situated in relatively cleaner and less corrosive environments—away from industrial areas.

Modern protective system has its origin around the middle of the 19th century. A good example of this system is provided by the Forth railway bridge in Scotland, which was opened to traffic in 1889. The paint system adopted was a red lead/linseed oil primer in the shop, followed by a second coat in the field. The steel was then provided with an additional coat of red oxide intermediate coat, followed by a coat of red oxide finishing coat [2]. The system bears a striking resemblance to the system which has been in use until recently in many bridges, or is being used in many other bridges even today.

Over the years, improvement in the coatings gave way to the use of aluminium and zinc in the coatings. Research in the area of protective coatings, particularly in polymer chemistry, has resulted in more sophisticated coating systems, thereby enhancing the service life of the system from 5-10 years to 15-30 years. No doubt, the costs have also gone up, but considering the Life Cycle Costing (LCC) analysis of the steel bridges (to be discussed in a later section of this chapter), the system with increased life expectancy is well worth the initial investment.

26.3.2 Surface Preparation

One of the important factors which influence the protection quality of any coating system is its capabilty to adhere to the underlying steel surface. To achieve this, it is necessary that the steel surface is prepared for the application of the protective coating system.

The extent to which the steel surface should be prepared varies, depending on the protective coating system to be used. The principal methods used for surface preparation are manual cleaning and blast cleaning. These methods have been discussed in Chapter 19.

26.3.3 Types of Coatings

Protective coating systems commonly in use in bridge structures can be broadly classified into two categories:

- Non-metallic, e.g., paint coatings
- Metallic, e.g., metal spraying and galvanising

Discussions in the following sections will broadly cover these categories of coatings.

26.3.4 Paint Coatings

General

Paints are essentially a mixture of film forming liquid called binder and solid particles called pigments. Often, a drying agent is also added. A variety of binders are commonly used, such as natural oil, resin, alkyd, epoxy ester, synthetic resins, etc. Pigments are dispensed in the binder and give the paint its colour, hardness, and corrosion resisting properties. The binder is an important part of most paints, because it provides properties to resist attack from the environment. Pigments, however, are no less important, as they also influence durability and the properties of the paint. They contribute to the hardness and abrasion resistance of the paint film and reduce its permeability.

Normally, a paint film has tiny pores called *pinholes*, caused by bursting of air bubbles trapped in the paint during application. These holes are invisible to the naked eye. If only a single coat of paint is applied, there is a possibility of oxygen and moisture from the atmosphere to be transmitted through these tiny holes to the underlying steel surface and cause corrosion. Application of several coats of paint obviates this possibility by substantially reducing the chance of pinholes being formed at identical locations in the different coats.

Paint systems

Usually, a number of coats are applied to constitute a particular paint system. These are:

(a) Primary coat: Primary coat or primer is the foundation of the paint system. This is the initial coat applied to the virgin surface of steel immediately after the surface is prepared. The first primary coat must adhere well to the surface of the steel. In effect, the primer film prevents moisture and oxygen from reaching the steel, neutralises the penetrants, and thus prevents corrosion of the steel surface. Sometimes, a second coat of primer is applied over the first one, in order to improve corrosion prevention quality of the system.

(b) Top coat: This is applied over the primary coat(s) and should have the properties of resisting external environment such as air, moisture, ionic material, etc. It should resist abrasion, in addition to being of smooth finish and aesthetically pleasing colour.

The primary coat and the top coat differ substantially in function as well as in chemical composition, and therefore, due care should be taken while choosing the combination in order that they can perform effectively.

(c) Intermediate coat: Often, an intermediate coat is used between the primary coat and the top coat, to provide additional protection against penetration of moisture and air, as well as good adhesion between the coats. Different tints should be used if more than one intermediate coat is used, so that proper coverage can be ensured.

Protection mechanism of paint films

In any coating system, paint films protect the steel through the use of three general mechanisms:

(a) Providing barrier: In this mechanism, the underlying steel surface is insulated from the effects of environment, viz., air, moisture, and ionic materials. Virtually, this mechanism is involved in all the coating systems. Some common types of barrier system of coatings are: coal tar enamels, low-build vinyl lacquers, epoxy and aliphatic urethanes, coal tar epoxies, etc.

(b) Inhibition: In this mechanism, corrosion is stopped through a process of chemical inhibition to prevent deterioration of steel caused by air and moisture. Red lead is an example of an inhibitive primer, widely used in steel construction work. There is, however, a general concern over the toxic nature of this primer, and all lead based primers are increasingly being discouraged from being used. Recent research and development work has produced a variety of less toxic inhibitive paints.

(c) Galvanic action: In this mechanism, the electrode potential of the metal is controlled so that the metal becomes immune or passive. Zinc is the most common material used for this mechanism. Zinc has an advantage, in that it is relatively less toxic than lead.

Types of paints

Paints commonly used in bridge steelwork are briefly discussed below.

(a) Conventional paints: These include paints based on drying oils (e.g., red lead), alkyd resin, modified alkyd resin, phenolic varnish, epoxy ester, etc. They dry in air, and the film is formed by absorbing oxygen from the atmosphere. They have limited solvent as well as chemical resistance, and are, therefore, suitable for bridges situated in less aggressive environments [4, 5].

(b) Chemical resistant paints: For comparatively more aggressive environments, chemical resistant paints are used. These are generally available in two varieties, viz., one-pack paints and two-pack paints.

One-pack paints: These paints are of ready-to-use quality and can be applied directly on the steel surface after necessary surface preparation. Examples of one-pack chemical resistant paints are: chlorinated rubber, vinyl, etc. These paints form a film by solvent evaporation and involves no oxidative process. They can be applied as moderately thick films. The films, however, remain relatively soft and have poor solvent resistance, but good chemical resistance. These paints can be repainted without difficulty. Bituminous paints also dry by solvent evaporation, and have moderate chemical resistance. Comparative impermeability to the ingress of water is their significant advantage. Their main drawback is their inability to withstand the detrimental effects of oil. Their other drawback is that they are available in black and dark colours only [4, 5].

Two-pack paints: These paints are available in two separate containers, one containing the base, and the other the curing agent. These two components have to be mixed together immediately before use, and are workable thereafter only upto a specified period of time. In other words, once mixed, these materials have limited 'pot life', by which time the mixed coating must be applied. They dry as a result of chemical reaction between the components, producing a hard film which is resistant to abrasion. Examples of this type of paint are : two-pack epoxies and urethanes. They need proper surface preparation by blast cleaning. They are generally applied by spraying, and require skilled workmen for their application. Application of such paints must be carried out strictly in accordance with the guidelines provided by the manufacturers. While epoxy paints rapidly become dull when exposed to strong sunlight, the urethane paints retain their gloss for a longer period. Another example of two-pack paint system is organic zinc-rich paint. This paint is mainly used as primer, but can be used also for a complete paint system at suitable thickness. This paint can provide galvanic protection to steel, and is often called 'cold galvanising'. This paint also requires high quality of surface preparation [1, 4, 5].

26.3.5 Metallic Coatings

Metallic coatings used for steel bridges can be broadly classified into two categories:

- Metal spraying
- Hot-dip galvanising

Metal spraying

In this process, metal (normally zinc or aluminium) is sprayed directly from a gun on to the steel surface. The metal, in wire or powder form, is fed through a nozzle with a stream of air or gas and is melted by a heat source (oxy-gas flame or electric arc). The melted globules of the metal are blown on to the previously blast cleaned steel surface by the compressed air jet or gas. A coating consisting of a large number of these globules is formed on the steel surface. Initially the coating is porous as it is made up of a large number of distinct globules which are not always melted together. The pores are subsequently sealed by allowing the metal coating (globules) to corrode and block the pores by the corrosion products. Often metal spray coatings are further protected by subsequent applications of paint coatings.

In the oxy-gas flame system, there is no metallurgical bonding between the globules and the steel surface. However, in the case of arc spraying, since it operates at a much higher temperature, some insipient welding may occur between the globules and the steel surface, resulting in improved adhesion. Consequently, in metal spraying process, adhesion between the coating and the steel is considered to be essentially mechanical in nature. Therefore some form of mechanical key is required for its functioning. To achieve this, blast cleaning of the steel surface with angular grits is recommended. Normally, aluminium requires a comparatively cleaner surface, for which cleaning to Swedish Standard SA3 is specified. For zinc coating, a slightly lower standard (SA 2-1/2) is commonly considered satisfactory. Coating thickness varies from 100-250 microns for aluminium, and 75-400 microns for zinc [1, 4].

Hot-dip galvanising

Hot-dip galvanising is an alternative method of providing metallic coating on structural steelwork. While it has diverse applications in other *fields*, its use in steel bridges has been somewhat limited. Although this method was used in the cables of Brooklyn bridge in New York as early as in 1883, its use in modern steel bridges has been limited, mainly to railings rather than the primary components.

For galvanising of structural steelwork, zinc is the metal most commonly used. It has an advantage over other metals (like aluminium, tin, lead) because of its low melting point (420° Centigrade) and the nature of the zinc layer formed on steel surface. Galvanising procedure involves the following stages of operation:

(a) Cleaning: Steelwork is cleaned of grease and oil, often, with warm caustic soda.

(b) Pickling: All rust and scale is cleaned off by immersing it in a bath containing acid (usually a diluted solution of hydrochloric or sulphuric acid). This process is termed 'pickling'. Blast cleaning may be employed before pickling to remove scales and roughen the surface.

(c) Fluxing: The clean steelwork is then fluxed with ammonium chloride in a pre-fluxing bath.

(d) Coating: The steelwork is then dipped into a bath of molten zinc (440-470° Centigrade), when a reaction between zinc and iron (from steel) occurs and a series of zinc-iron alloys are formed on its surface, along with a layer of pure zinc on top of the alloy layers. The degree of the alloying depends on many factors including the composition of the steel, temperature of the bath and duration of immersion of the steelwork in the bath.

Depending on the size of the bath, larger components of steelwork can be galvanised by 'double-dipping' in the bath, provided that its size is less than twice that of the bath.

In galvanised coatings, the amount of zinc in the coating is the measure of its protective quality and is normally denoted in weight (g/sq. cm.) rather than thickness (microns), as in other coatings. This can, however, be converted into thickness, which for structural steelwork, is in the range of 75-125 microns. With blast cleaning before pickling, the thickness can be increased to 175 microns [1].

In galvanised steelwork, steel is protected by metal coating by providing a barrier against atmospheric effects and at the same time offering cathodic protection against corrosion. A galvanised steel surface needs little or no maintenance by way of painting. Inspite of these advantages, however, galvanising process suffers from a number of drawbacks. The main drawback is that it is more expensive than most conventional coating systems. Also, the limited size of the bath makes its use in long bridge members rather awkward. Possibility of warping of fabricated components from zinc is an additional hazard. Furthermore, environmental regulations in many countries restrict toxic coating systems using zinc.

26.3.6   Selection Criteria of Coating System

Selection process of a coating system normally involves examination of a number of factors, and arriving at a compromise choice on their relative importance. Some of the well recognised factors are discussed below.

Environment

Environmental conditions are of fundamental importance in the selection process of a coating system. A particular environment may be aggressive on some types of coatings, but may not be on others. Metal coatings normally react with the atmosphere and corrode; but the rate of corrosion varies from one environment to another. Similarly, some paint coatings are more susceptible to deteriorate in some types of environment, compared to others. Also, ultra-violet rays from the sun may cause breakdown of organic coatings. For selection of a coating system, therefore, the following aspects need particular consideration:

(a) Environmental conditions: polluted or non-polluted, saline or non-saline.

(b) Proximity of water: chances of intermittent splash of water, or parts of steel fully immersed in water; presence of harmful salts (e.g., sulphates, chlorides, etc.), or fungi or bacteria in the water. Also, likelihood of impact or abrasion from water-borne debris.

Access for maintenance

Bridges situated in remote areas, such as in mountainous regions, where access for maintenance is both difficult and expensive, a more durable coating system may have to be selected, even if the initial cost is more. Similarly, where repainting at regular intervals is feasible, it may be preferable to select a less resistant and comparatively inexpensive coating system.

Availability of the coating system

It is necessary that the facilities required for specialised application of a particular coating system are easily avaiable. Delay in arranging such facilities might affect the overall progress of the project.

Ease of application

This aspect is particularly important where there is shortage of skilled workmen. In such cases, it may be preferable to choose coating systems which are comparatively straightforward to apply, without requiring specialist operators (for say, blast cleaning). This would preclude problems not only at the time of construction, but also during future repainting.

Hardness

Soft films on the steel components of a bridge are comparatively easily damaged, particularly during handling, transportation and erection. Therefore, a reasonably hard film is required for such structural steelwork. For some paints, development of adequate hardness is related to drying time. For example, red lead paint takes a long time to dry and produce a hard film. Modern polymer coatings dry comparatively quicker and yet provide acceptable hard films. However, hard films may suffer from brittleness, which is likely to reduce its service life.

Abrasion resistance

Resistance to abrasion and erosion is an important factor when a bridge is situated in an environment where sand, particles of coke, and other similar substances are blown about by wind. In general, two pack coatings are reasonably resistant to abrasion.

Economy

For financial evaluation of the protection system, not only the initial costs, but also the future maintenance costs should be considered. For this purpose, total life cycle costs of some *prima facie* suitable systems should be calculated and compared.

In this method, the future costs for a particular suitable system are brought together into a single value, often by applying the traditional present value concept on a base date corresponding to the initial costs, and added to the initial investment to give the net present value (NPV) of the system. Similar values of other suitable systems are also calculated and the results are compared for financial evaluation. It has often been found that, in an aggressive environment, a special corrosion resistant painting system, which would initially cost more, but would last longer than a traditional painting system, may prove to be more economical over a period of time, if analysed with Life Cycle Costing system [3].

Other factors

Other factors which also need consideration for the selection of a protective coating include:

- Appearance (after final coat) *vis-a-vis* the surroundings

- Time required to apply the coating system
- Past performance of coatings on existing bridges in the region

26.3.7 Factors Influencing Protective Quality of Coatings

Selection of appropriate coating system with correct specifications is, however, not the only criterion for satisfactory performance of the system. Failures can arise not only from incorrect use of basically sound material, but also due to some other extraneous factors which influence the protective quality of coatings. Some of these which are relevant to steel bridges are discussed in the following paragraphs.

Adhesion to steel surface

A reasonably good adhesion of the coatings to steel is a basic requirement of any protective system. For this purpose, steel surface must be clean and free of any loose material. It is also necessary that the steel surface is sufficiently roughened to provide mechanical key for the coatings to adhere to it.

Blast cleaned surfaces provide superior adhesive quality, because this process increases the area of contact as well as potential contact points of bonding with the substrate by removing dirt, mill scale, etc.

Another issue which should be guarded against, is moisture on the steel surface. This may lower the adhesive quality of the coating.

Inter-coat adhesion

A coating system normally consists of a number of coats, applied successively one coat over another. It is, therefore, necessary that these individual coats are compatible and adhere soundly to each other.

Procedures of application

Procedures of application are very important for the life and effectiveness of any coating system. This is important even for conventional oil-drying type of coatings. For many of the modern two-pack materials, temperature, humidity, and ventilation are particularly important, and should be considered. Inadequate storage, incorrect nozzle size for spraying, or poor mixing may lead to dry films with inadequate protective properties. Thickness of individual coatings should be closely monitored. Some of the other sources of deficiencies are:

- Use of material after the expiry of the shelf life
- Use of wrong or excessive thinner

- Inadequate mixing of two-pack materials

Permeability

It is obvious that an impermeable coating would provide maximum protection to steelwork. The method of application of the paint may increase permeability. Permeability may also be caused by insufficient binder in the film leading to creation of interstices or pores which allow water to penetrate the paint film.

Blistering

Blistering in a coated film may occur in two ways. The first type occurs within the coating itself caused by pressure of water or solvents trapped within or under the paint film. The other type is caused by corrosion of the underlying steel substrate arising from soluble iron corrosion products at the interface of metal and paint. Proper cleaning of the steel and careful application of the coat are imperative to avoid blistering.

Micro-organisms

Bacteria and other micro-organisms present in the environment may cause problems to certain coated films.

Ageing

On exposure to environment, all coated films suffer deterioration to some extent. One form of deterioration is the external change in appearance, viz., loss of gloss or fading of colour. This can be attributed to ultraviolet rays and pollution in the air. However, this deterioration may not necessarily reduce the protective properties of the film seriously. Nevertheless, ageing of the film may influence its protective properties and consequently it may eventually fail, necessitating repair or re-application of the coating.

Design and detailing

Design and detailing of the bridge structure have considerable influence on the performance of the coating. Plate girder bridges with large flat areas are comparatively easy to coat. Truss bridges, on the other hand, are more complex with joints and connections. Similarly, box sections are easier to paint compared to compound sections made up with angles and lacings. Access for application of coatings is also important for the performance of the protective film. Welds often present problems if the slags are not cleaned properly, thereby affecting adhesion of the paint to the metal surface. Also, in case of welds with considerable roughness, it may be difficult to achieve satisfactory and uniform thickness of the applied coat.

Handling, storage and transport of steelwork

Inadequate handling during loading, unloading or erection may damage portions of the priming coat on the steelwork. Also, inadequate storage at site, or during long voyages may impair the priming coat. Subsequent patch repairwork may not be good enough to restore the coat to its original standard.

26.4 PROTECTION OF CABLES

Corrosion affects all steel products in varying degrees, and cables used in cable supported bridges are no exception. Therefore, it is essential that an appropriate protection system be employed for ensuring their reasonable durability.

The traditional method of protecting the main cables of suspension bridges was by coating the steel with a red lead paste, wrapping the cables with galvanised, annealed wires, and applying red lead paint. This system was used in Brooklyn bridge with considerable success. However, in some bridges the results were not encouraging. An inherent defect in the system is that, with the stretching and shortening of the cables under live loads and temperature changes, separation of adjacent turns of the wire rapping may occur, leading to cracking of the paint and ultimate ingress of water into the cable surface. In early cable stayed bridges, stays (locked coil or structural strands) were usually protected against corrosion by galvanising and painting. These also did not always prove successful.

An alternative method is to provide protection to the wires by applying various thicknesses of zinc coatings, depending on the location of the wire in the cable and the atmospheric exposure. Some Standards (e.g., ASTM) specify the various thicknesses of zinc coatings to be applied. Cable manufacturers may also have their own standards for protection of the cables for particular atmospheric conditions. A current system is to cover the cable with a sheathing of steel or polyethylene tube and inject cement grout in the sheathing to fill up the space between the cable and the sheathing. The alkaline properties of the grout provide the corrosion protection. Another system is to coat the individual wire or strand with an epoxy material [7].

In view of the diverse methods available for corrosion protection of the cables, the designer would do well to carefully and critically study the different options available before deciding on the protection system to be adopted.

26.5   CONCLUDING REMARKS

In the preceding discussions the causes of corrosion of steel bridge super-structure and the various types of protective coating systems have been briefly discussed. It is hoped, that these discussions would serve as an indicator of the seriousness of the problem of corrosion and the need for protection against it, particularly in relation to the large number of steel bridges, existing as well as future bridges, around the world. The key to ensuring durability of these bridges lies in recognising the corrosive nature of the environment, and specifying the appropriate system with due regard to future maintenance needs. In this excercise, LCC method should help in comparing the overall cost of different protective systems and choosing the most appropriate and economical system.

REFERENCES

[1] Chandler, K.A. and Bayliss, D.A. 1985, *Corrosion Protection of Steel Structures*, Elsevier Applied Science Publishers, Barking, UK.

[2] Tonias, D.E. 1995, *Bridge Engineering*, McGraw-Hill Inc, New York, USA.

[3] Ghosh, U.K. 2000, *Repair and Rehabilitation of Steel Bridges*, A.A. Balkema, Rotterdam Netherlands.

[4] *Corrosion Protection of Structural Steels used in Building and Bridges*, Institute for Steel Development and Growth, Calcutta, India, 2001.

[5] *Standard Specifications and Code of Practice for Road Bridges (IRC: 24-2001 Section V, Steel Road Bridges)*, The Indian Roads Congress, New Delhi, India, 2001.

[6] Podolni, W. (Jr.) 1994, *Cable-suspended bridges*. In: Structural Steel Designer's Handbook, Brockenbrough, R.L. and Merritt, F.S. (eds.) McGraw-Hill Inc, New York, USA.

[7] Podolny, W. (Jr.) and Scalzi, J.B. 1986, *Construction and Design of Cable Stayed Bridges*, John Wiley & Sons, New York, USA.

Chapter 27

Post-construction Maintenance

27.1 INTRODUCTION

Construction of bridge structures involve heavy intital investments, and consequently they are invaluable assets to the community. However, they are permanently exposed to the adverse effects of environmental conditions. It is imperative that the bridge structures are protected from deterioration and are maintained properly to provide safe and uninterrupted traffic flow throughout the design life span of these high-investment assets. In order to achieve this objective, a well planned and monitored inspection and maintenance system is required. Such a system would ensure inspection and recording of the current state of the structure, and would provide feedback information to the concerned authorities about the present problems as well as potential sources of future troubles for taking timely corrective actions.

This chapter provides guidelines for preventive maintenance activities needed for steel bridges. While each situation requires careful thought and judgment, these guidelines should be helpful in achieving the overall objective of adequate maintenance of the structure.

Some of the important activities which comprise the maintenance system of steel superstructures are:

- Maintenance documentation
- Inspection and recording
- Replacement of loose rivets
- Minor repair work
- Cleaning and greasing of bearings
- Maintenance repainting

27.2 MAINTENANCE DOCUMENTATION

The foremost requirement of an effective maintenance system is to prepare a methodically entered current record for each bridge under the system. These records may be stored in an easily retrievable computerised data bank and should include the following :

- Number or name of the bridge
- Year of construction
- Location, giving distance from the nearest town or village, and indicating the position on a small scale map, if possible
- Type of the structure, e.g., plate girder, type of truss, riveted or welded, etc.
- General dimensions, e.g., overall length, width, number and lengths of spans, skew angle, if any, depth, width of girders, etc.
- Services carried on the bridge
- Details or types of components, such as bearings, expansion joints, parapets, guard rails, etc.

Additionally, the records should include the present live load capacity of the bridge supported by design calculations.

Apart from the above information, the records should provide full history of each structure, including 'as made' drawings as well as details of all repair or rehabilitation work undertaken previously, and the behaviour of the structure thereafter.

All data should be systematically preserved in a suitable format, so that information for a particular bridge can be easily retrieved. Such retrieved information would be very useful as a background for any inspection. Important features arising out of such inspection should, in turn, be stored in the records for future reference.

27.3 INSPECTION AND RECORDING

27.3.1 Objectives

The primary objective of inspection is to search for and identify defects in the structure in order that the maintenance repair can be undertaken. However, inspection should not be confined to this objective only. It should also include searching for and identifying potential problem areas, so that timely action can be initiated to avoid actual damage to the structure in the future.

27.3.2 Frequency

Normally, every bridge is subjected to two types of inspection, viz., 'General Inspection' and 'Principal Inspection'. In general inspection, certain items of each bridge are inspected at definite intervals, irrespective of the fact whether anything alarming has taken place or not. This inspection is carried out by visual observations from ground or deck level, or from existing walkways or platforms. Principal inspection, on the other hand, is detailed inspection of each and every component of the superstructure. This activity may need suitable temporary arrangements for access. Many authorities require that the general and principal inspections are carried out at least once every two years and six years respectively [2]. However, the frequency and the extent of inspection depend largely on several parameters such as age of the structure, traffic characteristics, history of deteriorating conditions, strategic importance, environment, etc. The person in charge of formulating the inspection programmes of a particular bridge, may consider these parameters for deciding on the specific frequency of inspection.

'Special Inspections' are required for a bridge when a particular problem needs special investigation, such as damage due to accident, flood, loss of camber, etc. These are undertaken on the basis of reports of unusual occurrences, as well as on routine inspection reports.

27.3.3 Timing

Inspection should be carried out in those periods of the year when the weather offers the most desirable condition for exhaustive and in-depth inspection. As for example, structures requiring high climbing should not be inspected during the seasons when high wind or extremes of temperatures are expected to prevail. On the other hand, the most suitable seasons for inspecting the bearings or the expansion joints would be during the times of extreme temperatures, when movements in these items can be noticed easily.

27.3.4 Personnel

A bridge inspector is normally called upon to face a diverse and complex nature of problems. Consequently, his judgment is often required for proper evaluation of the findings at site. To this end, a bridge inspector should ideally be a person closely associated with bridge work, and should have reasonable knowledge in material and structural behaviour under actual loading conditions. Additionally, he should have close acquaintance with the design and construction features of the bridge, and be capable of putting his findings in clear language and by simple sketches in the report. He should be conscientious and painstaking and should have developed an insight for observing deterioration of materials due to corrosion, weathering, fatigue, etc. He should be able to recognise any such structural deficiency, assess its seriousness, and initiate preventive maintenance programme to keep the structure in a safe condition, if required, by imposing restrictions on loading or speed, or any other measure deemed necessary.

It is hardly possible for an individual to have all the requisite qualifications and experiences of an ideal inspector. It is, therefore, often preferable to form an inspection team of engineers and technicians, with experience and knowledge of diverse areas such as structural design, construction, maintenance, emergency repair, etc. Additionally, bridge inspection often requires climbing on rough, steep and slippery terrain, working at heights, or working for days together, and not all persons are physically capable of doing so. This aspect also needs consideration while forming a team for inspection. Assistance of specialist engineers/agencies may have to be sought when regular staff is not available, particularly in the case of special structures such as movable bridges (involving mechanical and electrical/electronic equipment to be inspected), or unusually large structures [3].

27.3.5 Tools

Proper tools are required for carrying out satisfactory and comprehensive inspection of a steel bridge. For routine inspection, some of the most useful tools are a 2 m pocket tape, a 30 m steel tape, a chipping hammer, paint scraper, flat bladed screw driver, pocket knife, wire brush, plumb bob, vernier or jaw-type callipers, small level, steel straight edge, and rivet testing hammer. Other items, such as camera, flashlight, inspection mirror, magnifying glass, a set of crayons, nylon chord, gloves, and first aid kit are also useful during inspection. A clip board is essential for taking notes. Additional special inspection tools are surveying instruments, non-destructive testing equipment and dye penetrant kit.

27.3.6 Areas to be Inspected

Common defects which are noticed in a steel bridge are corrosion, cracks, overstress and collision damage. Of these, corrosion is the most common defect. Loss of sectional area due to corrosion should be ascertained during inspection by measuring the net thickness and the same should be recorded as percentage loss of the original cross section. Cracks usually originate at connection details, ends of welds, locations of heavy corrosion, and then propagate across the section. In painted bridges, areas with broken paint, accompanied by rust stainings, are suspect areas, and should be carefully inspected for cracks. Areas covered with debris and dirt attract accumulation of water and are prone to suffer from corrosion and cracking. Debris and dirt should be cleared and the area checked for such defects. Overstressing can be identified from inelastic elongation (yielding), or decrease in cross section (necking) in tension members and buckling in compression members. Locations of such defects should be accurately recorded, and the cause of the defects investigated. Vehicular collision damages can be distortion of shape, cracking, etc. Locations of such damage should be noted and repair work initiated. In case of major damage, restrictions in speed and loading may have to be imposed till the damage is repaired.

While all the components of a bridge need to be inspected, there are certain areas where serious defects are most likely to occur. Inspection should, therefore, be particularly focussed on these areas. Some of these areas are:

- Camber of truss bridges
- Members with high stresses or subjected to reversal of stresses
- Members and connections which are prone to waterlogging due to inadequate drainage system
- Connections which are not easily accessible to painting, such as troughs in ballasted deck system, areas adjoining interfaces between concrete deck slabs and supporting steel girders, areas behind the bends of joggled stiffeners, inner faces of small box sections, etc.
- Notches in flange/web at the ends of stringers and cross girders of deck system (for fatigue cracks)
- Surfaces of the windward side of a bridge situated near the sea
- Areas where waste water is discharged from passing trains in railway bridges, and areas directly below drainage spouts or weep holes
- Areas which are prone to corrosion due to emission of steam blasts or smoke from locomotives
- Areas (particularly, underside of bottom chords or bottom flanges) where dew collects or where water from the river touches the steelwork during floods
- Areas which are subjected to alternate wetting and drying, such as debris, underside of timber sleepers in railway bridges

- Fatigue prone areas of welded girders such as: ends of cover plates in flanges with multiple plates, butt welded connections of flange plates of different thicknesses, welded splices, ends of stiffeners, fillet welds across the direction of tensile stress, etc.
- Corroded cleats, and rivets/bolts in joints
- Web members of through and semi-through bridges, particularly the lower members, for damage due to collision with vehicular traffic
- Compression members of truss bridges, compression flanges and webs of plate girders for deformation due to buckling
- Inordinate vibration or excessive deflection during passage of heavy load

27.3.7 Testing

There are several non-destructive testing methods available for inspection of steel bridges. Some of these are:

- Liquid penetrant test
- Ultrasonic test
- Magnetic particle test
- Radiograpy

These inspection systems have already been described in Chapter 20, and are therefore not repeated here.

27.3.8 Report

Inspection report is an important document and is used primarily to identify and assess the present requirement of repair and maintenance of a particular bridge. Also, past inspection reports can give an idea about the history of the bridge, particularly about the repairs carried out previously, which can be compared with their present condition. It is, therefore, necessary that successive inspection reports in respect of every bridge are carefully maintained chronologically as a bridge history file. Some authorities require that the findings of any bridge inspection are recorded in standard inspection report formats.

The inspector and the structural engineer responsible for deciding on any action are often not the same person. Therefore, it is through the inspection report that the structural engineer has to 'see' the condition of the bridge and develop his stategy for maintenance. Thus, the effectiveness of any inspection depends largely on how accurately and clearly the entries have been made in the inspection report. The inspector should, therefore, ensure that the entries in the report are specific, detailed, quantitative, and complete. Also, deficiencies must be clearly described to enable any future inspector to make a comparison of the condition. Taking recourse to simple sketches often proves to be more informative than writing notes. Clear and sharp photographs are also very useful documents to support the bridge inspection reports. It is advisable to include in the photograph a clearly marked scale or an easily recognisable item for easy comprehension of the scale of the detail. Inclusion of a plumb bob to show the vertical line would also be helpful.

Inspection report should include the inspector's reasonable interpretation of the findings (defects), their causes, as also the inspector's recommendations for remedial measures (repairs), or alternative actions, such as imposition of speed or load restrictions or a further special inspection. In case of recommendation for repairs, the report should provide a brief description of the proposed repair work including specifications and an estimated bill of quantities.

27.3.9 Safety

Since inspection work is carried out when a bridge is in service, safety aspect becomes closely related with the arrangements for controlling the traffic during this operation. Traffic control, therefore, needs careful planning. During inspection, safety of the users of the bridge and of the local public assume prime importance. No less important is the safety of the inspection crew, and this must also be adequately addressed in this context.

Members of the inspection team must be aware of and comply with the safety rules of the authorities (such as, Railways, Public Works Departments, etc.), within whose reserve the particular bridge comes. These safety rules normally make it mandatory for every member of the team to wear safety vests, helmets, strong boots, full sleeved shirts, full pants, etc., to ensure his physical safety. The clothings should also be appropriate to protect against poisonous snakes, etc., as also from the effects of climatic conditions. Special equipment such as safety belts, safety goggles, life jackets, gloves, etc., should also be used, as necessary.

Inspectors should be careful about potential hazardous environments in the vicinity of the bridge. Examples are: proximity of high voltage power system, movement of railways, defective or loose ladder rungs, walkways, etc. Also, safe use of access vehicles, such as bucket snoopers, lifting platforms, telescopic hoists, etc., should be reviewed prior to the inspection operation.

27.4 LOOSE AND CORRODED RIVETS

Many of the existing bridges are of riveted construction. Rivets which are subjected to vibrations (at locations of bracings, trolley refuge, etc.) are prone to get loose. Often, some rivets are damaged due to corrosion. If the corrosion continues from one side of the connecting plates to the other side, the rivet gets loose and the rivet head on the other side also gets corroded. This has been shown in Fig. 27.1. Such loose rivets should be replaced during maintenance work. The normal practice is that, if 50% or more of the rivet head gets damaged, the rivet should be replaced.

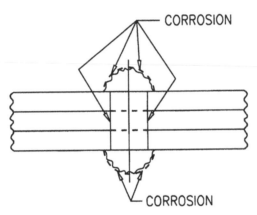

Fig. 27.1 Corrosion in rivet

Looseness of rivets is tested by placing the left hand index finger on one side of the rivet head and then striking the rivet head on the opposite side with a testing hammer. Testing hammer is a light hammer, typically weighing 110 gm with a 300 mm long handle. For a loose rivet, vibration movement is felt by the left hand index finger. Experienced inspectors are quite adept in detecting loose rivets by this method.

For replacement of rivets, the damaged rivets are to be removed first. This work requires utmost care and is done in stages. First, the head is sheared by using a pneumatic rivet breaker or 'buster', and then the rivet shank is driven out with a pneumatic punch. Alternatively, the rivet head may be flame cut about 2 mm above the base metal, and the shank may then be driven out by means of a pneumatic punch. Rivet shank may also be removed by drilling in both these cases. Whichever method is employed, care should be taken to ensure that the parent metal is not damaged. The pneumatic rivet breaker machine imparts heavy shocks to adjoining rivets in groups, which may cause loosening of adjoining sound rivets. Therefore, when only a few rivets at scattered locations in a group are required to be replaced, it is a common practice to flame cut the rivet head, and drill the shank out, thereby avoiding the possibility of loosening of adjoining rivets.

Once a rivet is removed, it should be replaced immediately by a bolt. Use of high strength close tolerance bolts is generally preferable over other types of bolts. Since close tolerance bolts can be made to fit tightly (slip free) to match the existing rivet holes, their behaviour pattern becomes somewhat identical to that of the adjoining existing rivets, thereby ensuring satisfactory transmission of forces.

27.5 MINOR REPAIRS

Based on the inspection report, repair work of minor nature (like isolated corrosion damages of members and cracks) is normally attended to during routine maintenance work. Minor repair works of this type are discussed in this section.

27.5.1 Corrosion

The scheme of repair of a member damaged by corrosion depends largely on the extent of corrosion. If the loss of thickness of the material is such that the net remaining cross sectional area is well within the safety limit, the usual action is to clean the area of any rust and then apply the primary coat of paint followed by the finishing coats. If, however, the net cross sectional area is less than the permissible limit, additional steel material is required to be added. A typical repair scheme for a corrosion damaged top flange of a riveted girder is shown in Fig. 27.2. In this case, a corrosion plate is provided over the top flange and secured to the top flange by bolts, using the holes of the existing rivets. Similarly, web plates damaged due to corrosion can be repaired by fixing corrosion plates of adequate thickness, preferably on both sides of the web, and securing by bolts. Figure 27.3 shows a few examples of repair of corroded web plates.

27.5.2 Cracks

Cracks are very often found in locations damaged by corrosion. These are normally caused due to overstressing. Cracks also originate in other areas,

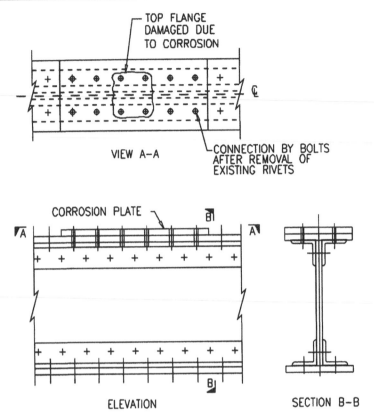

Fig. 27.2 Typical repair work of corroded flange plates of riveted girders

such as connections with incorrect details, roots of flange angles, etc. Fatigue cracks are common where stress concentration due to sharp corners and abrupt changes of cross section exist. Cracks occurring in isolated locations can be repaired by the following method:

First, a smooth hole (12-25 mm diameter) is drilled at about 20 mm beyond the tip of the crack along its assumed line of further progress, to arrest propagation of the crack. Then, splice plates or splice angles are fixed by an adequate number of bolts on either side of the crack. This is a very common solution for isolated cases of cracks. Figure 27.4 illustrates a typical repair work of this nature. However, when too many cracks are noticed in a single member, the damaged member may have to be replaced by an identical member. This case may have to be considered as a separate work outside the scope of routine maintenance and, therefore, needs to be highlighted in the inspection report.

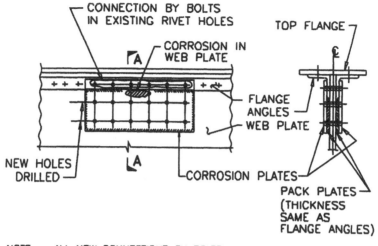

NOTE :- ALL NEW CONNECTIONS BY BOLTS

CORROSION NEAR TOP FLANGE

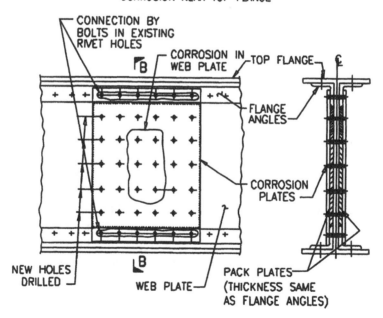

Fig. 27.3 Typical repair work of corroded web plates of riveted girders

Welded girders suffer from some typical cracks. One such crack is shown in Fig. 27.5a. Normally, these cracks originate at the end of the weld

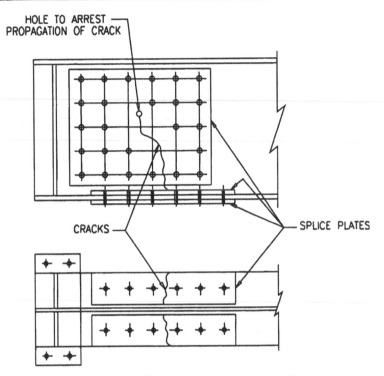

Fig. 27.4 Typical repair work of crack

and propagate along the web plate. These cracks are caused by residual stress concentration due to welding process at the time of fabrication, or subsequent cyclic out-of-plane movement and high bending stresses in the area of the unstiffened gap in the web, leading to fatigue condition and subsequent crack. These cracks can be repaired by first drilling smooth holes (12-25 mm diameter), about 20 mm beyond the tips of the crack and then gouging out the cracked portion and depositing weld metal in its place, followed by removal of excess metal by grinding. Finally, the damaged portion is strengthened by fixing a built-up T member on the other face of the web plate by bolts, as shown in Fig. 27.5b.

An example of typical fatigue crack is often found at the ends of additional cover plate welded to the bottom flange of rolled or built-up beam. The cracks first originate at the toes of the welds and slowly propagate through the flange. Here, concentration of stress due to abrupt change in the cross sectional area of the flange is the primary cause of the cracks. In such a case, first smooth holes (12-25 mm diameter) are drilled at the tips of the cracks to arrest further propagation of the cracks. The cracked portions are

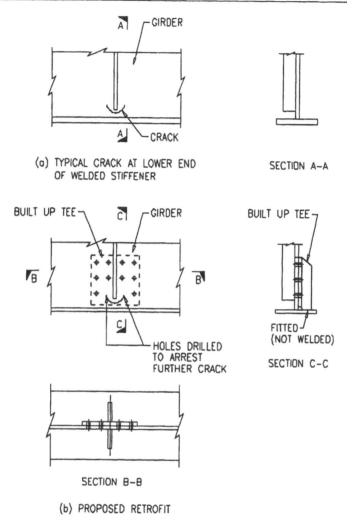

Fig. 27.5 Typical crack in welded girder and repair work

then gouged out and weld metals deposited in their place. The surfaces should be ground to a smooth finish. Finally, splices similar to usual flange splice should be fixed by bolts.

27.6 BEARINGS AND BED BLOCKS

Most bearings have limited life span and normally do not outlast a steel bridge. However, careful inspection and periodic maintenance can extend their service life considerably. Bearings may be broadly classified into two categories, viz., (1) metal bearings, and (2) elastomeric bearings. Various types of bearings used in steel bridges have been described in Chapter 14. However, regardless of their type, it is important to keep bridge bearings free from dirt, debris and excessive moisture. All bearings should be examined to ascertain whether they are functioning properly or not. Changes in other parts of the structure, such as settlement of foundations or tilt in the piers or abutments, may have an adverse effect on the functioning of bearings.

27.6.1   Metal Bearings

Some of the common problems associated with metal bearings and their remedial solutions are discussed below:

Corrosion

As discussed earlier, corrosion and rusting in steel are mostly caused by water. Dust and debris normally absorb and retain moisture. Hence, debris should not be allowed to accumulate at the locations of bearings.

Corrosion on contact surfaces increases the coefficient of friction, thereby impeding movements of sliding plates or rollers, and rendering them ineffective. Often, the smaller components, such as tooth bars, or pins get corroded and resist the movement of the bearings. Once they get jammed or 'frozen', bearings may induce additional forces to the structure, particularly in the substructure, causing damage around the concrete bed block. Minor repair work, such as replacement of damaged components, may be done during routine maintenance. Where components do not show any major loss of section due to corrosion, they may need only cleaning *in situ* and greasing. Any major defect necessitating replacement of major components (such as rollers or sliding plates), may be considered as an extra job and must be recorded in the inspection report, indicating thereon the inspector's recommendations for rehabilitation.

Misalignment

Misalignment is normally caused by excessive vibration of the superstructure due to live loads, error in initial setting of the bearings, or defective fabrication. This condition may restrict the movement of the superstructure and induce additional forces in the bridge structure. The defect should be recorded in the inspection report.

If the misalignment is not caused by defective fabrication, it can be corrected during maintenance work by first relieving the load from the bearing by jacking up the superstructure, introducing temporary props to support it. Then the bearing components are to be re-set with correct alignment, giving due consideration to the temperature effect for the inclination of the rollers.

Tilting

Tilting of bearings may be attributed to the movement of the substructure or the superstructure. In either case, the defect must be recorded in the inspection report. Re-setting of the tilted bearings can be done in a manner similar to that of misalignment of bearings.

Replacement

Since frozen or non-functioning bearings can induce additional forces in the members of the bridge structure, it is necessary for the inspector to carefully study the distresses in the bearings, and recommend corrective measures. In case some bearings are found to be beyond repair, replacement option may have to be recommended. The following points need to be considered in the event of replacement of bearings:

- The height of the new bearings should match the existing available height.
- The bearings must satisfy the functional requirements, viz., the horizontal and/or rotational movements, as necessary.
- In case the existing metal bearings are replaced by elastomeric bearings, the horizontal force on the substructure may change. Adequacy of the substructure should be checked in such an event.

Oiling and greasing

Like other steel elements in the bridge structure, metal bearings also need some form of protective system to contain corrosion. This may be in the form of painting, which needs to be attended to during the maintenance activity. The maintenance programme should also include periodic lubrication and greasing of the movable surfaces of the metal bearings.

Some authorities recommend movable bearings to be encased in oil baths, and keeping them submerged in specified brands of lubricating oil. For such bearings, the maintenance programme should include inspection of the level of the oil in the bath, and keeping it to the specified level. The seal of the oil bath should be checked for possible leakage, which must be rectified, if found defective.

27.6.2 Elastomeric Bearings

The common defects found in elastomeric bearings are:

- Excessive bulging due to bearings being too 'tall'
- Splitting, tearing, or shearing due to excessive shearing for a long period of time
- Failure of bond

Like metal bearings, elastomeric bearings also require periodic inspection and maintenance. The accumulated dirt and debris in the vicinity of these bearings should be cleared. The integrity of the bearings should be checked for the above defects. If necessary, they should be replaced by jacking up the superstructure in a manner similar to that of replacement of metal bearings.

27.6.3 Bed Blocks

The loads from the bearings are transmitted and distributed to the underlying substructure *via* the concrete bed blocks. Restriction of the free movement of the superstructure (due to malfunctioning of the bearings) may result in defects such as, development of cracks in the bed blocks, or shearing/bending/loosening of holding down bolts. Hence the bed blocks should be properly inspected for these defects, and the cause of the defect established. Damaged bed blocks should be repaired, or if necessary, replaced using stronger concrete mix. For this purpose, the superstructure has to be first jacked up to relieve the dead load, and supported on temporary props. The bearing should be removed to enable the bed block to be repaired or replaced. The bearings should then be re-installed, but only after the concrete has hardened.

27.7 MAINTENANCE REPAINTING

27.7.1 General

Maintenance repainting of a steel bridge is done to protect it from corrosion and thereby to ensure that the structure can perform its intended functions over a specified period of time, without the need of structural repair.

27.7.2 Common Problems of Repainting

As part of the maintenance programme, many of the world's existing steel bridges are given coats of protective paint at regular intervals. It has often been noticed that the locations which are easily accessible receive the paint coat regularly, whereas other areas which are not so easily accessible do not receive proper attention. This results in the less accessible areas of a bridge being corroded earlier than those which are accessible easily.

Another problem area is the lack of proper cleaning of the surface before application of the paint coat. Unless the rust already developed on a steel surface is properly cleaned, and the dirt removed, the rust is likely to reappear within a short period of time after the repainting is done. In case the loss of area due to corrosion is beyond the permissible limit, it will be necessary to add corrosion plate to make up the loss of area and only then the repainting should be carried out.

Some members are critical for the overall stability of the structure. These are called fracture critical members. Loss of cross-sectional area due to corrosion in such members should be avoided. A system of monitoring the condition of these members should be introduced, so that these areas receive special attention and repainting needs are promptly attended to.

Another aspect that needs looking into is the excessive thickness of the coatings on a member. While thick coatings may appear to provide more protection to the steel surface, it may, in fact, be detrimental, as it tends to induce cracking and flaking, needing removal of such coating. This aspect needs consideration during maintenance repainting stage.

The quality and workmanship of the initial painting system largely influence the efficiency of the maintenance painting of an existing bridge. If the original painting system was inadequate for the service conditions or the workmanship was not upto the desired level, efficient repainting work becomes rather difficult. Also, steel bridges suffering from long spells of neglect, or inadequate maintenance, may pose problems while carrying out subsequent maintenance repainting work, as extensive cleaning and/or structural repair work may be needed prior to repainting. Normally, local damaged/repaired areas are provided first with the primary coat, and then the finishing coats, prior to the application of final coats of repainting.

27.7.3 Coating Defects

Coating defects which are generally encountered in an existing bridge are briefly discussed in this section.

Blistering

Blistering may occur within the coating itself due to presence of water or solvents trapped within or under the paint film, causing swelling. It may also be caused by local accumulation of iron corrosion products at the interface of the paint and the metal.

Flaking

When the top coat peels off from the underlying coat, the phenomenon is termed flaking. The main reason of flaking is loss of adhesion. Some of the causes of poor adhesion are:

- Presence of loose or powdery materials, rust, dust, etc., on the surface
- Presence of contamination (such as grease, oil, silicone, etc.) which prevents proper bonding between the paint and the underlying surface
- Surface too smooth to provide mechanical bonding with the paint

Cracking

Crackings are not always visible by naked eyes and often require use of magnifying glass to detect them. Surface hair cracking may be limited to the top coat only. However, some crackings penetrate beyond the top coat, causing failure of the protective system. Crackings are generally caused when the internal stress in a paint film increases beyond its tensile strength. The capacity of a paint film against cracking depends on a number of factors such as, its condition during application and curing, plasticisation, age, etc.

Blooming or blushing

Blooming or blushing is identified as a local loss of gloss and dullness of colour. It is generally caused by condensation, or high humidity during curing period. This defect is not generally considered to be detrimental to the coating, but may reduce its adhesion to the subsequent overcoats. It can normally be removed by application of an appropriate solvent.

Pinholes

Pinholes are tiny holes in the paint film caused by bursting of air bubbles trapped in the paint during application, which the adjoining wet paint is unable to cover before the setting of the paint.

27.7.4 Selection of Protective System

Normally, in a bridge structure, maintenance repainting is done by applying the same type of protective paint as the existing one, provided, of course, its performance has been satisfactory. In case, the existing system is not satisfactory, or needs upgrading, due to change in the environment or other reasons, a new system may have to be selected. In such an event, it will be necessary to consider the following aspects:

Compatibility

The most important aspect to consider is the compatibility of the material of the proposed system with the coating already existing on the structure, in order that the new coat adheres to the existing one for a long period. If the existing coat has hardened during service, it may require abrasive treatment for providing mechanical keys on the surface to hold the new system.

Other factors

Other criteria to be considered for the selection of a new coating system are environment, availability of the coating system, ease of application, economy, etc. These have been discussed in Chapter 26, and are not repeated here.

27.7.5 Surface Preparation Prior to Repainting

As in the case of a new bridge structure, surface preparation prior to repainting an existing bridge is very important to make the repaint coating to be effective and durable. Unless the surface is properly cleaned and made free from rust or other chemicals, corrosion under the new coating is likely to start again. This has been a major problem for maintenance of a large number of existing bridges.

Surface preparation for maintenance repainting is normally done in localised areas, and for this purpose the following guidelines may be followed:

(a) Where the existing system does not show any defect such as blistering or flaking, no special treatment is required. The surface should only be thoroughly washed, and dried before the maintenance painting is applied.

(b) Where the existing system shows only minor blistering or flaking locally, with no rusting, the loose paint coatings are to be removed first, and then the surface should be thoroughly washed and dried before the maintenance painting is applied.

(c) Where the existing system shows major damage of the painting system with or without rusting of the steel, it will be necessary to clean the area thoroughly by power tools or flame cleaning, or other appropriate methods to remove all the rust from the steel, so as to provide a firm base for the future coat. The full protective system, viz., the primary coat, and the finishing coats should be applied on the cleaned surface, to complete the repair work. The surface is now ready to receive the normal maintenance paint.

27.8 CONCLUDING REMARKS

In 1921, J.A.L. Waddell, in his *Economics of Bridgework* stated : The life of a metal bridge that is scientifically designed, honestly and carefully built and not seriously overloaded, if properly maintained, is indefinitely long. This statement is true, but the criteria required for attaining an 'indefinitely long' life are hardly fully achieved in real life conditions. Nevertheless, the criteria mentioned by Waddell includes the words 'if properly maintained' and these words carry great importance. Although many authorities consider 100 - 120 years to be a reasonable useful life of a steel bridge, modern methods of inspection, maintenance, timely repair and rehabilitation work, including application of appropriate repainting system, would certainly go a long way in extending the life of steel bridges even further.

REFERENCES

[1] *Manual for Maintenance Inspection of Bridges*, 1974, American Association of State Highway and Transport Officials (AASHTO), Washington, D.C., USA.

[2] *Bridge Inspection Guide*, 1983, Her Majesty's Stationery Office, London, UK.

[3] Ghosh, U.K. 2000, *Repair and Rehabilitation of Steel Bridges*, A.A. Balkema, Rotterdam, Netherlands.

Index

Printed and bound by CPI Group (UK) Ltd, Croydon, CR0 4YY

01/11/2024

01782599-0008